PHYSICS IN MOLECULAR BIOLOGY

Tools developed by statistical physicists are of increasing importance in the analysis of complex biological systems. *Physics in Molecular Biology* discusses how physics can be used in modeling life. It begins by summarizing important biological concepts, emphasizing how they differ from the systems normally studied in physics. A variety of subjects, ranging from the properties of single molecules to the dynamics of macro-evolution, are studied in terms of simple mathematical models. The main focus of the book is on genes and proteins and how they build interactive systems. The discussion develops from simple to complex phenomena, and from small-scale to large-scale interactions.

This book will inspire advanced undergraduates and graduates of physics to approach biological subjects from a physicist's point of view. It requires no background knowledge of biology, but a familiarity with basic concepts from physics, such as forces, energy, and entropy is necessary.

KIM SNEPPEN is Professor of Biophysics at the Nordic Institute for Theoretical Physics (NORDITA) and Associate Professor at the Niels Bohr Institute, Copenhagen. After gaining his Ph.D. from the University of Copenhagen, he has been research associate at Princeton University, Assistant Professor at NORDITA and Professor of Physics at the Norwegian University of Science and Technology. He has lectured and developed courses on the physics of biological systems and mathematical biology, and has also organized several workshops and summer schools on this area. Professor Sneppen is a theorist whose research interests include cooperativity in complex systems, the dynamics and structure of biological networks, evolutionary patterns in the fossil record, and the cooperative behavior of genetic switches and heat shock response of living cells.

GIOVANNI ZOCCHI is Assistant Professor of Physics at the University of California, Los Angeles. Following his undergraduate education at the Università di Pisa and Scuola Normale Superiore, Pisa, he obtained his Ph.D. in Physics from the University of Chicago. He has worked in diverse areas of complex system physics at the Ecole Normale Supérieure, Paris, and the Niels Bohr Institute, Copenhagen, before his current position at UCLA. He has taught courses on introductory biophysics at the Niels Bohr Institute and currently teaches biophysics at advanced undergraduate and graduate level. Professor Zocchi is an experimentalist; his present research is focused on understanding and controlling conformational changes in proteins and DNA.

# PHYSICS IN MOLECULAR BIOLOGY

KIM SNEPPEN & GIOVANNI ZOCCHI

CAMBRIDGE UNIVERSITY PRESS
Cambridge, New York, Melbourne, Madrid, Cape Town,
Singapore, São Paulo, Delhi, Mexico City

Cambridge University Press
The Edinburgh Building, Cambridge CB2 8RU, UK

Published in the United States of America by Cambridge University Press, New York

www.cambridge.org
Information on this title: www.cambridge.org/9780521844192



First published 2005
Reprinted 2006

*A catalogue record for this publication is available from the British Library*

*Library of Congress Cataloguing in Publication Data*

ISBN 978-0-521-84419-2 Hardback

# Contents

# Preface

This book was initiated as lecture notes to a course in biological physics at Copenhagen University in 1998–1999. In this connection, Chapters 1–5 were developed as a collaboration between Kim Sneppen and Giovanni Zocchi. Later chapters were developed by Kim Sneppen in connection to courses taught at the Norwegian University of Science and Technology at Trondheim (2001) and at Nordita and the Niels Bohr Institute in 2002 and 2003.

A book like this very much relies on feedback from students and collaborators. Particular thanks go to Jacob Bock Axelsen, Audun Bakk, Tom Kristian Bardøl, Jesper Borg, Petter Holme, Alexandru Nicolaeu, Martin Rosvall, Karin Stibius, Guido Tiana and Ala Trusina. In addition, much of the content of the book is the result of collaborations that have been published previously in scientific journals. Thus we would very much like to thank:

- Jesper Borg, Mogens Høgh Jensen and Guido Tiana for collaborations on polymer collapse modeling;
- Terry Hwa, E. Marinari and Lee-han Tang for collaborations on DNA melting;
- Audun Bakk, Jacob Bock, Poul Dommersness, Alex Hansen and Mogens Høgh Jensen for collaborations on protein folding models and models of discrete ratchets;
- Deborah Kuchnir Fygenson and Albert Libchaber for collaborations on nucleation of microtubules and inspiration;
- Erik Aurell, Kristoffer Bæk, Stanley Brown, Harwey Eisen and Sine Svenningsen for collaborations on $\lambda$-phage modeling and experiments;
- Ian Dodd, Barry Egan and Keith Shearwin for collaborations on modeling the 186 phage;
- Jacob Bock, Mogens Høgh Jensen, Sergei Maslov, Petter Minnhagen, Martin Rosvall, Guido Tiana and Ala Trusina for ongoing collaborations on the properties of molecular networks, and modeling features of complex networks;
- Sergei Maslov and Kasper Astrup Eriksen on collaborations on large-scale patterns of evolution within protein paralogs in yeast;
- Per Bak and Stefan Bornholdt for collaborations on macro-evolutionary models, quantifications of large-scale evolution, and modeling evolution of robust Boolean networks.

Kim Sneppen is particularly grateful for the hospitality of KITP at the University of California, Santa Barbara, where material for part of this book was collected during long visits at programs on Physics in Biological Systems in winter/spring 2001 and 2003.

I, Kim, thank my infinitely wise and beautiful wife Simone for patience and love throughout this work, and my children Ida, Thor, Eva and Albert for putting life into the right perspective.

# Introduction

This book covers some subjects that we find inspiring when teaching physics students about biology. The book presents a selection of topics centered around the physics/biology/chemistry of genes. The focus is on topics that have inspired mathematical modeling approaches. The presentation is rather condensed, and demands some familiarity with statistical physics from the reader. However, we attempted to make the book complete in the sense that it explains all presented models and equations in sufficient detail to be self-contained. We imagine it as a textbook for the third or fourth years of a physics undergraduate course.

Throughout the book, in particular in the introductions to the chapters, we have expressed basic biology ideas in a very simplified form. These statements are meant for the physics student who is approaching the biological subject for the first time. Biology textbooks are necessarily more descriptive than physics books. Our simplified statements are meant to reduce this difference in style between the two disciplines. As a consequence, the expert may well find some statements objectionable from the point of view of accuracy and completeness. We hope, however, that none is misleading. One should think of these parts as first-order approximations to the more complicated and complete descriptions that molecular biology textbooks offer. On the other hand, the physical reasoning that follows the simplified presentation of the biological system is detailed and complete.

The book is not comprehensive. Large and important areas of biological physics are not discussed at all. In particular we have not ventured into membrane physics and transport across membranes, signal transmission along neurons and sensory perception, to mention a few examples. While there are already excellent books and reviews on all these subjects, the reason for our limited choice of topics is more ambitious. The basic physics ideas that are relevant for molecular biology can be learned on a few specific examples of biological systems. The examples were chosen because we find them particularly suited to illustrate the physics.

We have chosen to place the focus on genes, DNA, RNA and proteins, and in particular how these build a functional system in the form of the $\lambda$-phage switch. We further elaborate with some larger-scale examples of molecular networks and with a short overview of current models of biological evolution. The overall plan of the book is to proceed from simple systems toward more complex ones, and from small-scale to large-scale dynamics of biological systems.

Chapter 1 gives some impression of important ideas in biology. To be more precise, the chapter summarizes those concepts which, we think, strike a physicist who approaches the field, either because they have no counterpart in physics, or, on the contrary, because they are all too familiar. The chapter grew out of discussions with biologists, and we normally use it as a first introductory lecture when we give the course. Of the subsequent chapters, we regard Chapter 7 on the $\lambda$-phage in *E. coli* as especially central: it deals with the interplay between elements introduced earlier in the book, and it contains a lot of the physics reasoning that the book is meant to teach.

In Chapter 2 we describe the physics of polymer conformations, emphasizing the interplay between energy and entropy and examining both the behavior of extended polymers and how compact configurations may be reached. In the next chapters we introduce and discuss the most important biological polymers: DNA, RNA and proteins. Although the covalent bonds forming the polymer backbone have binding energies $\Delta G > 1$ eV, the form and function of these biomolecules is associated to the much weaker forces perpendicular to the polymer backbone. These interactions are of order $k_B T$, and it is the combined effect of many of these forces that forms the functional biomolecule. In Chapters 3–5 we characterize the stability of DNA, RNA and proteins, with emphasis on the cooperativity responsible for this stability.

Biological molecules can be used for various types of computations. Chapter 3 includes a section on DNA computation and DNA manipulation in the laboratory. This is in part a continuation of Chapter 2 (reptation), and also an introduction to the computational aspects of molecular replication (the PCR reaction). Chapters 4–6, on the other hand, focus on proteins and protein folding and thus the functional aspects are left to subsequent chapters. In this book we have addressed in considerable detail one of these aspects, namely how a protein may control the production of another protein (Chapter 7). As we explain in Chapter 7, genetic control involves mechanisms associated to both equilibrium statistical mechanics and to the timescales involved in complex formation and disruption. Topics in this chapter include a discussion of cooperativity, of target location by diffusion, of timescales in a cell and of stability of expressed genetic states.

Chapter 7 also forms a microscopic foundation for the large-scale properties of molecular networks, which we discuss in Chapter 8. Chapter 8 thus continues the subject of genetic regulation and molecular networks, in part by venturing into the heat shock mechanism. This shows that protein folding is also a control mechanism in a living cell, and it introduces a type of genetic regulation that was not treated in the previous chapter: $\sigma$ sub-units of RNAp, which control the expression of larger classes of genes. Chapter 8 also discusses the larger-scale properties of genetic regulatory networks, introducing a few recent physics attempts at modeling these.

Chapter 9 discusses evolution, with emphasis on the interplay between randomness and selection from the smallest to the largest scales. The chapter introduces concepts such as neutral evolution, hill climbers and co-evolution, and uses these concepts to discuss questions related in part to the concept of punctuated equilibrium, and in part to the origin of life in the form of autocatalytic networks. Thus Chapter 9 introduces some simple models that allow us to discuss the nature of the history leading to the emergence of life, and in particular aims at stressing the importance of interactions and stochastic events on all scales of the biological hierarchy.

In the Appendix we have a short introduction to statistical mechanics, including the fluctuation–dissipation theorem and the Kramers escape problem; it is meant to render the book self-contained from the point of view of the physics.

# 1

# What is special about living matter?

Kim Sneppen & Giovanni Zocchi

Life is self-reproducing, persistent (we are $\sim 4 \times 10^9$ years old), complex (of the order of 1000 different molecules make up even the simplest cell), "more" than the sum of its parts (arbitrarily dividing an organism kills it), it harvests energy and it evolves. Essential processes for life take place from the scale of a single water molecule to balancing the atmosphere of the planet. In this book we will discuss the modeling and physics associated, in particular, to the molecules of life and how together they form something that can work as a living cell. First we briefly review some basic concepts of living systems, with emphasis on what makes biological systems so different from the systems that one normally studies in physics.

Conceptually, molecular biology has provided us with a few fundamental/universal mechanisms that apply over and over. Some concepts, like evolution, do not have counterparts in physics. Others, like the role of stochastic processes, are, on the contrary, quite familiar to a physicist.

(1) **Biology is the result of a historical process.** This means that it is not possible to "explain" a biological system by applying a few fundamental laws in the same way that is done in physics. A hydrogen atom could not be different from what it is, based on what we know of the laws of nature, but an *E. coli* cell could. In evolution, it is much easier to modify existing mechanisms than to invent new ones. Thus on evolutionary timescales nearly everything comes about by cut and paste of modules that are already working. We will end the book with a chapter dedicated to evolutionary concepts and models.

(2) **The molecules of life are polymers.** At the molecular scale, life is made of polymers: DNA, RNA and proteins. Even membranes are built of molecules with large aspect ratios. Perhaps mechanics at the nano-scale can work only with polymers, molecules that are kept together by strong forces along their backbone, while having the property of forming specific structures by utilizing the much weaker forces perpendicular to the backbone. In molecular biology we witness nano-mechanics at work with polymers. We will discuss polymers in Chapter 2, and thereby introduce concepts necessary

for understanding DNA (Chapter 3), proteins (Chapters 4–5) and polymers in action (Chapter 6).

(3) **Genetic code.** Information is maintained on a one-dimensional, double-stranded DNA molecule, which will be discussed in Chapter 3. Thus the one-dimensional nature of the information mirrors the one-dimensional nature of the polymers that make life work. The DNA strands open for *copying* and *transcribing*, by separating the double-stranded DNA into two single strands of DNA that each carry the full information. The copying is done by DNA polymerase using the complementarity of base pairs. Similarly the genetic code is read by RNA polymerase and ribosomes that again use the matching of complementary base pairs to translate codons into amino acids. This is usually summarized in terms of the central dogma

$$\text{DNA} \rightarrow \text{RNA} \rightarrow \text{protein} \tag{1.1}$$

This is highly simplified: proteins modify other proteins, and most importantly proteins provide both positive and negative feedback on all the arrows in (1.1). If one has only DNA in a test tube, nothing happens. One needs proteins to get DNA → RNA, etc. Then Eq. (1.1) should be supplemented at least by an arrow from protein to DNA. Thus it is not always clear where the start of this loop is, and the whole scheme has to be extended to the complicated molecular networks discussed in Chapter 8.

(4) **Computation.** A living cell is an incredible information-processing machine: an *E. coli* transcribes about $5 \times 10^6$ genes during 1/2 h, i.e. about 10 Gb/h of information. All this within a 1 $\mu m^3$ cell, coded by about $5 \times 10^6$ base pairs. The information density far outnumbers that in any computer chip, and even a million *E. coli* occupy

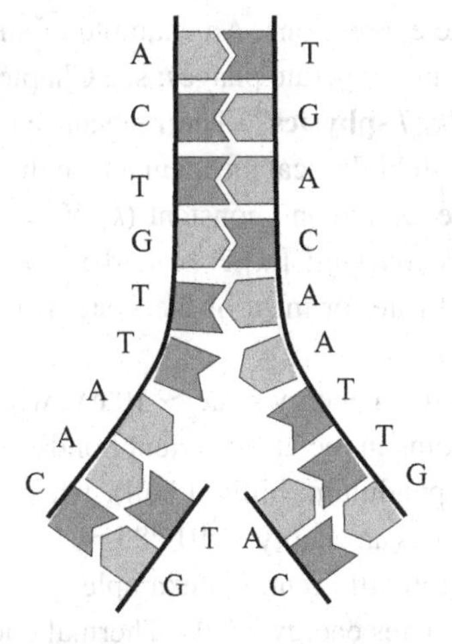


Figure 1.1. Information in life is maintained one-dimensionally through a double-stranded polymer called DNA. Each polymer strand in the DNA contains exactly the same information, coded in form of a sequence of four different base pairs. Duplication occurs by separating the strands and copying each one. This interplay between memory and replication opened 4 billion years of complex history.

much less space than a modern CPU, thus beating PCs on computation speed as well. The levels of computation in a living system increase when one goes to eukaryotes and especially to multi-cellular organisms (where each cell must have encoded awareness of its social context also). The simplest organisms (e.g. the prokaryote *M. pneumonomiae* with 677 genes) can manage essentially without transcription control. Larger genome size prokaryotes typically need a number of control units that grow with the square of number of genes. We discuss modeling of processes within living cells in Chapter 7 and, to some extent, also in Chapters 6 and 8.

(5a) **Life is modular.** It is build of parts that are build of parts, on a wide range of scales. This facilitates robustness: if a process doesn't work, there are alternative routes to replace it. Molecular-scale examples include the secondary, tertiary and quaternary structures of proteins (complexes of proteins); they may include network modules, such as sub-cellular domains, that each facilitate an appropriate response to external stimuli. Most importantly, the minimum independent living module is the cell.

(5b) **Life is NOT modular.** Life is more than the sum of its parts. Removing a single protein species often leads to death for an organism. Another observation is that the number of regulatory proteins for prokaryotes increases with the square of the number of proteins that should be regulated. Thus regulatory networks are an integrated system, and not modular in any simple way. This is the subject for the chapter on networks.

(6) **Stochastic processes play an essential role** from molecules to cells; in particular, they include mechanisms driven by Brownian noise, trial-and-error strategies, and the individuality of genetically identical cells owing to their finite number of molecules. An example of a trial-and-error mechanism is microtubule growth, attachment and collapse (see Chapter 6). Individuality of cells has been explored by individual cell measurements of gene expression, and variability of cell fate has been associated with fluctuations in gene expressions. An example of such stochasticity includes the lysis–lysogeny decision in temperate phages; see Chapter 7.

(7) **Biological physics is "$k_B T$-physics".** The relevant energy scale for the molecular interactions that control all biological mechanisms in the cell is $k_B T$, where $T$ is room temperature and $k$ is the Boltzmann constant ($k_B N_A = R$, where $N_A$ is Avogadro's number and $R$ is the gas constant; $1\ k_B T = 4.14 \times 10^{-14}$ ergs $= 0.62$ kcal/mole at $T = 300\ K$). This is not true for most of the systems described in a typical physics curriculum, for example:

- the hydrogen atom, with an energy scale $\sim$ 10 eV, whereas $k_B T_{\text{room}} \simeq 1/40$ eV;
- binding energies of atoms in metals; covalent bonds: energy $\sim$ 1 eV;
- macroscopic objects (pendulum, billiard ball), where even a 1 mg object moving with a speed of 1 cm/s has an energy $\sim 10^{-10}$ J $\sim 10^9$ eV (1 eV $= 1.602 \times 10^{-19}$ J).

The approach is therefore different. For example, in the solid state one starts with a given structure and calculates energy levels. Thermal energy may be relevant to kick carriers in the conduction band, but $k_B T$ is not on the brink of destroying the ordered structure.

Soft-matter systems often self-assemble in a variety of structures (e.g. amphiphilic molecules in water form micelles, bilayers, vesicles, etc.; polypeptide chains fold to form globular proteins). These ordered structures exist in a fight against the disruptive effect of thermal motion. The quantity that describes the disruptive effect of thermal motion is the entropy $S$, a measure of microscopic disorder that we review in the Appendix. So for these systems energy and entropy are both equally important, and one generally considers a free energy $F = E - TS$. The language and formalism of thermodynamics are effective tools in describing these systems. For example: free-energy differences are just as "real" as energy differences; therefore entropic effects can result in actual forces, as we discuss in Chapter 2.

## Further reading

Berg, H. C. (1993). *Random Walks in Biology*. Princeton: Princeton University Press.
Boal, D. H. (2002). *Mechanics of the Cell*. Cambridge University Press.
Bray, D. (2001). *Cell Movements: From Molecules to Motility*. Garland Publishing.
Crick, F. H. C. (1962). The genetic code. *Sci. Amer.* **207**, 66–74; *Sci. Amer.* **215**, 55–62.
Eigen, M. (1992). *Steps Towards Life*. Oxford University Press.
Godsell, D. (1992). *The Machinery of Life*. Springer Verlag.
Gould, S. J. (1991). *Wonderful Life, The Burgess Shale and the Nature of History*. Penguin.
Howard, J. (2001). *Mechanics of Motor Proteins and the Cytoskeleton*. Sinauer Associates.
Kauffman, S. (1993). *The Origins of Order*. Oxford University Press.
Lovelock, J. (1990). *The Ages of Gaia*. Bantam Books/W. W. Norton and Company Inc.
Pollack, G. H. (2001). *Cells, Gels and the Engines of Life*. Ebner & Sons Publishers.
Ptashne, M. & Gann, A. (2001). *Genes & Signals*. Cold Spring Harbor Laboratory.
Raup, D. (1992). *Extinction: Bad Genes or Bad Luck*? Princeton University Press.
Schrödinger, E. (1944). *What is Life*? Cambridge University Press.

# 2

# Polymer physics

Kim Sneppen & Giovanni Zocchi

Living cells consist of a wide variety of molecular machines that perform work and localize this work to the proper place at the proper time. The basic design idea of these nano-machines is based on a one-dimensional backbone, a polymer. That is, these nano-machines are not made of cogwheels and other rigid assemblies of covalently interlocked atoms, but rather are based on soft materials in the form of polymers – i.e. one-dimensional strings. In fact most of the macromolecules in life are polymers. Along a polymer there is strong covalent bonding, whereas possible bonds perpendicular to the polymer backbone are much weaker. Thereby, the covalent backbone serves as a scaffold for weaker specific bonds. This opens up the possibility (1) to self-assemble into a specific functional three-dimensional structure, (2) to allow the machine parts to interact while maintaining their identity, and (3) to allow large deformations. All three properties are necessary ingredients for parts of a machine on the nano-scale. In this chapter we review the general properties of polymers, and thus hope to familiarize the reader with this basic design idea of macromolecules.

Almost everything around us in our daily life is made of polymers. But despite the variety, all the basic properties can be discussed in terms of a few ideas. Some of these properties are astounding: consider a metal wire and a rubber band. The metal wire can be stretched about 2% before it breaks; its elasticity comes from small displacements of the atoms around a quadratic energy minimum. The rubber band, on the other hand, can easily be stretched by a factor of 4. Clearly its elasticity must be based on an entirely different effect (it is in fact based on entropy); see also Fig. 2.1.

Polymers are long one-dimensional molecules that consist of the repetition of one or a few units (the "monomers") bound together with covalent bonds. You can think of beads on a string. Figure 2.2 shows three examples; the first two are synthetic polymers, the third represents the primary structure of proteins. What is radically different between these molecules and all others is that the number of

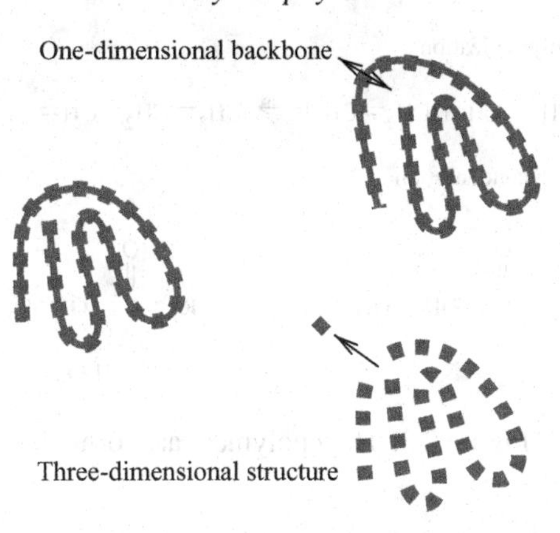


Figure 2.1. Illustration of the self-healing properties of a device with a one-dimensional backbone. Thermal or other fluctuations may dislodge a single element, but if attached to a backbone it typically will move back into the correct position (from Hansen & Sneppen, 2004).

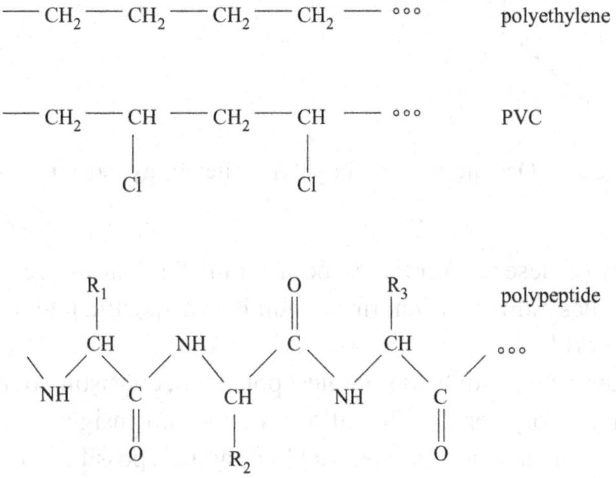

— $CH_2$ — $CH_2$ — $CH_2$ — $CH_2$ — ooo   polyethylene

— $CH_2$ — CH — $CH_2$ — CH — ooo   PVC
(with Cl attached to each CH)

polypeptide: NH — CH($R_1$) — C(=O) — NH — CH($R_2$) — C(=O) — NH — CH($R_3$) — C(=O) — ooo

Figure 2.2. Examples of polymers.

monomers, $N$, is large, typically $N \sim 10^2 - 10^4$ (but note that for DNA $N$ can be $\sim 10^8$). The single most dramatic consequence is that the molecule becomes flexible. We normally think of the relative motion of atoms within a small molecule, say $CO_2$, in terms of vibrational modes. A polymer, however, can actually bend like a string! There are more consequences. Perpendicular to the strong (covalent) forces along the one-dimensional backbone, weaker forces may come into play; forces that would be insignificant if the atoms were not brought together by the backbone bonds. But given that the backbone forces these monomers together, the cooperative

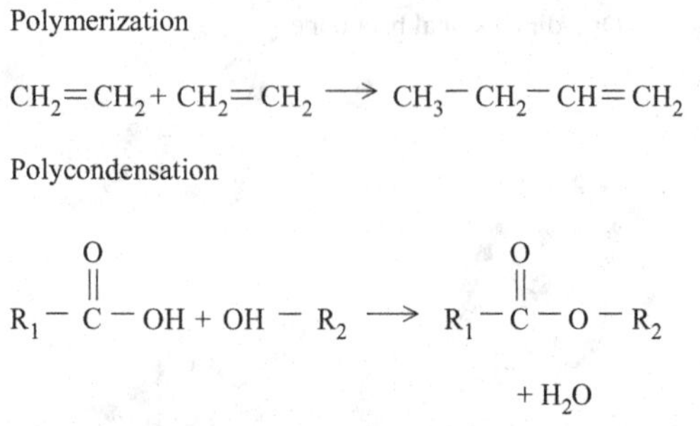

Polymerization

$$CH_2{=}CH_2 + CH_2{=}CH_2 \longrightarrow CH_3{-}CH_2{-}CH{=}CH_2$$

Polycondensation

$$R_1 - \overset{\overset{\displaystyle O}{\|}}{C} - OH + OH - R_2 \longrightarrow R_1 - \overset{\overset{\displaystyle O}{\|}}{C} - O - R_2 + H_2O$$

Figure 2.3. How polymers are formed.

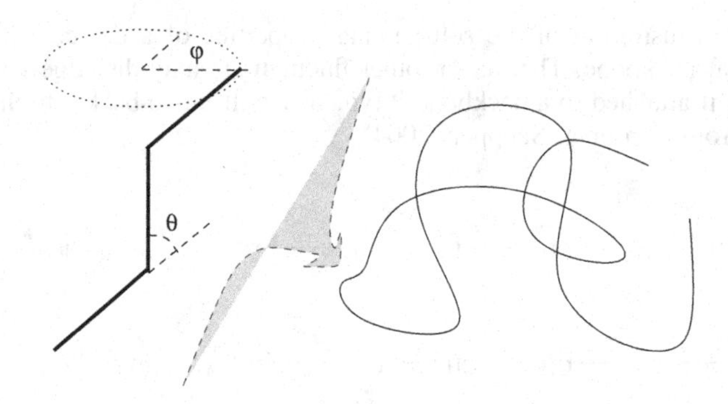


Figure 2.4. One mechanism for polymer flexibility: bond rotations.

binding of many of these weaker forces, both within the same molecule and between different molecules, allows the enormous number of specific interactions found in the biological world.

In this chapter we will study the simplest polymers, consisting of many identical monomers ("homopolymers"). This allows us to gain insight into the interplay between the one-dimensional polymer backbone and the possible three-dimensional conformations of the molecule.

Polymers are formed by polymerization (e.g. polyethylene) or by polycondensation (e.g. polypeptides); see Fig. 2.3. The single most important characteristic of polymers is that they are flexible. The simplest mechanism for their flexibility comes from rotations around single bonds. Figure 2.4 shows three links of, say, a polyethylene chain; the C atoms are at the vertices and the segments depict the C–C bonds. The bond angle $\theta$ is fixed, determined by the orbital structure of the carbon, but $\phi$ rotations are allowed. As a result, on a scale much larger than the monomer size a snapshot of the polymer chain may look as depicted on the right in the figure, i.e. a coil with random conformation. For other polymers, for example double-stranded DNA, the chemical structure does not allow bond rotations as in

Fig. 2.4. These polymers are generally stiffer (for example, double-strand DNA is much stiffer than single-strand DNA), but still flexible at large enough scales; this flexibility is similar to the bending of a beam.

In the next sections we will study some basic properties of polymer conformation, such as how the size scales with length in a good solvent and polymer collapse in a bad solvent, with simple models based on the random walk.

*Question*

(1) Discuss the information needed to assemble a machine sequentially in one dimension, and compare it with that needed to assemble it directly in three dimensions. If there are 20 different building blocks, how many different neighborhoods would there be if each building block is assigned a position on a three-dimensional cubic lattice?

## Persistence length of a polymer

In order to quantify the stiffness, one introduces a length scale called the persistence length $l_p$ of the polymer. Operationally $l_p$ is associated with the decay length of correlations of directionality along the chain. Referring to Fig. 2.5, if $\mathbf{e}(x)$ is the unit vector tangent to the chain at some position $x$ along the chain (i.e. $x$ is the arclength), one considers the *correlation function* $C(x, y) = \langle e(x) \cdot e(y) \rangle$. Here the $\langle \rangle$ refer to an ensemble averaging, that is an average over many different copies of the polymer. The averaging can be replaced by a time averaging, and thus can be done experimentally by measuring over long time over the same polymer. For a very long homopolymer the correlation is solely a function of the distance $|x - y|$; and the correlation function decays exponentially with distance, with a characteristic scale $l_p$

$$C(|x - y|) = \langle e(x) \cdot e(y) \rangle \propto \exp(-|x - y|/l_p) \tag{2.1}$$

(see Question 4 on p.18). The physical meaning of $l_p$ is that if one walks along the chain, after a distance of order $l_p$ the direction where the chain is pointing is essentially uncorrelated to the direction at the starting point. An equivalent statement is that the persistence length counts how short a segment can bend considerably (e.g. in a circle) by a fluctuation of order $k_B T$. Clearly this is a measure of the stiffness of the molecule. For example, consider a simple mechanical model in which the molecule behaves like an elastic beam; this is appropriate for double-stranded DNA, for example. The work per unit length necessary to bend a beam through a curvature $1/R$ is:

$$\frac{F}{l} = \frac{B}{R^2} \tag{2.2}$$

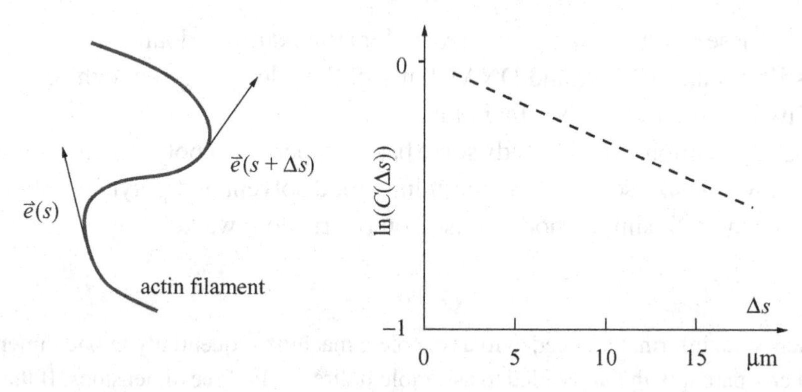


Figure 2.5. Measurement of the persistence length of actin (Ott *et al.*, 1993). One takes the average of the cosines of the angle between the tangent vectors separated by the contour length $\Delta s$. This will decrease exponentially with contour length, with a characteristic length called the persistence length of the polymer.

(quadratic in the curvature); $B$ is called the bending modulus. Thus a fluctuation $F \sim k_B T$ will bend a length $l_p$ into a circle ($R \sim l_p$) for:

$$l_p \sim \frac{B}{k_B T} \tag{2.3}$$

which shows the relation of the persistence length $l_p$ to the elastic parameter $B$.

For very flexible polymers the persistence length is normally inferred from the size of the coil determined in a scattering experiment. For stiffer polymers, however, it is sometimes possible to measure the persistence length directly from the relation in Eq. (2.1); this has been done in the case of polymerized actin; see Fig. 2.5 (Ott *et al.*, 1993).

Actin is a polymer made of polymers. The actin monomer is a 375 residue globular protein; it polymerizes in a two-stranded helix of 8 nm diameter that can be many micrometers (microns) long. Such actin filaments form part of the cytoskeleton, and are also a component of the muscle contraction system. Because the polymer is so long and stiff, its contour can be visualized by fluorescence microscopy. It is remarkable that one can thus directly "see" a single molecule! In such experiments one observes, at any given time, a conformation of the molecule as depicted in Fig. 2.5. From the images one constructs the correlation function (Eq. (2.1)), averaged over an ensemble of conformations. The plot of $\ln(\langle e(x) \cdot e(y) \rangle)$ vs $|x - y|$ is linear (Fig. 2.5), and from the slope one obtains a value for $l_p$, which for actin is $l_p \approx 17$ μm. Polymer stiffness varies over quite a range; for comparison, the persistence length is ∼10 μm for polymerized actin, ∼100 nm for double-strand DNA, and ∼1 nm for single-strand DNA; see also Figs. 2.6 and 2.7.

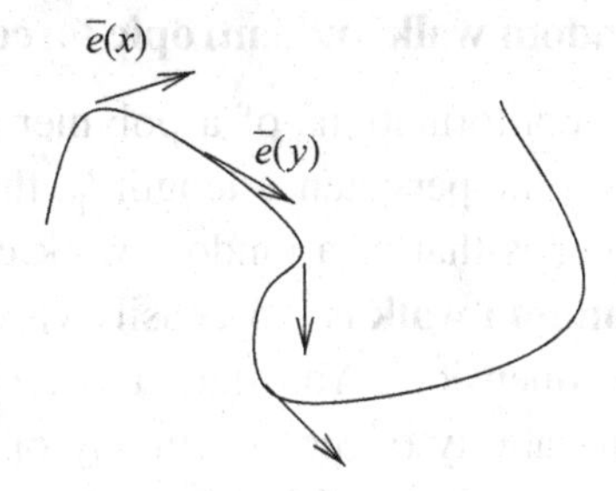


Figure 2.6. Definition of the persistence length of a polymer, with unit vectors at position $x$ and position $y$ along the backbone indicated. When the distance $x-y$ between two unit vectors is increased, the directions of the vectors become uncorrelated. The scale over which this happens is the persistence length of the polymer.

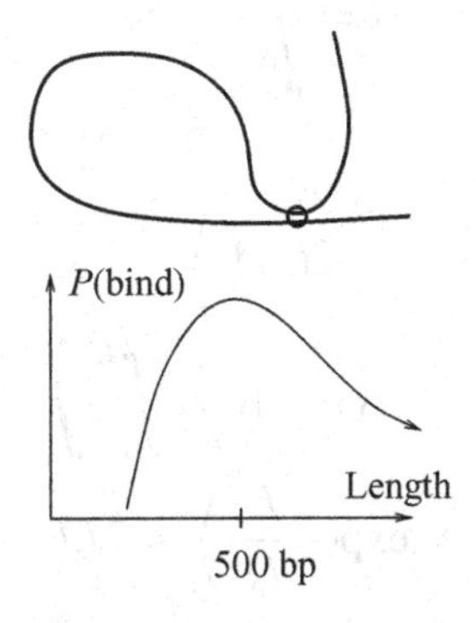


Figure 2.7. DNA has a persistence length of about 200 base pairs. DNA loops in a living cell can result in non-local control of gene expression through DNA binding proteins, which bind simultaneously to distant binding sites along the DNA. The figure illustrates that there is an optimal loop size for such binding. For shorter loops, significant energy is used to bend the DNA; for longer loops the entropy loss becomes increasingly prohibitive. The effective concentration of sites is maximal for a separation of about 500 base pairs (where it corresponds to an equivalent concentration of 100 nM (Mossing & Record, 1986).

In summary, a polymer is stiff at length scales $L \ll l_{\mathrm{p}}$, and flexible at length scales $L \gg l_{\mathrm{p}}$.

What are the different states of polymeric matter? In this book we encounter mainly polymers in solution, whereas the one-component system (consisting only of the polymer) can be in the liquid (polymer melt) or solid state. The solid can be a crystal or, more commonly, a glass, or something in between. But there are other states of matter that are peculiar to polymers. One can have a cross-linked polymer melt (polymer network), for example rubber. The long polymer molecules can bundle together to form fibers. Finally, a cross-linked polymer in solution forms a gel (for example, jelly).

**Random walk and entropic forces**

We now come back to the conformations of a polymer chain. For length scales that are large compared with the persistence length $l_p$, the simplest picture of the conformation of the polymer is that of a random walk of step size given by the Kuhn length $l_k \approx 2l_p$. A **random walk** is most easily visualized on a lattice, let us say a square lattice in two dimensions. You start at a site and walk for $N$ steps; at each step there is equal probability of moving in any of the four directions. The question we want to answer is: what is the average size of the walk (the average end-to-end distance, EED)? This will also be, in the ideal chain approximation, the size of a polymer chain of $N$ links.

To investigate the random walk properties of a chain we express the EED as:

$$\mathbf{R} = \int_0^L \vec{e}(s)\mathrm{d}s \tag{2.4}$$

and can then calculate

$$\langle R^2 \rangle = \left\langle \int_0^L \vec{e}(t)\mathrm{d}t \cdot \int_0^L \vec{e}(s)\mathrm{d}s \right\rangle \tag{2.5}$$

$$= \int_0^L \int_0^L \langle \vec{e}(t) \cdot \vec{e}(s) \rangle \mathrm{d}t\mathrm{d}s = \int_0^L \int_0^{L-s} \exp(-t/l_p)\mathrm{d}t\mathrm{d}s \tag{2.6}$$

$$= 2l_p \left( \frac{L}{l_p} - 1 + \exp(-\frac{L}{l_p}) \right) \approx 2Ll_p \quad \text{for} \quad L \gg l_p \tag{2.7}$$

Alternatively one may consider a random walk on a square lattice with step length $l_k$. In that case the end point is given by

$$\mathbf{R} = \sum_{i=1}^{N} \mathbf{l}_i \tag{2.8}$$

where $\{\mathbf{l}_i\colon i = 1, 2, \ldots, N\}$ is the realization of the walk (each $l_i$ of length $l_k$). If we average over an ensemble of realizations, obviously $\langle \mathbf{R} \rangle = \mathbf{0}$ because of the symmetry, but the typical extension of the walk $\langle |\mathbf{R}|^2 \rangle^{1/2}$ is given by

$$\langle |\mathbf{R}|^2 \rangle = \left\langle \sum_{i,j} \mathbf{l}_i \cdot \mathbf{l}_j \right\rangle = \sum_{i=1}^{N} \langle |\mathbf{l}_i|^2 \rangle = N l_k^2 = L l_k \tag{2.9}$$

where we have used $\langle \mathbf{l}_i \cdot \mathbf{l}_j \rangle = 0$ for $i \neq j$ because the steps are uncorrelated. Comparing Eq. (2.7) with Eq. (2.9) we see that one can define the Kuhn length

$$l_k = \langle R^2 \rangle / L \approx 2l_p \tag{2.10}$$

as the characteristic step length for a random walk along the polymer chain.

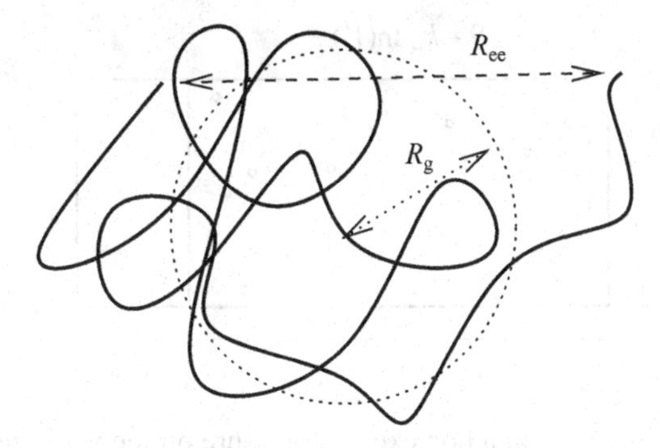


Figure 2.8. Illustration of polymer with end-to-end distance $R = R_{ee}$ and radius $R_g$ indicated. In the standard treatment we will ignore this quantitative difference. However, one should keep in mind that the measure of polymer extension $R$ involves the entire polymer, and not only its ends.

Thus for a polymer of contour length $L \gg l_k$ (and thus also $\gg l_p$) the simplest description of the conformation is in terms of a random walk of step size $l_k = 2l_p$ and with an end-to-end distance EED:

$$R = l_k\sqrt{L/l_k} \tag{2.11}$$

For the chain tracing a random walk, the end-to-end distance $R = \sqrt{\langle R^2 \rangle}$ differs from the radius of gyration $R_g = \langle (1/N)\sum_i (\mathbf{r_i} - \langle \mathbf{r} \rangle)^2 \rangle^{1/2}$ by a constant factor

$$R = \sqrt{6}\, R_g \tag{2.12}$$

The difference is illustrated in Fig. 2.8.

As an example, consider the single DNA molecule of the bacterium *E. coli*. The molecule is about $5 \times 10^6$ base pairs (bp) long; the persistence length of DNA is about $l_p = 200$ base pairs (~60 nm) (see Fig. 2.7). The Kuhn length is 400 bp (or 120 nm) and thus we obtain $R = 60 \times \sqrt{5 \times 10^6/400}$ nm ~6 μm. Thus DNA inside the bacterium, which has a volume 1 μm³, is much more condensed than it would be outside the cell.

The random walk picture is only true until self-avoidance of the polymer begins to count. This happens when the polymer becomes long. If we assume that each monomer is a hard sphere of radius $b$ then the polymer may be viewed as a system composed of $N = L/l_k$ reduced monomer units, each with an excluded volume $v = l_k b^2$. In the next section we will investigate the behavior of such polymers. We will do this by using entropy $S \propto \ln(\text{number of states})$, and the fact that entropy is a source for expansion $X$ and thus for a force $\propto dS/dX$:

$$\text{Entropy} \rightarrow \text{Force} \tag{2.13}$$

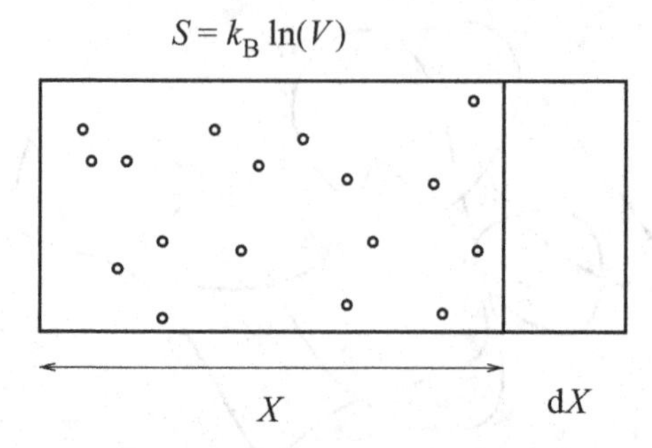


Figure 2.9. The ideal gas in a box exerts a pressure on the walls because entropy increases with the volume.

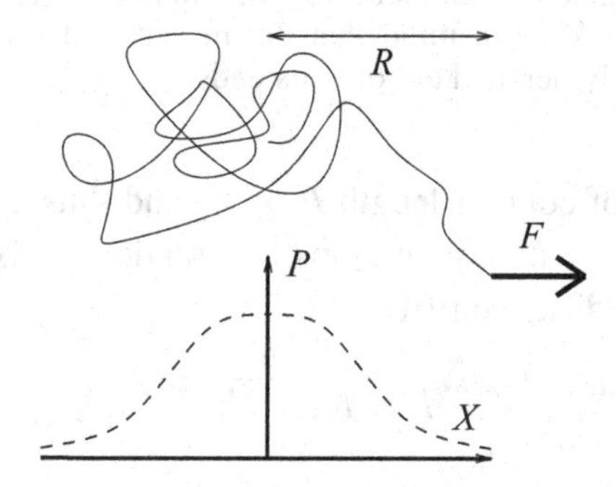


Figure 2.10. Number of confirmations $\mathcal{N}$ of a polymer as function of end-to-end distance $X$. Note that $\mathcal{N} \propto P(X)$, where $P$ is the probability. When the polymer is stretched to distance $X$ it therefore exerts a force $= -T(\mathrm{d}S/\mathrm{d}X) = T(\mathrm{d}\ln(P)/\mathrm{d}X) \propto -k_B T \cdot (3X/Nl_k^2)$ on its surroundings. Thus the polymer acts as a spring with spring constant $k_{\text{spring}} = 3k_B T/(Ll_k)$. Typical entropic spring constants are a fraction of a pico Newton per nanometer (pN/nm).

Entropic forces are familiar from simple systems such as an ideal gas confined in a volume (Fig. 2.9). The force exerted by the gas on the container is purely entropic, because the energy of the ideal gas does not depend on the volume. In fact, an ideal gas with $N$ particles in the volume $V$ has entropy $S = \text{constant} + k_B\, N \ln(V)$. Thus it exerts a force on its surroundings given by

$$\text{Force} = -\frac{\mathrm{d}F}{\mathrm{d}X} = T\frac{\mathrm{d}S}{\mathrm{d}X} = TA\frac{\mathrm{d}S}{\mathrm{d}V} = k_B\, TAN/V = pA \tag{2.14}$$

where $p$ is the ideal gas pressure and $A$ is the area that encloses it; $F = \text{const.} - TS$ is the free energy. Similarly we will see that a polymer behaves as an entropic coil, which tends to be extended to a size that maximizes the number of microstates of the polymer. The essence of the argument, which we present in more detail in the next section, is as follows.

Consider the conformation of a polymer coil as a random walk (see Fig. 2.10). The probability distribution for the end-to-end distance (EED) **R** is Gaussian, because

**R** is the sum of many random variables. Thus

$$P(\mathbf{R}) \sim \mathrm{e}^{-\frac{3R^2}{2\langle R^2\rangle}} \tag{2.15}$$

and we have already seen that $\langle R^2\rangle = Nl_\mathrm{k}^2$. The probability $P(\mathbf{R})$ is proportional to the number of states $\Gamma$ with end-to-end distance $\mathbf{R}$: $\Gamma(\mathbf{R}) \propto P(\mathbf{R})$, and the entropy is $S = k_\mathrm{B} \ln \Gamma$. Therefore

$$S(\mathbf{R}) = -k_\mathrm{B}\frac{3R^2}{2Nl_\mathrm{k}^2} + \text{constant} \tag{2.16}$$

So even if the energy $E$ is completely independent of conformation we still obtain a free energy that depends (quadratically) on $R$:

$$F(\mathbf{R}) = k_\mathrm{B}T\frac{3R^2}{2Nl_\mathrm{k}^2} \tag{2.17}$$

This means that the polymer chain behaves like a spring (an "entropic" spring):

$$\text{restoring force} = -\frac{\mathrm{d}F}{\mathrm{d}R} = -\frac{3k_\mathrm{B}T}{Nl_\mathrm{k}^2}R \tag{2.18}$$

$$\text{spring constant} = \frac{3k_\mathrm{B}T}{Nl_\mathrm{k}^2} \tag{2.19}$$

This effect is the reason for the incredible elasticity of rubber, and as we see from Eq. (2.19), it is temperature dependent (the spring becoming stiffer at higher temperature!).

A single polymer molecule makes a very soft spring. Consider, for example, a polystyrene coil with $N \sim 10^3$, $l_\mathrm{k} = 1$ nm; the size of the coil is $\langle R^2\rangle^{1/2} = l_\mathrm{k}\, N^{1/2} \sim 30$ nm and the spring constant is $3k_\mathrm{B}T/(Nl_\mathrm{k}^2) \sim 10^{-2}$ pN/nm, since

$$\frac{k_\mathrm{B}T}{1\ \text{nm}} \approx 4\ \text{pN} \quad (\text{at room temperature } T = 300\ \text{K}) \tag{2.20}$$

The entropic elasticity of a single coil can be measured directly through micromechanical experiments (Fig. 2.11).

### *Questions*

(1) Assume that the 5 000 000 base pairs of a long DNA molecule homogeneously occupy a spherical cell volume of 1 $\mu\text{m}^3$. One base pair has a longitudinal extension of about 3.5 Å, and the persistence length is about 200 base pairs.

(a) What would be the root mean square (r.m.s.) radius for the DNA outside a cell? What is the effect of excluded volume on this scale?

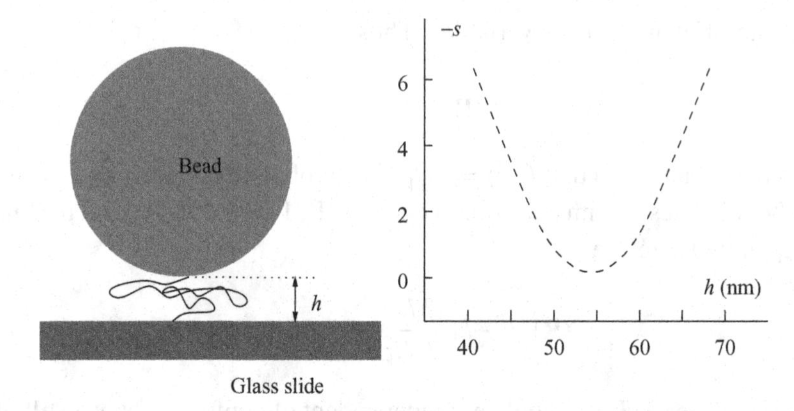


Figure 2.11. Polymer entropic elasticity experiment by Jensenius & Zocchi (1997). A micron-size bead is tethered to a surface by a single polymer coil. The interaction potential between the bead and the surface is measured by analyzing the vertical Brownian motion of the bead. After subtracting the contributions from electrostatic and van der Waals forces (i.e. the measured potential in the absence of the tether), one obtains a parabolic potential, which represents the entropic spring due to the polymer.

(b) What is the average distance between nearby DNA segments inside the cell?
(c) DNA has a radius of about 1 nm. If everything is disordered what is the average distance between nearby DNA intersections in the cell?

(2) Polymers in water can be charged, because of the dissociation of certain groups, e.g carboxyl ($-COOH \leftrightarrow COO^- + H^+$) and amino groups ($-NH_2 + H^+ \leftrightarrow NH_3^+$) in proteins. From the electrostatic energy of the dissociated pairs, explain why this happens only in water.
(3) Suppose there are two preferred conformations (energy minima) for the bond rotation $\phi$ in Fig. 2.4, with an energy difference $\epsilon$. If $a$ is the monomer size, give an expression for the persistence length in terms of $\epsilon$ and $a$.
(4) Consider the following model of a polymer in two dimensions:
   - independent segments of length $a$,
   - successive segments make a fixed angle $\gamma$ with each other (either $+\gamma$ or $-\gamma$ with equal probability).

   Thus, $\gamma$ is the bond angle and rotations by $\pi$ around the bonds are allowed. Show that the angular correlation function $\langle \mathbf{n}(0) \cdot \mathbf{n}(s) \rangle$ decays exponentially along the chain and calculate the persistence length ($\mathbf{n}(s)$ is the unit vector tangent to the polymer at position $s$).

## Homopolymer: scaling and collapse

Here we give a somewhat more sophisticated modeling of polymer conformations that takes into account monomer–monomer interactions. This allows us to obtain the Flory scaling (which describes the behavior of a real polymer in a good solvent quite well), and a description of the polymer collapse transition in bad solvents.

At high temperatures or in good solvents homopolymers behave essentially like a random walk. However, this is not entirely true, because a polymer conformation is not allowed to cross the same space point twice. This is called the excluded volume effect, and it accounts for the fact that the polymer effectively occupies a larger volume (is "swollen") compared with a random walk. We will now repeat Flory's classic argument for polymer scaling, in a way that allows us to generalize to the case where there is an attractive interaction between the monomers. Subsequently we present the classical theory according to DeGennes for polymer collapse.

The physics here is to count the number of states of a self-avoiding polymer, as a function of its extension $R$, and then to find the value of $R$ where the number of states is maximal. The number of states as function of $R$ is a Gaussian, $\exp(-R^2/N)$, supplemented with a reduction factor of order $(1 - v/R^3)^N$ because self-crossing is suppressed ($v$ is a microscopic volume). The famous Flory scaling then follows from maximizing this product, whereas the discussion of collapse follows when one includes a two-body attraction described by an energy term $\propto N^2/R^3$.

First consider a random walk in one dimension (see Fig. 2.12). The number of walks with $N_+$ steps to the right and $N_-$ steps to the left is

$$\mathcal{N}_1 = \frac{N!}{N_+!N_-!} \tag{2.21}$$

where $N = N_+ + N_-$ is the total number of steps and the resulting net number of steps to the right $x = N_+ - N_-$. Thus the number of states of a polymer of $N$ units

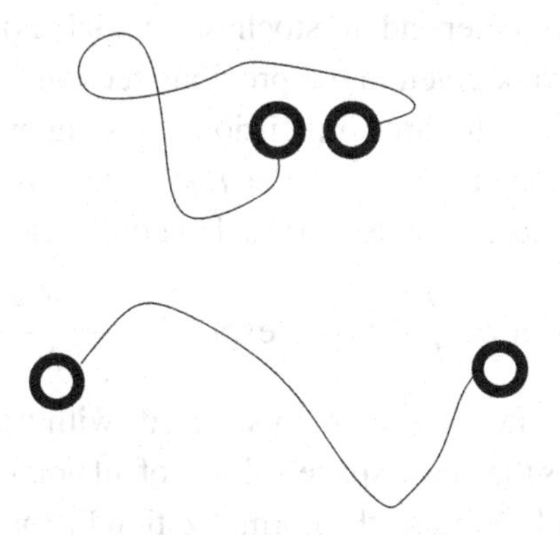

Figure 2.12. The number of ways to go from A to B, given that we must take, say, $N$ steps, decreases greatly as the distance between A and B increases. This is why homopolymers acts as entropic springs, with a tendency to keep end-to-end distances to a minimum in one dimension.

that is stretched to a position $x$ in one dimension is

$$\mathcal{N}_1 = \frac{N!}{((N-x)/2)!((N+x)/2)!} \tag{2.22}$$

Using Sterling's formula ($n! = (n/\mathrm{e})^n$ up to factors of order $\sqrt{n}$) this is rewritten as

$$\mathcal{N}_1 = \frac{(2N)^N}{(N-x)^{(N-x)/2}(N+x)^{(N+x)/2}} \tag{2.23}$$

Rewriting again by dividing with $N^N$ in both nominator and denominator, we obtain

$$\mathcal{N}_1 = \frac{2^N}{(1-x/N)^{(N-x)/2}(1+x/N)^{(N+x)/2}} \tag{2.24}$$

Accordingly

$$\ln(\mathcal{N}_1) = N\left(\ln(2) - \frac{1}{2}(1-x/N)\ln(1-x/N) - \frac{1}{2}(1+x/N)\ln(1+x/N)\right) \tag{2.25}$$

which approximately (to first order in $x/N$, using $\ln(1+x) = x - x^2/2$) is

$$\ln(\mathcal{N}_1) = N(\ln(2) - \frac{1}{2}(\frac{x}{N})^2) \tag{2.26}$$

or, in fact, the standard Gaussian approximation gives for the number of states associated with the end-to-end distance $x$:

$$\mathcal{N}_1 = 2^N \exp(-\frac{x^2}{2N}) \tag{2.27}$$

This formula means that the probability distribution for the EED is a Gaussian of width $\sqrt{N}$ (i.e. $\langle x^2 \rangle = N$). This also follows from the Central Limit Theorem (the EED is the sum of many independent stochastic variables) and the calculation of the size of the random walk given in the previous section.

If we now want to extend to three dimensions keeping overall end-to-end length at position $x = R_x/l_k$ along the $x$-axis, $y = R_y/l_k$ along the $y$-axis and $z = R_z/l_k$ along the $z$-axis, the number of states for such a configuration is

$$\mathcal{N}(R)\mathrm{d}R \propto \int \mathrm{d}x\mathrm{d}y\mathrm{d}z \exp(-\frac{(x^2+y^2+z^2)}{2N}) \tag{2.28}$$

where we integrate over all volume elements $\mathrm{d}x\mathrm{d}y\mathrm{d}z$ within the radius $R$ to $R + \mathrm{d}R$. Thus when we re-express the phase space volume of all points within $R$ and $R + \mathrm{d}R$, include the Kuhn length $l_k$ directly, the normalization factors, and take into account the coordination number $C$ and polymer size $\sqrt{N}$, we have

$$\mathcal{N}_{\text{free}}(R) = \frac{4\pi C^N}{(2\pi N)^{3/2}}(R/l_k)^2 \exp(-\frac{3(R/l_k)^2}{2N}) \tag{2.29}$$

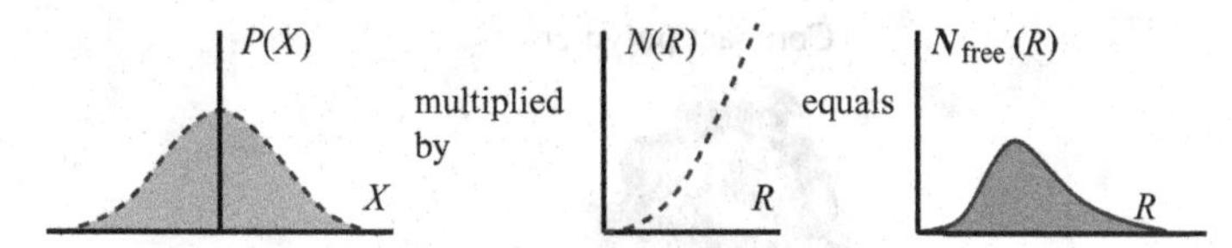


Figure 2.13. The number of states to extend a polymer to an end-to-end distance $R$ is given by the probability of reaching a specific point at distance $R$ multiplied by number of points at distance $R$. This is how we obtain Eq. (2.29).

with $R^2 = R_x^2 + R_y^2 + R_z^2$; see Fig. 2.13. The factor 3 in $3R^2$ takes into account that of the $N$ possible moves along the polymer, $N/3$ are available for movement in each direction. For simplicity in the following we write $N$ for $L/l_k$. $C$ is the coordination number; that means the number of possible turns of the polymer at each step along the polymer. For one dimension $C = 2$, whereas for a three-dimensional cubic lattice we would normally set $C = 6$.

All this was for the ideal polymer chain on a square lattice (where we chose one of three dimensions to move in for each step along the chain, giving $C = 6$ possible moves). Now we include corrections due to self-avoidance. Assume that the free chain has $C$ options for each subsequent link. With self-avoidance this is immediately reduced to $C - 1$. Further, following Flory, we count the number of available spots for the polymer as we lay it down, counting at each stage the average occupation on the lattice:

- the first element can be anywhere, and the acceptance probability is 1;
- the second element can be everywhere except where the first was: the acceptance probability is $1 - v/V$, where $v$ is the excluded volume per monomer;
- the third element can be everywhere except at the positions of the first two, and the acceptance probability is $1 - 2v/V$; and so forth.

This leads to the following overall phase space reduction factor for a polymer with $N$ monomers each filling a volume $v$, the overall polymer being confined into a volume $V$:

$$\chi = 1(1 - \frac{v}{V})(1 - \frac{2v}{V})(1 - \frac{3v}{V}) \cdots (1 - \frac{(N-1)v}{V}) \tag{2.30}$$

which should then be multiplied by the above $\mathcal{N}$, with volume $V = R^3$. Let us rewrite $\chi$ by multiplying and dividing by $V^N$, $v^N$:

$$\chi = \frac{v^{N-1}}{V^{N-1}} \frac{V}{v} (\frac{V}{v} - 1)(\frac{V}{v} - 2)(\frac{V}{v} - 3) \cdots (\frac{V}{v} - (N-1)) \tag{2.31}$$

Compact polymer

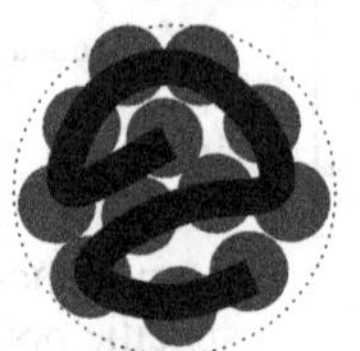

Number of states = $((C-1)/\mathrm{e})^N$

Figure 2.14. The number of different conformations of a compact polymer is $((C-1)/e)^N$, where $N$ is the number of independent segments and $C-1$ is the maximum number of new states that each consecutive segment can be in.

Because $V/v \geq N$ we can also write this as

$$\chi = \frac{(V/v)!v^N}{(V/v-N)!V^N} = \mathrm{e}^{-N}\left(\frac{V/v}{V/v-N}\right)^{V/v-N} \tag{2.32}$$

where we have used Sterling's formula in the last equality.

For a **compact polymer**, that is where the $N$ monomers each of volume $v$ exactly fill the volume $V$, i.e. $V/v = N$ (and $R \propto N^{1/3}$), Eq. (2.31) reduces to $\chi = \mathrm{e}^{-N}$; see Fig. 2.14. The total possible overlapping or non-overlapping configurations are

$$\mathcal{N}_{\text{free}} = (C-1)^{N-1} \tag{2.33}$$

which reflects the fact that we first take one monomer, and then add subsequent ones in any of the remaining $C-1$ directions. Thus this relation has been obtained from direct counting without regard to the end-to-end distance. Ignoring small corrections due to differences between $N$ and $N-1$, we can now count the number of states of a compact polymer:

$$\mathcal{N}(\text{compact polymer}) = \mathcal{N}_{\text{free}}\ \chi = \left(\frac{C-1}{\mathrm{e}}\right)^N \tag{2.34}$$

which was derived by Flory counting the number of possible states of a compact polymer of length $N$ (in units of the Kuhn length).

For a **non-compact** polymer, $V \gg Nv$ and the sum of all monomers' hard core volumes fills only a small part of the total volume occupied by the polymer. Then Eq. (2.32) gives

$$\chi = \mathrm{e}^{-N}(1-\frac{Nv}{V})^{N-V/v} \sim \exp(-v\frac{N^2}{R^3}) \tag{2.35}$$

where in the last equality we have used $V = R^3$, and $R$ must now take the meaning of the radius of gyration of the polymer; see Fig. 2.15. Notice that, on the contrary, the first term of our free energy, $\propto \exp(-R^2)$, was deduced for the end-to-end

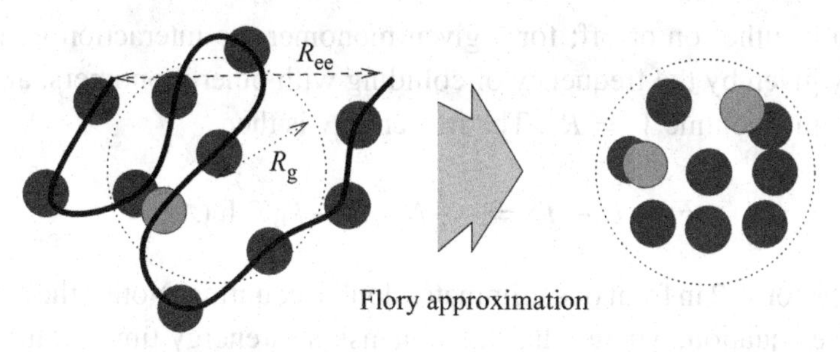


Figure 2.15. The Flory estimate of the number of self-crossings of a polymer. The polymer is subdivided into elements, each with volume $v$ given by the Kuhn length (multiplied by the cross section of the polymer). Each of these elements is assigned an independent position within a volume $V$ set by the end-to-end distance $R$ (or rather the radius of gyration). The probability of avoiding overlap is calculated from noting that each element has the chance $(1 - Nv/V)$ of avoiding overlapping with other elements. The probability that none of the elements overlaps is then $(1 - Nv/V)^N \sim e^{-N^2 v/V}$.

distance. If we simply assume that the end-to-end distance is representative of the radius of gyration there is no problem, but that is not always true.

The equilibrium size of the coil is now found by maximizing the product of the Gaussian chain with the penalty for self-exclusion for a polymer confined to be within the end-to-end distance:

$$\mathcal{N} \propto \frac{R^2}{l_k^2} \cdot \exp(-\frac{3(R/l_k)^2}{2N}) \cdot \exp(-v\frac{N^2}{R^3}) \tag{2.36}$$

Here we ignore prefactors that do not depend on $R$ because they do not contribute to the derivative below. Setting

$$\mathrm{d}\mathcal{N}/\mathrm{d}R = 0 \Rightarrow \frac{2}{R} - \frac{3R}{Nl_k^2} + 3\frac{vN^2}{R^4} = 0 \tag{2.37}$$

we can examine the terms for large $N$. Assuming the last two terms cancel each other we obtain $R \propto N^{3/5}$, which is the Flory scaling. This result is consistent for large $N$, because it implies that the first term, $2/R \propto N^{-3/5}$, decreases faster toward zero than the other two terms, both $\propto N^{-2/5}$. We will now derive this result by expressing everything in terms of free energies, thereby in addition opening a discussion of polymer collapse.

We parametrize the monomer–monomer interaction by the quantity $\epsilon$:

$$\epsilon = -\int \mathrm{d}^3 x U(x) \tag{2.38}$$

where $U$ is the two-body short-range interaction potential (negative $U \rightarrow$ positive $\epsilon$ means attraction). This allows a description in which the monomer–monomer interaction is either on or off; for a given monomer the interaction is on, with a probability given by the frequency of colliding with other monomers, at a density $\rho = N/V$, in a volume $V = R^3$. The free energy is then

$$F = E - TS = -\frac{1}{2}N\ \rho\ \epsilon - k_B T\ \ln(\mathcal{N}) \tag{2.39}$$

where the factor $1/2$ in front of $\epsilon$ eliminates double counting. Notice the dimensions in the above equation, where $\epsilon$ has the dimension of energy times volume. Again writing only the $R$-dependent parts we obtain

$$\frac{F}{k_B T} = -\frac{\epsilon}{2k_B T}\frac{N^2}{R^3} \tag{2.40}$$

$$+\frac{3}{2}\frac{R^2}{N l_k^2} - 2\ln(\frac{R}{l_k}) + (\frac{R^3}{v} - N)\ln(1 - \frac{vN}{R^3}) \tag{2.41}$$

where we remind the reader that $\epsilon > 0$ means attraction.

For $R^3 \gg vN$ the formula can be rewritten (removing terms that do not depend on $R$) as

$$\frac{F}{k_B T} = +\frac{1}{2}(1 - \frac{\epsilon}{v k_B T})v\frac{N^2}{R^3} + \frac{3}{2}\frac{R^2}{N l_k^2} + \frac{v^2}{6}\frac{N^3}{R^6} - 2\ln(\frac{R}{l_k}) \tag{2.42}$$

where we expand the logarithm to third order (since we keep three-body interaction terms): $\ln(1 + x) = x - x^2/2 + x^3/3$, i.e. $\ln(1 - (Nv/V)) = -((Nv/V) - \frac{1}{2}(Nv/V)^2 - \frac{1}{3}(Nv/V)^3)$. Figure 2.16 illustrates the different terms in this expansion. Notice that (1) the mean field excluded volume acts as a positive (repulsive) interaction term between a monomer and the rest of the polymer, proportional to $N \cdot N/R^3$; and (2) the Flory expansion also provides a repulsive three-body potential where each monomer interacts with intersections of the polymer with itself (with density $N^2/R^6$).

Now let us consider the lessons to be drawn from the above mean field expression (mean field because the repulsion is treated by ignoring those correlations between monomers that should exist because of their position along the polymer). In all cases we will compare the leading term that tends to expand the polymer (positive pressure outwards, or $dF/dR < 0$) with the leading term that favors a compact polymer.

(1) **Flory scaling.** The expression $\delta = 1 - (\epsilon/k_B T v)$ quantifies the net two-body interaction between monomers. When it is positive the hard core repulsion wins; when it is negative the attraction wins. At $\delta = 0$ we are at the theta ($\Theta$) point. For $\delta > 0$ one obtains Flory scaling by differentiation of Eq. (2.42) and subsequently we consider the balance between the self-avoidance term $\delta \cdot (N^2/R^3)$ and the Gaussian limit on

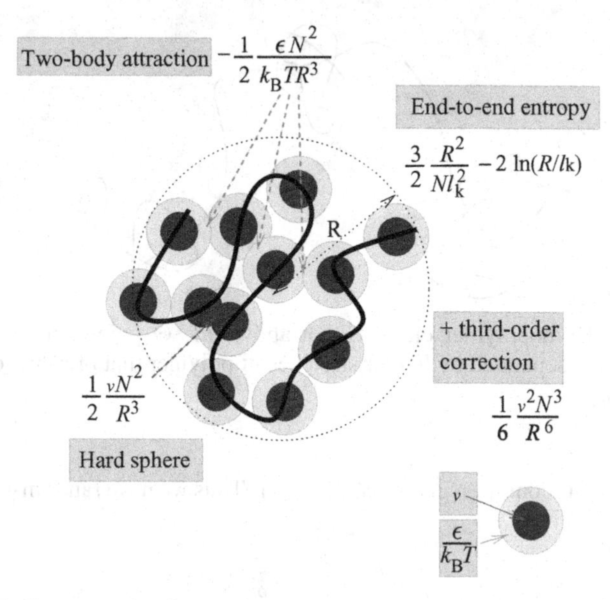


Figure 2.16. Free energies for a polymer with attraction between its elements. The polymer is subdivided into elements each of size $v$ that are assigned a hard core repulsion, and an attraction (indicated by the light-shaded region around the dark spheres). The third-order correction is a repulsive (positive) term corresponding to simultaneous overlap of three spheres of radius $v$. In the bottom-right corner we illustrate the two ingredients in polymer collapse, the hard sphere volume $v$ and the effective size of the attraction "volume" $\epsilon/(k_B T)$.

expansion $-R^2/N$: $R \propto N^{3/(2+d)}$. This will be valid for large $N$, where for $\delta > 0$, the hard core will always dominate. Simulations of large polymers in three dimensions give $R \propto N^{0.58\pm0.005}$, whereas simulations in two dimensions are indistinguishable from the Flory scaling result (Li *et al.*, 1995). The Flory scaling describes a polymer that is considerably swollen with respect to the "ideal" random coil result ($R \propto N^{1/2}$). This is not a small difference: for example, for $N = 100$ we have $(100)^{0.5} = 10$, but $(100)^{0.6} \approx 16$! This is the case of a polymer in a good solvent ($\epsilon$ small, or even $\epsilon < 0$); from the expression for $\delta$ we see that solvent quality can also be controlled by temperature (high $T \rightarrow$ good solvent).

(2) **Theta point**. For $\delta = 1 - (\epsilon/k_B T v) = 0$ the balance is determined by the phase space expansion term $2\ln(R)$ against the $R^2/N$ term. This gives the random-walk scaling for the polymer

$$R \propto N^{1/2} \tag{2.43}$$

The temperature at which this happens is called the $\Theta$ point. For finite but small $\delta \geq 0$ one has random-walk scaling out to a certain scale. In fact as long as $\delta N^2/R^3 < R^2/(N l_k^2)$ the self-repulsion can be ignored and the differential of the $R^2$ term should balance the differential of the logarithmic term. As long as this is the case,

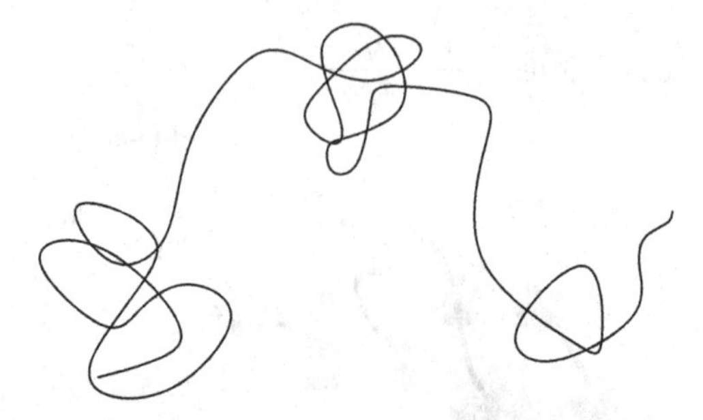

Figure 2.17. Homopolymer close to, but above, the $\Theta$ point. Small regions of random coil are separated by longer stretches of polymer that prevents overlap on larger scales.

$R \sim \sqrt{N}$ and this condition gives $\delta N^{2-3/2} < 1$. Thus we have random-walk scaling for

$$N < n^* = \frac{1}{\delta^2} \tag{2.44}$$

For $N > n^*$, Flory scaling takes over, giving a picture of compact random-walk blobs, separated by stretches of self-avoiding polymer, see Fig. 2.17.

(3) **Homopolymer collapse**. For $\delta < 0$ attraction dominates until the hard core repulsion stops the collapse. At this point we have a compact polymer, and the term $Nv/R^3$ is of order 1, i.e. when the monomers are hard packed. Then $R = N^{1/3}$, making the collapsed state distinctly different from the random walk coil at $\delta = 0$. A similar investigation in two dimensions ($d = 2$) would give $R = N^{1/2}$, which is the same scaling as for the random walk (because the random walk is space filling in two dimensions, and accordingly there is no transition from random coil to collapsed state). Thus a transition exists only for $d > 2$. However, whether the transition is gradual or sharp depends on the coefficient of the $N^3/R^6$ term (DeGennes, 1985). In fact if there was no such hard core repulsion, Eq. (2.42) would predict infinite negative $F$ for $R \to 0$, and thus complete collapse. Following DeGennes we use

$$r = \frac{R}{\sqrt{N} l_k} \tag{2.45}$$

and the free energy for $d = 3$ is rewritten (with $v' = v/l_k^3$) as

$$\frac{F}{k_B T} = +\frac{1}{2}(1 - \frac{\epsilon}{k_B T v'}) \cdot v' \cdot \frac{\sqrt{N}}{r^3} + \frac{v'^2}{6r^6} + \frac{3}{2}r^2 - 2\ln(r) \tag{2.46}$$

or

$$\frac{F}{k_B T} = \frac{3}{2}r^2 - 2\ln(r) + \frac{W_1(T)}{2} \cdot \frac{\sqrt{N}}{r^3} + \frac{W_2}{2}\frac{1}{r^6} \tag{2.47}$$

where $W_1 = W_1(T) = 1 - \epsilon/(k_B T v') \propto T - \Theta$, for $T$ close to the theta point $\Theta = \epsilon/v' = \epsilon l_k^3/v$ and where

$$W_2 = \frac{v'^2}{3} = \frac{1}{3}\left(\frac{v}{l_k^3}\right)^3 \tag{2.48}$$

accounts for the hard core repulsion (all three-body interactions).

Equation (2.47) also follows from an expansion in powers of the monomer density, where the first two terms are entropic, the third term represents monomer–monomer interactions, and the fourth term represents a three-body interaction.

For any given temperature $T$ (value of $W_1$) the free energy $F$ has at least one minimum as a function of polymer extension $r$. If, for a given temperature, we set the derivative $\mathrm{d}F/\mathrm{d}r = 0$ (equivalent of setting the pressure equal to zero), we obtain a possible equilibrium value of $r$ as

$$3r - \frac{2}{r} - \frac{3\sqrt{N}}{2}\frac{W_1}{r^4} - \frac{3W_2}{r^7} = 0 \tag{2.49}$$

This has unique solutions for most values of $W_1$, but not all! When there are two solutions the system can be in two local minima at the corresponding temperature. In the left-hand part of Fig. 2.18 one sees the existence of such degenerate solutions as a local minimum at a dilute state (large $r$) and at a high-density state for some temperatures around the $\Theta$ temperature where $W_1 \sim 0$. When there are two solutions, a temperature change may induce a sudden shift between one optimum and the other at an entirely different density. At this point the **first** derivative of $F$ may change discontinuously, and the system then experiences a **first**-order phase transition.

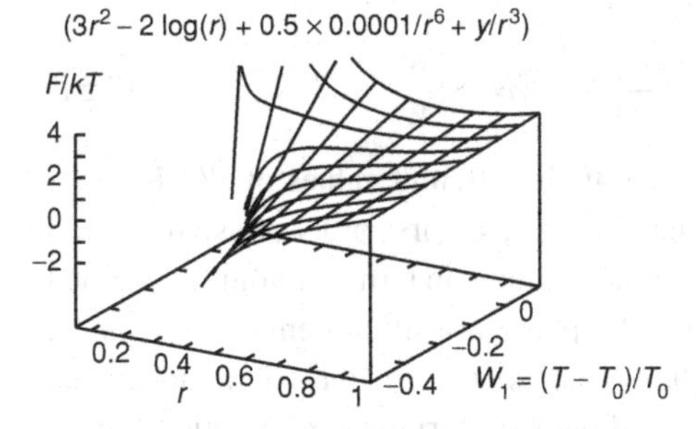


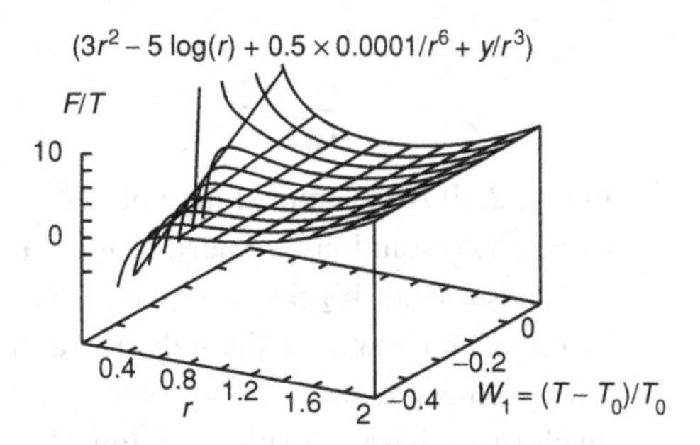


Figure 2.18. The free energy of a homopolymer as a function of radius ($\alpha$) and temperature away from the $\Theta$ point. The figure on the right shows the effect of increased logarithmic suppression of small end-to-end distances. Because there is a barrier due to $\ln(R)$ versus $W_1/R^3$ an increased prefactor for $\ln(R)$ makes an even higher barrier. The prefactor 2 was from the assumption that we identified the radius with end-to-end distance, which is not strictly correct.

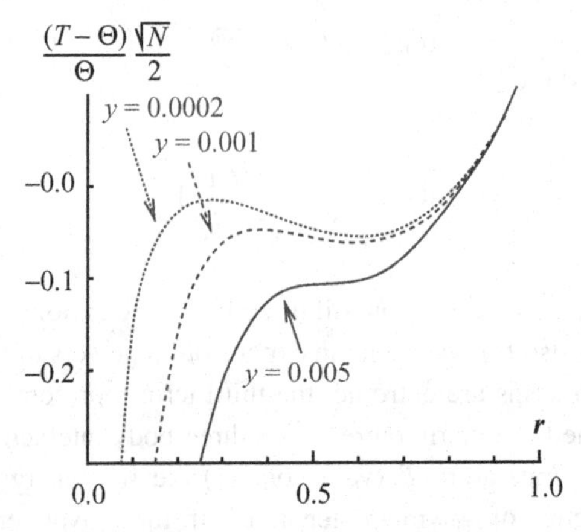


Figure 2.19. Pressure $P = \mathrm{d}F/\mathrm{d}V \propto \mathrm{d}F/\mathrm{d}r = 0$ curves for a homopolymer in the deGennes mean field approach (plotting $x \propto (T - \Theta)$ versus extension $r$ using Eq. (2.50)). The different curves correspond to different values of the hard core repulsion $y$ (i.e. $W_2$). For small $y$ there are three solutions for each $W_1$ and accordingly the system tends to be either at the metastable solution at high $r$ or the metastable solution at low $r$. In that case, the system undergoes a first-order transition at a temperature $T$ close to (but slightly below) the $\Theta$ point.

In Fig. 2.19 we examine the paths in $r$, $W_1$ that correspond to the local thermodynamic optimum. This is done by rewriting Eq. (2.49) in the form

$$x = r^5 - \frac{2}{3}r^3 - \frac{y}{r^3} \tag{2.50}$$

where $y$ $(=W_2)$ contains the three-body interactions, and $x$ contains the two-body effects ($x = W_1\sqrt{N}/2$) and thereby the temperature dependence ($x \propto T - \Theta$). From this we learn that the collapse ($r \ll 1$) happens to a state where $x = -y/r^3$ with $R = N^{1/3}$. In this densely packed state the bulk energy given by the last two terms of Eq. (2.47) is

$$\frac{F_{\text{bulk}}}{k_B T} = \frac{x}{N} + \frac{y}{2N^2} = -\frac{y}{2N^2} = -(N/8)\frac{W_1^2}{W_2} \tag{2.51}$$

In Fig. 2.19 we examined the optimal paths for three different values of $W_2$. In order to have two solutions we need to have a small $W_2$. Thus a first-order phase transition demands a small $W_2 \propto v^2$. As $v$ includes excluded volume in terms of volume spanned by one Kuhn length of the polymer, a first-order phase transition demands persistent thin polymers. However, for proteins with their rather small size, one should not expect much of a phase transition. In Fig. 2.20 we show equilibrium configurations of an $N = 500$ homopolymer simulated by Metropolis sampling at different temperatures around the transition temperature.

The precise value of $W_2$ cannot be derived from this very simplified approach. In particular we emphasize that the nature of the transition is closely linked to the prefactor to the log; that is, our calculation of translational entropy by counting only the end-to-end distance of the polymer. This is probably not realistic, as other segments of the

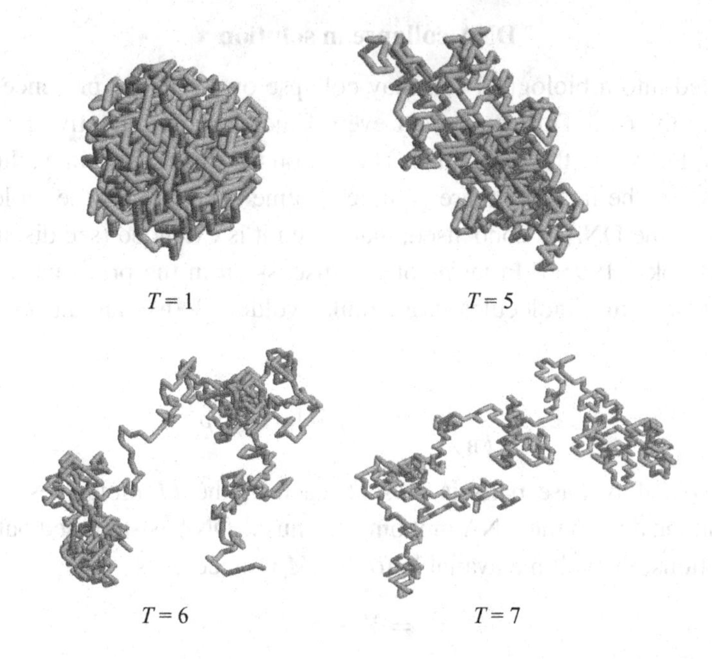


Figure 2.20. Simulation of homopolymer equilibrium configurations at four different temperatures, by J. Borg (personal communication). One sees that the change is fairly gradual with temperature, reflecting a soft transition. The simulation was done with a volume per monomer equal to the distance between the monomers, properties which are reminiscent of amino acids in proteins. The size of the homopolymer was $N = 500$.

polymer also contribute. In the right-hand panel of Fig. 2.18 we show the rather large increase in barrier that one obtains by using a factor 5 in front of $\ln(r)$ in Eq. (2.47), instead of the factor 2.

For experiments on homopolymer collapse see Wu & Wong (1998). They observed a transition with a change in radius of factor 4 when using poly($N$-isopropylacrylamide) with $N$ of order $10^5$. Sun *et al.* (1980) found that a polymer of length 30 does not collapse, one of length 1000 collapses its radius by a factor 3, whereas a polymer of length 250 000 shows a factor 6 reduction in radius. The sharpness of the transition was seen to increase with length.

## *Questions*

(1) Assume that a polymer is confined inside a volume of radius $R$. Argue for the scaling of the number of times that the polymer touches its boundary as $R$ decreases far below $\sqrt{N}$.

(2) What would be an appropriate correction to the free energy of a collapsing polymer if one assumes that the decreased size of the global extension is parametrized in terms of a polymer broken up into uncorrelated polymer segments?

(3) Repeat the DeGennes mean field model for collapse with the hereby modified collapse free energy.

**DNA collapse in solution**

DNA injected into a biological cell may collapse owing to the presence of other molecules in the cell. This may occur even if there are no attractive interactions between the DNA and the molecules. The reason for this resides in excluded volume effects, i.e. the fact that there is more volume available for the molecules in solution when the DNA is condensed, than when it is extended (see discussion by Walter & Brooks (1995)). In terms of our discussion in the previous section the presence of $M$ solute molecules in a cellular volume $V$ has an entropy $S$ given by

$$\exp\left(\frac{S}{k_{\mathrm{B}}}\right) \propto V^M/M! \propto 1/\rho^M \tag{2.52}$$

When a DNA of $N$ base pairs is present, each of the $M$ molecules cannot be within a volume $\delta$ from the DNA monomers. Thus if DNA is stretched out without self-interactions, the volume available for the $M$ molecules is

$$V' = V - N\delta \tag{2.53}$$

which is somewhat smaller than $V$. Here $\delta \sim \pi(r_{\mathrm{DNA}} + r_{\mathrm{protein}})^2 l_{\mathrm{k}}$. Thus the presence of DNA is diminishing the entropy of the solution. Some of this entropy can be regained by letting the DNA bend upon itself, thereby making excluded volumes overlap in the intersections (Fig. 2.21). With a radius $R$ for the DNA one obtains an excluded volume

$$V' = V - N\delta + \frac{N^2}{R^3}\delta^2 \tag{2.54}$$

where $N(N\delta/R^3)$ counts the number of intersections of DNA of length $N$, with the volume $N\delta$ in the volume $R^3$. Note that in the intersection between two volumes each of size $\delta$ randomly placed within volume $R^3$ there is an expected

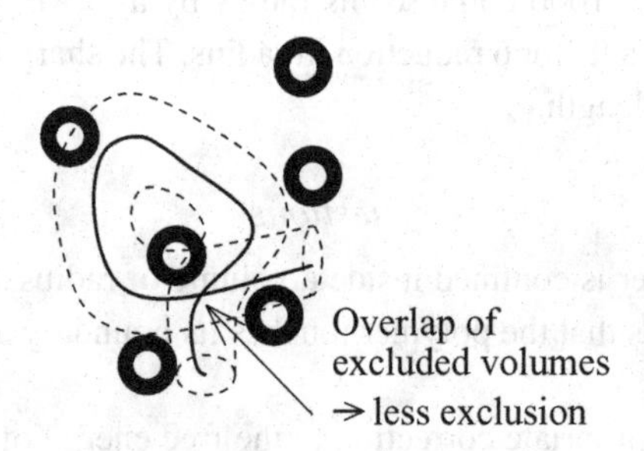


Figure 2.21. The excluded volume effect effectively squeezes the polymer into a more compact state, thereby decreasing the reduction of volume for solute molecules around the polymer.

average overlap of $\delta^2/R^3$. All units are in terms of the Kuhn length $l_k = 2l_p$ of the DNA. Each of these intersections has volume $\delta$, giving the overlapping volume $(N^2/R^3)\delta^2$.

In the following discussion we will ignore the $N\delta$ term because it is much smaller than $V$. That is, $N\delta$ is the volume spanned by the 1 mm DNA in an *E. coli* cell, with a radius of about 5 nm, giving a volume 1000 μm · π(5 nm)$^2$~0.1 μm$^3$ << the $V = 1$ μm$^3$ volume of an *E. coli* cell. Notice that $N\delta$ is compared with $V$ and not the $R$-dependent term $N\delta^2/R^3$. This is because we separate the terms that depend on $R$ from the terms that have no $R$ dependence.

The added free energy due to the excluded volume will be

$$\begin{aligned} F &\sim -k_B T \ln(V'^M/M!) \\ &\sim -k_B T M \ln\left(V\,(1+\frac{N^2\delta^2}{VR^3})\right) + k_B T M \ln(M) - k_B T M \ln(e) \end{aligned} \tag{2.55}$$

Expanding the log we obtain the polymer-dependent part

$$F_{\text{solute}} \sim -k_B T \frac{(N\delta)^2}{R^3}\rho \tag{2.56}$$

where $N^2\delta^2/R^3$ counts the overlapping excluded volume, which multiplied by $\rho = M/V$ then counts the number of doubly excluded solute molecules. Notice that this free energy acts as an attractive potential between the DNA monomers, i.e. a positive $\epsilon$. Thus solute molecules act as a condensing force on the DNA.

To estimate the size of this attractive potential we must compare it with the repulsive potential from the excluded volume of the polymer with itself (from the first term in Eq. (2.42)):

$$F_{\text{self-avoidance}} = \frac{k_B T}{2} \cdot v \cdot \frac{N^2}{R^3} \tag{2.57}$$

where $N$ is counted in units of the Kuhn length, and $v$ is the volume of one Kuhn length: for DNA, $v \sim 120$ nm · π · 1 nm$^2$ = 400 nm$^3$. To estimate the relative importance of attraction to self-avoidance we calculate the ratio of the attractive to the repulsive term

$$\text{ratio} = \frac{F_{\text{solute}}}{F_{\text{self-avoidance}}} = 2\frac{\delta}{v}\delta\rho \sim 60 \tag{2.58}$$

where we use $\delta/v \sim (r_{\text{protein}} + r_{\text{DNA}})^2/r_{\text{DNA}}^2 \sim 9$ (the radius of DNA is 1 nm, and of a protein typically 2 nm); further $\delta\rho$ is the number of proteins inside a volume of length 120 nm and radius $r_{\text{protein}} + r_{\text{DNA}} \sim 3$ nm, i.e. $\delta \approx 4000$ nm$^3$. The density of proteins inside *E. coli* is of order $2 \times 10^6$ in a volume of 1 μm$^3$, $\rho = 2 \times 10^{-3}/\text{nm}^3$, giving $\rho\delta \sim 10$. Thus the attraction due to solutes is larger than the self-repulsion,

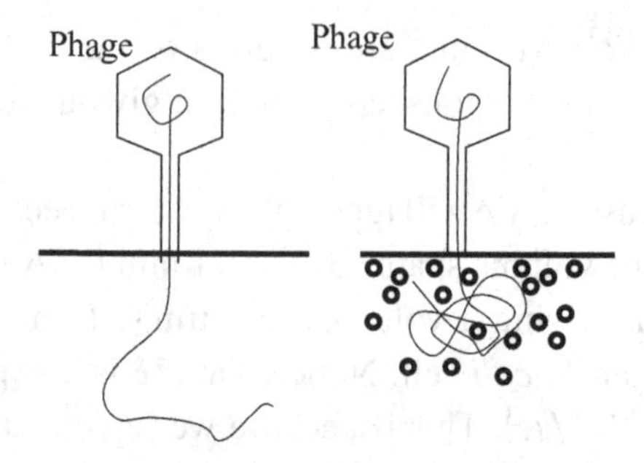


Figure 2.22. Visualization of the excluded volume effects imposed on injected DNA into a bacterial cell where about 20%–30% of the volume is occupied by proteins, thus presenting a huge excluded volume pressure on the injected DNA. It is possible that excluded volume also plays a role in the condensation of bacterial DNA into about 10% of the bacterial volume.

and in principle provides enough pressure on the DNA to collapse it into a dense phase. In *E. coli* the DNA is indeed confined to a volume far smaller than the Gaussian random chain expected at the $\Theta$ point. In fact, it is collapsed to about a tenth of the cell volume, that is with an overall extension of about 0.5 μm. For comparison, the $\Theta$ point end-to-end radius was about 25 μm.

Finally, we stress that factors other than osmotic pressure contribute to confine the *E. coli* DNA. There are specific sites on the DNA bound to the cell membrane, there is DNA supercoiling, and in *E. coli* there are certain histone-like proteins that bind unspecifically to the DNA and bend it. Thus the relevance in vivo of the above approach is speculative, and at most relevant to DNA injected into the living cell by phages, for example. This is shown in Fig. 2.22.

*Questions*

(1) Estimate the free energy associated with self-avoidance and the free energy associated with the excluded volume of proteins for *E. coli* DNA confined to a radius of 0.3 μm inside an *E. coli* cell.
(2) Discuss two chromosomes of DNA, each with 5 000 000 bp confined to the same cell volume of 1 $\mu m^3$ in a 30% solution of proteins. Will they segregate or mix?

## Helix–coil transition

The helix–coil transition is one of the simplest transitions observed for (some) polymers. It is a transition between a random coil state similar to the self-avoiding polymer described in Chapter 1, and a helix state where subsequent monomers make a short turn stabilized by monomer–monomer binding. For a polypeptide

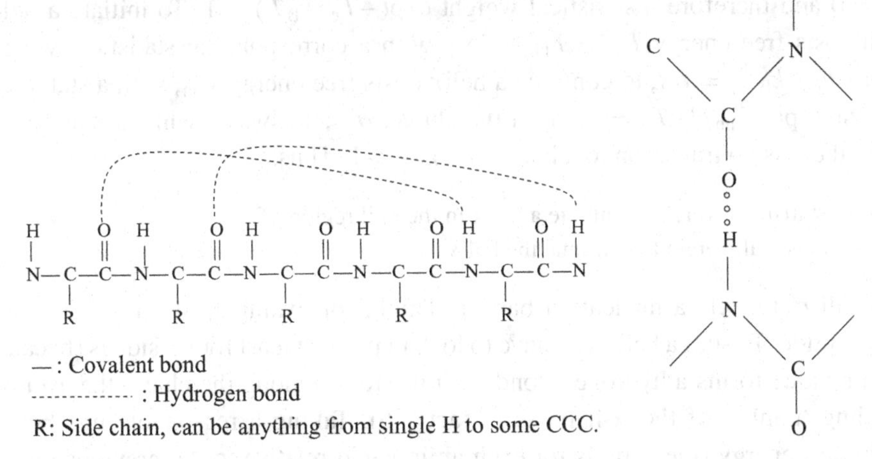


Figure 2.23. Left: peptide backbone with possible hydrogen bonds indicated by dashed lines. Hydrogen bonds appear between a hydrogen donor (N) and an acceptor (O). The first backbone represents an energy gain of only one unit, but an entropy loss of order of four units. Subsequent hydrogen bonds each cost only one entropy unit. Right: geometry of the hydrogen bonds formed between the peptide backbone of a polymer. The donor and acceptor atoms are on opposite sides of the polymer, and perpendicular to the backbone. Thereby a helix is favored. As discussed later, the constraints imposed by H-bonds also leads to a competing state, the so-called β-sheets.

chain, the helix state is characterized by hydrogen bonds between monomers that are only a few amino acids away along the polymer chain (see Fig. 2.23). There is an entropic cost in forming a helix turn, and the overall balance between a helix state with high binding energy and a coil state with high entropy makes a transition possible.

Helices are not seen only in homopolymers, but are also important structural elements of proteins; α-helices were predicted by Pauling & Corey (1951). Their structure is closely associated to hydrogen bonds between amino acids; see Chapter 4.

Experimentally the helix content is measured by using the fact that the helix state absorbs polarized light differently from the coil state. Circular dichroism (CD) spectra are different for the random coil, α-helix or β-sheet. A detailed account of this technique is given by Cantor & Schimmel (1980).

The helix–coil transition can be parametrized in terms of two quantities (Zimm, 1959; Scheraga, 1973): an entropy-cost term $\sigma$ that counts the loss in conformational entropy to form the first bond, and an energy-gain term that counts the energy for each subsequent addition of a bond. In more detail, let us normalize the statistical weights relative to the coil (c) state, i.e. we assign to it a free energy

$F_c = 0$ and therefore a statistical weight $\exp(-F_c/k_B T) = 1$. To initiate a helix then costs free energy $F_{ch} = F_{hh} - TS_{ch}$ with a corresponding statistical weight $\exp(-F_{ch}/k_B T) = \sigma s$; to continue a helix costs free energy $F_{hh}$, with a statistical weight $\exp(-F_{hh}/k_B T) = s$. From the above, $\sigma < 1$ always, while $s$ can be $<1$ (then there is no transition to a helix state), or $>1$. Thus:

- $\sigma s$ = statistical weight to initiate a helix in the coil region,
- $s$ = statistical weight to continue the helix.

A small $\sigma$ reflects a nucleation barrier. Think, for example, of an $\alpha$-helix in a polypeptide. To start a helix you have to lock in place at least four residues (because each residue forms a hydrogen bond with the fourth along the chain, the average winding number of the $\alpha$-helix is in fact 3.6). Taking a reasonable number of minimum-energy orientations for each amino acid relative to the previous one to be $\sim$6, a naive estimate of the conformational entropy loss is

$$\Delta S/k_B \sim -3.6 \ln(6) \Rightarrow \sigma = e^{\Delta S/k_B} \approx 10^{-3} \quad (2.59)$$

To relate $\sigma$ and $s$ to experimental data on the helix content in polymers at various temperatures we now review the helix–coil transition model. This will also allow us to introduce some tools from statistical mechanics.

Denoting a helix segment with "h" and a coil segment with "c", an example of a configuration of a polymer of length 30 is

cccccccchhhhhhhhhhccccchhhhhhh

The statistical weight of this segment is $\sigma^j s^k = \sigma^2 s^{17}$, where $j$ is the number of initiated helix segments (which is 2 in the above example), and $k$ is the number of helix monomers (17 in above example).

We will now solve the helix–coil system, and thereby illustrate the usefulness of partition sums, and also show how transition matrix methods are used. To solve the model we have to calculate the statistical weight for all possible sequences $\{s\}$ of h and c, and calculate the partition sum

$$Z = \sum_{\{\text{sequences } s\}} \exp(-\frac{F\{s\}}{k_B T}) \quad (2.60)$$

where we express the total partition sum in terms of a sum of coarse-grained partition sums $\exp(-F(s)/k_B T)$, each associated with one of the possible helix–coil states of the system.

The total partition sum $Z$ can be calculated explicitly by the so-called *transition matrix* method. This is based on setting up a recursion relation, which tells you the partition sum for the system of length $N+1$ if you know the partition sum for the system of length $N$. From $\sigma$ and $s$ introduced above, we can write down the statistical weights associated to the four possible transitions.

$$\begin{aligned}
&\mathrm{c} \rightarrow \mathrm{h}\text{: } \sigma s \\
&\mathrm{c} \rightarrow \mathrm{c}\text{: } 1 \\
&\mathrm{h} \rightarrow \mathrm{h}\text{: } s \\
&\mathrm{h} \rightarrow \mathrm{c}\text{: } 1
\end{aligned}$$

Given a system of size $N$, we denote as $Z_c(N)$ the statistical weight that the last state in the sequence is in the coil state. The statistical weight that the last state is in a helix state is denoted $Z_h(N)$. Then

$$Z(N) = Z_c(N) + Z_h(N) \tag{2.61}$$

is the total partition sum for the system of size $N$. Now the basic trick is to use the transition probabilities above and decompose the statistical weight of a system of size $N + 1$ as

$$Z_c(N+1) = Z_c(N) + Z_h(N) \tag{2.62}$$
$$Z_h(N+1) = s\sigma\ Z_c(N) + s\ Z_h(N) \tag{2.63}$$

The first equation reflects that the coil state at position $N + 1$ can be obtained from either a helix or a coil state at position $N$. In any case there is no cost, because the coil state is the reference state. The second equation reflects that either we have to continue a helix from position $N$, or we have to initiate a new helix at position $N + 1$. The above set of recursion relations is called a transfer matrix relation, because it can be written as

$$\begin{pmatrix} Z_c(N+1) \\ Z_h(N+1) \end{pmatrix} = \mathbf{M} \begin{pmatrix} Z_c(N) \\ Z_h(N) \end{pmatrix} \tag{2.64}$$

with a transfer matrix

$$\mathbf{M} = \begin{pmatrix} 1 & 1 \\ \sigma s & s \end{pmatrix} \tag{2.65}$$

Because $Z_c(1) = 1$, $Z_h(1) = \sigma s$, and formally

$$\begin{pmatrix} 1 \\ \sigma s \end{pmatrix} = \mathbf{M} \begin{pmatrix} 1 \\ 0 \end{pmatrix} \tag{2.66}$$

we can write the recursion formula as

$$\begin{pmatrix} Z_c(N) \\ Z_h(N) \end{pmatrix} = \mathbf{M}^N \begin{pmatrix} 1 \\ 0 \end{pmatrix} \tag{2.67}$$

Now the basic idea is that we are interested in the large $N$ limit. Thus we are interested in the result after we have multiplied this matrix by itself many times. The overall product is then determined by the largest eigenvalue $\lambda_{\max}$, because $\lambda_{\max} > \lambda_{\min}$ implies that $\lambda_{\max}^N \gg \lambda_{\min}^N$. The eigenvalues $\lambda$ are determined by the equation

$$\det(\mathbf{M} - \lambda \mathbf{1}) = 0 \tag{2.68}$$

which in this case reads

$$(1 - \lambda) \cdot (s - \lambda) = \sigma s \tag{2.69}$$

or

$$\lambda = \frac{1}{2}\left(s + 1 \pm \sqrt{(1-s)^2 + 4\sigma s}\right) \tag{2.70}$$

Only the largest eigenvalue matters in the large $N$ limit, and the total statistical weight of all configurations is

$$Z(N) = Z_c(N) + Z_h(N) \sim \lambda_{\max}^N \tag{2.71}$$

The free energy is therefore

$$\begin{aligned} F &= -k_B T \ln(Z) = -k_B T N \ln(\lambda_{\max}) \\ &= -k_B T N \ \ln(\frac{1}{2}(1 + s + \sqrt{(1-s)^2 + 4\sigma s})) \end{aligned} \tag{2.72}$$

Now we ask what to use this total statistical weight $Z$ for? The free energy is anyway only defined relative to a reference state (here the coil state cccccc . . . cccc). To understand this is important, and it emphasizes a fundamental property of partition sums as so-called generating functions. The point is that the partition sum can also be expressed as

$$Z = \sum \Omega_{jk} \sigma^j s^k \tag{2.73}$$

where $j$ is the number of helix segments, and $k$ is the total number of monomers in the helix state; $\Omega_{jk}$ is the corresponding number of configurations. Now the quantity

$$\frac{\mathrm{d}\ln(Z)}{\mathrm{d}\ln(s)} = \frac{s}{Z}\frac{\mathrm{d}Z}{\mathrm{d}s} = \frac{\sum \Omega_{jk} k \sigma^j s^k}{\sum \Omega_{jk} \sigma^j s^k} = \langle k \rangle \tag{2.74}$$

determines the average of $k$. Thus the average helix content of the polymer is

$$\Theta = \frac{1}{N}\frac{\mathrm{d}\ln(Z)}{\mathrm{d}\ln(s)} = \frac{s}{2\lambda_{\max}}\left(1 + \frac{s + 2\sigma - 1}{\sqrt{(1-s)^2 + 4\sigma s}}\right) \tag{2.75}$$

Similarly the average frequency of helix initiation is

$$\langle j \rangle = \frac{1}{N} \frac{\mathrm{d}\ln(Z)}{\mathrm{d}\ln(\sigma)} = \frac{s\sigma}{\lambda_{\max}\sqrt{(1-s)^2 + 4\sigma s}} \tag{2.76}$$

These two equations represent the predictions of the helix–coil model.

Before discussing the temperature dependence of the parameters, let us stress the operational range for $s$ and $\sigma$. Note that $s$ counts the statistical weight of continuing a helix. If $s < 1$ a helix will tend to break up; if $s > 1$ helices tend to grow. In fact from the equations above you can see that for $s \ll 1$ then $\Theta \to 0$, while $s \gg 1$ implies $\Theta \to 1$.

The value of $\sigma$ is typically $\ll 1$, reflecting a nucleation threshold for initiating helices along the polymer. The smaller the value of $\sigma$, the longer the interval between subsequent helix segments, as we can see from $\langle j \rangle \propto \sigma$. In fact the helix–coil transition behaves somewhat as a first-order phase transition as long as $N < 1/\sigma$, where there is at most one helix segment in the system. However, when $N > 1/\sigma$, there will be several nucleation sites along the polymer, and the sharpness of the transition will not increase with further increase in $N$.

Experimentally the helix–coil transition can be induced by varying temperature. In terms of the parameters in the model, changing temperature changes $s$, while $\sigma$ remains fairly constant. Let us investigate the connection between $s$ and $T$ around the transition temperature $T_m$, i.e. for $s \approx 1$; remember that $\sigma \ll 1$, and consider the case $N < 1/\sigma$. Then we have a problem similar to a chemical equilibrium between h and c, and we can consider the corresponding equilibrium constant

$$K = \mathrm{e}^{-\Delta G/k_B T} \tag{2.77}$$

where $\Delta G$ is the free energy cost of transforming a coil into a helix. Assuming this equilibrium constant obeys a van't Hoff relation[1]

$$\frac{\mathrm{d}\ln(K)}{\mathrm{d}T} = \frac{\Delta H}{k_B T^2} \tag{2.78}$$

and noting that the equilibrium constant for adding one helical unit to a preexisting helix is $s$, we write

$$\frac{\mathrm{d}\ln(s)}{\mathrm{d}T} = \frac{\Delta H_1}{k_B T^2} \tag{2.79}$$

[1] Note that $\mathrm{d}\ln(K)/\mathrm{d}T = -(\mathrm{d}\Delta G/T)/k_B\mathrm{d}T = -\mathrm{d}/\mathrm{d}T((\Delta H/k_B T) - (\Delta S/k_B)) = \Delta H/k_B T^2$, where in the last equation we use $(\mathrm{d}\Delta H/T\mathrm{d}T) = \mathrm{d}\Delta S/\mathrm{d}T$. Both in fact equal $C_p/T$. Alternately the derivation follows directly from $\mathrm{d}\ln(Z)/\mathrm{d}\beta = -\langle E \rangle$.

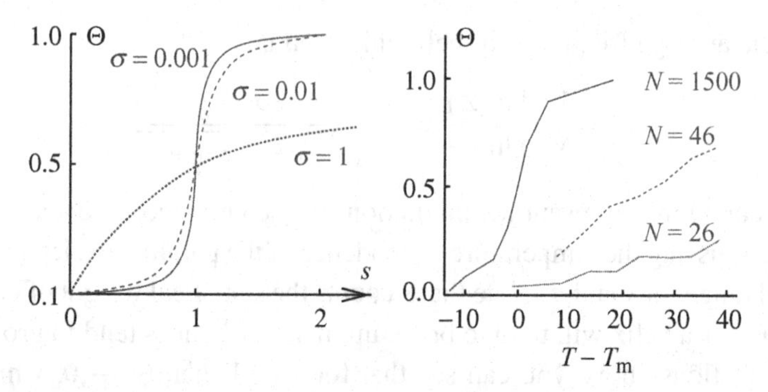


Figure 2.24. Left: model-predicted fraction of polymer in the helix state as a function of the control parameter, for different values of helix initiation costs (from Eq. (2.75)). For $\sigma = 1$ this fraction is linear around the transition, but on a wider scale behaves as $s/(1+s)$. For small $\sigma$ only very long helix segments contribute to $Z$, and therefore the helix fraction increases as $\sigma \to 0$ for $s > 1$. Right: the data used here are the temperature-induced coil-to-helix transition of the polypeptide poly-gamma-benzyl-L-glutamate in mixed dichloroacetic acid–ethylene dichloride solvent (Zimm *et al.*, 1959). Usually the helix fraction would decrease with $T$, but for this polymer the helix-stabilizing forces increase more with temperature than the contribution from the entropy.

where $\Delta H_1$ is the enthalpy cost of adding one helical unit (which can be measured experimentally), and where we know that $s = 1$ at the transition point $T_m$. Therefore:

$$\ln s(T) = \frac{1}{k_B} \Delta H_1 \int_{T_m}^{T} \frac{dT}{T^2} = (\Delta H_1) \frac{T - T_m}{k_B T T_m} \tag{2.80}$$

from where we find a roughly linear dependence of $s$ on $T$ around the transition temperature $T_m$:

$$s(T) = 1 + \epsilon \quad \text{with} \quad \epsilon \approx \frac{\Delta H_1}{k_B T_m^2}(T - T_m) \tag{2.81}$$

The value of $\sigma$ is constant, because it reflects an entropy cost that depends weakly on the absolute temperature $T$, and is not proportional to $T_m - T$. Figure 2.24 shows the sharpness of the transition for different values of $\sigma$.

Finally we give the experimental values of helix parameters for amino acids, as measured by Scherage (1973). Even for the best helix formers, $\sigma$ and $(s - 1)$ are surprisingly small. Helices of amino acids are essentially not stable by themselves, at least not for helix sizes that are characteristic in proteins (see Question 3 below).

| | | |
|---|---|---|
| Glycine | $\sigma = 0.00001$ | $s = 0.62$ |
| L-Alanine | $\sigma = 0.0008$ | $s = 1.06$ |
| L-Serine | $\sigma = 0.00075$ | $s = 0.79$ |
| L-Leucine | $\sigma = 0.0033$ | $s = 1.14$ |

One lesson from this section was in part methodological, an illustration of the use of transfer matrix and partition functions as generating functions. The other lesson was that the helix–coil transition, to a degree set by $\sigma$, is cooperative, but it is not a phase transition. Helix–coil parameters for amino acids demonstrate that secondary structures in proteins are not stable by themselves.

### *Questions*

(1) Estimate the conformational pressure of confining a random polymer of length $L$ in a box of size $X$, both measured in units of the Kuhn length.
(2) Discuss the $\Theta$ point transition in two dimensions and in four dimensions.
(3) Calculate minimal stable $\alpha$-helix of the four monomers above.
(4) Consider the formal steps that we glossed over in Eqs. (2.68)–(2.72). The total partition sum was

$$Z(N) = Z_c(N) + Z_h(N) = (1, 1)\begin{pmatrix} Z_c(N) \\ Z_h(N) \end{pmatrix} = (1, 1)\mathbf{M}^N \begin{pmatrix} 1 \\ 0 \end{pmatrix} \tag{2.82}$$

When diagonalizing $\mathbf{M}$ we apply the transformation $\mathbf{T} : \mathbf{M} \to \mathbf{T}^{-1}\mathbf{M}\mathbf{T} = \mathbf{\Lambda}$ such that

$$\mathbf{\Lambda} = \begin{pmatrix} \lambda_{\max} & 0 \\ 0 & \lambda_{\min} \end{pmatrix} \tag{2.83}$$

Where $\mathbf{M}^N = \mathbf{T}\mathbf{T}^{-1}\mathbf{M}\mathbf{T}\mathbf{T}^{-1} \ldots \mathbf{M}\mathbf{T}\mathbf{T}^{-1} = \mathbf{T}\Lambda^N\mathbf{T}^{-1}$ and thus

$$Z(N) = (1,\ 1)\,\mathbf{T}\begin{pmatrix} \lambda_{\max}^N & 0 \\ 0 & \lambda_{\min}^N \end{pmatrix}\mathbf{T}^{-1}\begin{pmatrix} 1 \\ 0 \end{pmatrix} \tag{2.84}$$

Calculate explicitly $\mathbf{T}$ in the above expression, and examine to what extent $Z(N) \approx \lambda_{\max}^N$.

## Collapse versus helix formation

For protein folding, both polymer collapse and the formation of helix structures are important. A model encompassing both effects may contain a spherically attractive potential (positive $\epsilon$) and a directed hydrogen-bonding potential associated to hydrogen donors and acceptors along the polymer chain. For the peptide backbone the donors (N) and acceptors (C=O) are directed opposite and perpendicular to the side chain.

We here adopt a lattice implementation of polymers that is widely used in the literature, originally introduced by Flory (1949), Ueda *et al.* (1975) and Lau & Dill (1989). The polymer is defined by a string of monomers placed on a three-dimensional cubic lattice. The energy of a configuration $\{\mathbf{r}_i\}$ is defined by the Hamiltonian

$$H_{\mathrm{VW}} = -\epsilon_{\mathrm{VW}} \sum_{i<j} \delta(|\mathbf{r}_i - \mathbf{r}_j| - 1) \tag{2.85}$$

where $\sum$ includes all monomer pairs, the $\delta$ function ensures contributions from nearest neighbors only and $\epsilon_{\mathrm{VW}}$ represents the strength of the spherically symmetric potential.

In addition to $H_{\mathrm{VW}}$, one may introduce an energy term associated with the directed interaction that hydrogen bonds would correspond to (Pitard *et al.*, 1997; Borg *et al.*, 2001). To each monomer $i$ we assign a spin $\mathbf{s}_i$ representing a hydrogen donor–acceptor pair. This can be easily pictured as a spin because of the opposite directions of the carbon–oxygen $C_i{=}O$ bond (H acceptor) and the nitrogen–hydrogen $N_{i+1}{-}H$ bond (H donor) on the peptide backbone. The spin is constrained to be perpendicular to the backbone. The hydrogen bond part of the Hamiltonian reads (Borg *et al.*, 2001):

$$H_{\mathrm{H}} = -\epsilon_{\mathrm{H}} \sum_{ij} \delta(\mathbf{s}_i \cdot \mathbf{s}_j - 1)\delta(|\mathbf{r}_i - \mathbf{r}_j| - 1), \tag{2.86}$$

where only $\mathbf{s}_i$ that are perpendicular to the backbone are allowed, and thus interactions along the backbone are automatically ignored. The Hamiltonian $H(\epsilon_{\mathrm{VW}}, \epsilon_{\mathrm{H}}) = H_{\mathrm{VW}} + H_{\mathrm{H}}$ specifies the energy of any homopolymer configuration including hydrogen bonds.

Figure 2.25 shows examples of structures obtained at low temperatures using H(0, 2) and H(1, 2), respectively. In the first case, with only hydrogen bonding

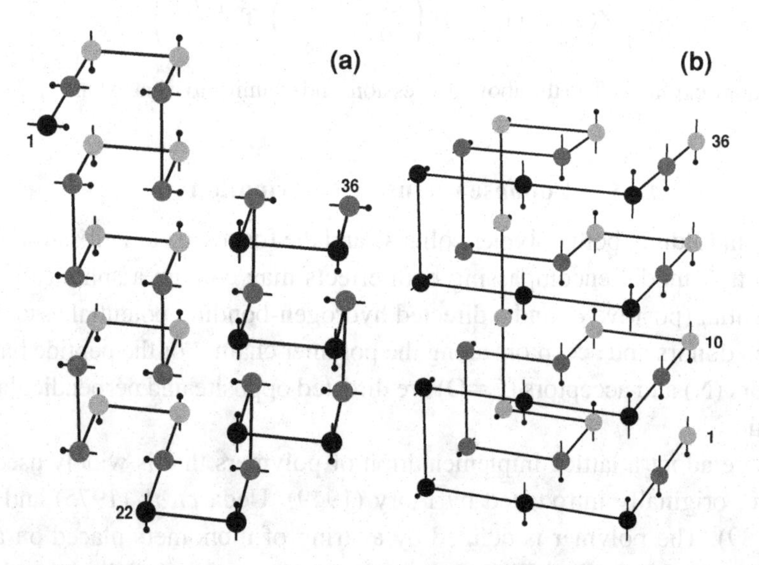


Figure 2.25. Collapsed states of a homopolymer with hydrogen bonds; the right-hand side is the ground state. Both folds displayed here reveal long-range order. In particular, (b) displays an up–down symmetry and an organization where sheets are on one side of the structure and the backbone connections between layers are concentrated on the opposite side. Figure from Borg *et al.* (2001).

energies, one can observe helix-like structures denoted p-helices (because they are caused by lattice constraints and are not true helices). For a given polymer topology one can quantify these helices by counting the number of bonds involved in them. Consider Fig. 2.25(a) where the monomers between 2 and 7 initiate a pseudo-helix, where neighbors 2–5 and neighbors 4–7 contribute. The p-helix continues until monomer 22, which breaks it because this monomer is not a neighbor to any members of a p-helix. A new pseudo-helix is initiated at monomer 25 and lasts throughout the chain.

Figure 2.25(b) displays the ground state of a polymer where fairly large attractive van der Waals interactions are included. In this case one can observe structures resembling β-sheets. The sheets can be either parallel or anti-parallel, as in natural proteins, in both cases quantified by identifying at least three pairs of consecutive neighbors (for a parallel sheet it would be $\{(i, j), (i+1, j+1), (i+2, j+2)\}$ and for an anti-parallel sheet it would be $\{(i, j), (i+1, j-1), (i+2, j-2)\}$). In Fig. 2.25(b), for example, monomer pairs (1, 10), (2, 9), (3, 8) and (4, 7) contribute to an anti-parallel sheet that gets broken at monomer 10. Monomers 6–13 also participate in an anti-parallel sheet with the layer above.

Both the folds displayed in Fig. 2.25 reveal long-range order. The key result is that spin interactions induce large-scale organization even in the case of homopolymer collapse.

Figure 2.26 shows the number of bonds involved in secondary structures ($I$), as function of total number of bonds ($N_{\mathrm{B}}$). The dependence on the number of bonds is obtained by thermal averaging as a function of temperature. The choice to use $N_{\mathrm{B}}$ rather than the temperature as a free variable is more convenient, because we are comparing systems with different energy scales. The three curves represent

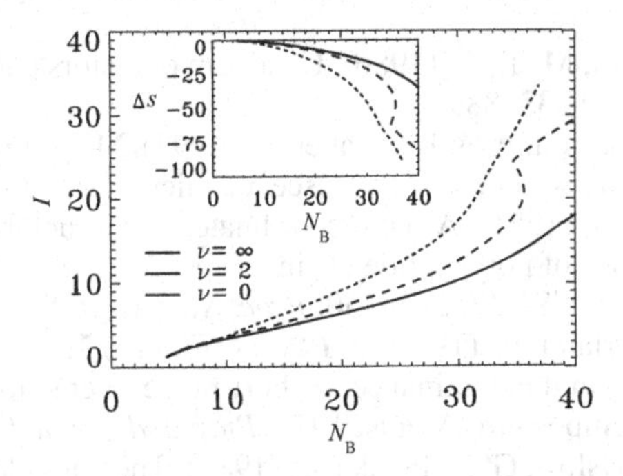


Figure 2.26. State of homopolymer as function of number of contacts, for various ratios of directed (hydrogen) to spherical symmetric (van der Waals) bindings. $I$ denotes structures that are associated to helix or sheet structures. $S$ is the total entropy of the chain given the constraint imposed by number of contacts $N_{\mathrm{B}}$. Figure from Borg *et al.* (2001).

the case where only van der Waals energy is present ($\nu \equiv \epsilon_H/\epsilon_{VW} = 0$, full line), where van der Waals and spin coupling energy are present ($\nu = 2$), and the case where only spin coupling is present ($\nu = \infty$). For any degree of compactness the amount of secondary structure increases with increasing spin coupling. In general, it also increases with compactness; however, a back bending for the $\nu = 2$ case is present at nearly maximal compactness. The back bending for $\nu = 2$ in both plots is the mark of a phase transition. The transition takes place at almost maximum compactness (which is associated with low temperature), where entropy is reduced abruptly while compactness changes to a minor extent. In fact, this transition distinguishes between two types of compact polymers, a phase in which helices are predominant, and an ordered, highly symmetric phase, rich in β-sheets. For real proteins, some are dominated by α-helices, whereas others are dominated by β-sheets.

## References

Cantor, C. R. & Schimmel, P. R. (1980). *Biophysical Chemistry*, Part III, Chapter 20. New York: Freeman.

DeGennes, P.-G. (1985). Kinetics of collapse for a flexible coil. *J. Phys. (France) Lett.* **46**, L639.

Flory, P. J. (1949). The configurations of a real polymer chain. *J. Chem. Phys.* **17**, 303.

Hansen, A. & Sneppen, K. (2004). Why nature uses polymers for mechanical nanodevices (in preparation).

Jensenius, H. & Zocchi, G. (1997). Measuring the spring constant of a single polymer chain. *Phys. Rev. Lett.* **79**, 5030.

Lau, K. F. & Dill, K. (1989). A lattice statistical mechanics model of the conformational and sequences spaces of proteins. *Macromolecules* **22**, 3986.

Li, B., Madras, N. & Sokal, A. D. (1995). Critical exponents, hyperscaling, and universal amplitude ratios for two- and three-dimensional self-avoiding walks. *J. Stat. Physics* **80**, 661–754.

Mossing, M. C. & Record, M. T., Jr. (1986). Upstream operators enhance repression of the lac promoter. *Science* **233**, 889.

Ott, A., Magnasco, M., Simon, A. & Libchaber, A. (1993). Measurement of the persistence length of filamentous Actin using fluorescence microscopy. *Phys. Rev. E* **48**, R1642.

Pauling, L. & Corey, R. B. (1951). Atomic coordinates and structure factors for two helical configurations of polypeptide chains. *Proc. Natl Acad. Sci. USA* **37**, 235–240; *Proc. Natl Acad. Sci. USA* **37**, 241–250; *Proc. Natl Acad. Sci. USA* **37**, 251–256.

Pitard, E., Garel, T. & Orland, H. (1997). *J. Phys. I (France)* **7**, 1201.

Scherage, H. A. (1973). On the dominance of short-range interactions in polypeptides and proteins. Prague Symposium, August 1972. *Pure and Applied Chemistry* **36**, 1.

Sun, S. T., Nishio, I., Swislow, G. & Tanaka, T. (1980). The coil–globule transition: radius of gyration of polystyrene in cyclohexane. *J. Chem Phys.* **73**, 5971–5975.

Ueda, Y., Taketomi, H. & Go, N. (1975). Studies on protein folding, unfolding and fluctuations by computer simulation. *Int. J. Peptide Prot. Res.* **7**, 445–449.

Walter, H. & Brooks, D. E. (1995). Phase separation in cytoplasm, due to macromolecular crowding, is the basis for microcompartmentation. *FEBS Lett.* **361**, 135–139.

Wu, C. & Wang, X. (1998). Globule to coil transition of a single homopolymer chain in solution. *Phys. Rev. Lett.* **80**, 4092.
Zimm, B. H. (1959). *Proc. Natl Acad. Sci. USA* **45**, 1601.
Zimm, B. H., Doty, P. & Iso, K. (1959). Determination of the parameters for helix formation in poly-gamma-benzyl-L-glutamate, *Proc. Natl Acad. Sci. USA* **45**, 1601–1607.

## Further reading

Beyreuther, K., Adler, K., Geisler, N. & Klemm, A. (1973). The amino acid sequence of lac repressor. *Proc. Natl Acad. Sci. USA* **70**, 3526.
Bloomfield, V. A. (1999). Statistical thermodynamics of helix coil transitions in biopolymers. *Am. J. Phys.* **67** (12), 1212–1215.
Borg, J., Jensen, M. H., Sneppen, K. & Tiana, G. (2001). Hydrogen bonds in polymer folding. *Phys. Rev. Letters* **86**, 1031–1033.
Bustamante, J. F., Marko, E. D., Siggia, E. & Smith, S. (1994). Entropic elasticity of lambda-phage DNA. *Science* **265**, 1599.
Cluzel, P., Chatenay, D. *et al.* (1996). DNA: an extensible molecule. *Science*, 264, 792.
Doi, M. & Edwards, S. F. (1986). *The Theory of Polymer Dynamics*. Oxford: Clarendon.
DeGennes, P.-G. (1991). *Scaling Concepts in Polymer Physics*. Ithaca: Cornell University Press.
Doty, P. & Yang, J. T. (1956). Polypeptides VII. Poly-gamma-benzyl-l-glutamate: the helix–coil transition in solution. *J. Am. Chem. Soc.* **78**, 498.
Flory, P. J. (1969). *Statistical Mechanics of Chain Molecules*. New York: Hansen Publishers.
Grosberg, A. Y. & Khoklov, A. R. (1994). *Statistical Physics of Macromolecules*. New York: AIP Press.
(1997). *Giant Molecules: Here and There and Everywhere*. New York: Academic Press.
Hiemenz, P. C. (1984). *Polymer Chemistry – The Basic Concepts*. New York: Marcel Dekker.
Lifshitz, I. M., Grosberg, A. Yu. & Khokhlov, A. R. (1978). Some problems of the statistical physics of polymer chains with volume interactions. *Rev. Mod. Phys.* **50**, 683.
Lifson, S. & Roig, A. (1961). On the helix–coil transition in polypeptides. *J. Chem. Phys.* **34**, 1963.
Poland, D. & Scheraga, H. A. (1970). *Theory of Helix Coil Transitions in Biopolymers*. New York: Academic.
Schellman, J. A. (1955). The stability of hydrogen bonded peptide structures in aqueous solution. *Compt. Rend. Trav. Lab. Carlsberg. Ser. Chim.* **29**, 230–259.
Smith, S. B., Finzi, L. & Bustamante, C. (1992). Direct mechanical measurements of the elasticity of single DNA-molecules by using magnetic beads. *Science*, **258**, 1122.
Smith, S. B., Cui, Y. & Bustamante, C. (1996). Overstretching B-DNA: the elastic response of individual double-stranded and single-stranded DNA. *Science* **271**, 795.
Zimm, B. H. & Bragg, J. K. (1959). Theory of the phase transition between helix and random coil in polypeptide chains. *J. Chem. Phys.* **31**, 526.
Zimm, B. H. (1960). Theory of 'melting' of the helical form in double chains of the DNA type. *J. Chem. Phys.* **33**, 1349–1356.

# 3

# DNA and RNA

Kim Sneppen & Giovanni Zocchi

## DNA

The molecular building blocks of life are DNA, RNA and proteins. DNA stores the information of the protein structure, RNA participates in the assembly of the proteins, and the proteins are the final devices that perform the tasks. Two other prominent classes of molecule are lipids (they form membranes and, thus, compartments), and sugars (they are the product of the photosynthesis; ultimately life depends on the energy from the Sun). Sub-units from these building blocks show up in different contexts. For example ATP is the ubiquitous energy storing molecule, but adenine (the A in ATP) is also one of the DNA bases; similarly, the sugar ribose is also a component of the DNA backbone.

Before entering into the details of the basic building units, let us list their respective sizes. We do this with respect to their information content. DNA is used as the main information storage molecule. The information resides in the sequence of four bases: A,T, C and G. Three subsequent bases form a codon, which codes for one of the 20 amino acids. The genetic code is shown in Table 3.1. The weight of a codon storage unit is larger for DNA (double stranded and 330 Da per nucleotide, 1 Da = 1 g/mol) than for RNA (single stranded) and proteins.

DNA: 2000 Da per codon,
RNA: 1000 Da per codon,
proteins: 110 Da per amino acid.

Thus whereas a 300 amino acid protein weighs 33 kg/mol, the stored information at the DNA level has a mass of 600 kg/mol.

The peculiar structure of biological macromolecules represents the existence of two distinct energy scales. The polymer backbone is held together by covalent (strong) bonds, i.e. energies of order ∼200 kcal/mol (∼ 10 eV). The secondary and tertiary structure, i.e. the DNA double helix (see Fig. 3.1), the folded form of a

Table 3.1. *The genetic code*

| 5′ end | U | C | A | G | 3′ end |
|---|---|---|---|---|---|
| U | Phe (F) | Ser (S) | Tyr (Y) | Cys (C) | U |
| U | Phe (F) | Ser (S) | Tyr (Y) | Cys (C) | C |
| U | Leu (L) | Ser (S) | STOP | STOP | A |
| U | Leu (L) | Ser (S) | STOP | Trp (W) | G |
| C | Leu (L) | Pro (P) | His (H) | Arg (R) | U |
| C | Leu (L) | Pro (P) | His (H) | Arg (R) | C |
| C | Leu (L) | Pro (P) | Gln (Q) | Arg (R) | A |
| C | Leu (L) | Pro (P) | Gln (Q) | Arg (R) | G |
| A | Ile (I) | Thr (T) | Asn (N) | Ser (S) | U |
| A | Ile (I) | Thr (T) | Asn (N) | Ser (S) | C |
| A | Ile (I) | Thr (T) | Lys (K) | Arg (R) | A |
| A | Met (M) | Thr (T) | Lys (K) | Arg (R) | G |
| G | Val (V) | Ala (A) | Asp (D) | Gly (G) | U |
| G | Val (V) | Ala (A) | Asp (D) | Gly (G) | C |
| G | Val (V) | Ala (A) | Glu (E) | Gly (G) | A |
| G | Val (V) | Ala (A) | Glu (E) | Gly (G) | G |

Each of the 64 different triplets read from the 5′ end to the 3′ end corresponds to one amino acid. AUG also codes for the start of translation. Amino acids are building block of proteins. Some properties of the corresponding amino acids can be seen in Table 4.1.

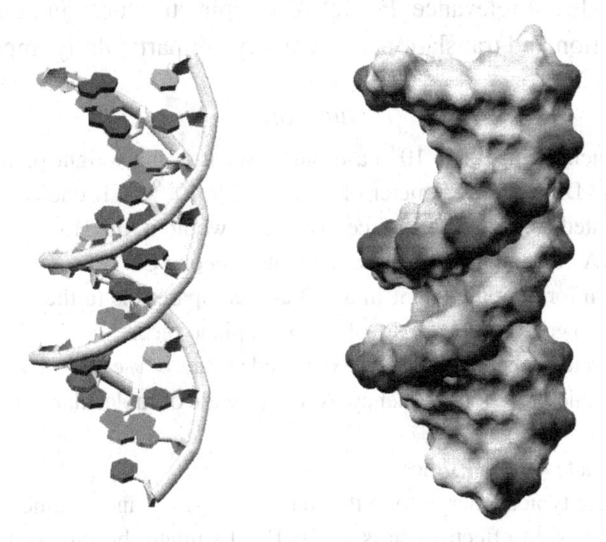

Figure 3.1. A little more than one turn of the DNA double helix. The left-hand panel shows the two strands with bases that match the complementary bases. The right-hand panel shows the DNA external surface that is accessible for a water molecule.

protein, etc., as well as molecular recognition processes, depend on weak bonds (e.g. hydrogen bonds), of order 1–2 kcal/mol (1/10 eV). As room temperature corresponds to an energy of 0.617 kcal/mol, the structure of a large functional biomolecule depends on many such weak bonds. Functional interactions of large biomolecules are likewise mediated by many such weak bonds, in order to reach a somewhat universal strength of order 10–20 kcal/mol. The understanding of these weak interactions is complicated by the fact that all biological processes take place in water, and interaction with water is a key uncertainty in all attempts at exact modeling. We will see that the main forces responsible for the functional behavior of biological macromolecules are associated with hydrogen bonds, or the absence of hydrogen bonds. That is, a main driving force for polymer collapse is hydrophobicity associated with deficiency in hydrogen bonds compared with the surrounding water, whereas the internal order is mostly associated with hydrogen bonds within the polymer.

In this chapter we will briefly introduce the structure of DNA and RNA, and then go on to discuss their physical properties such as rigidity, the melting transition of DNA, and RNA folding. These aspects, as well as the mechanism for diffusion of DNA through a gel, which is called reptation and forms the basis for gel electrophoresis, can be described in terms of simple statistical mechanics models. Further, these processes are technologically important. However, the reader may keep in mind that the biologically relevant properties are much more related to the structure of DNA and RNA at the conditions in the living cell. For DNA, supercoiling is also of relevance. For RNA, hairpin structures and their interaction with the translation and translational machinery are particularly important.

## *Questions*

(1) The human genome has $3 \times 10^9$ base pairs. What is the weight of one copy of the human DNA? DNA has a diameter of 2 nm $= 2 \times 10^{-9}$ m. If one copy of the human genome is plated densely on a surface, what area would it fill? How many species with the same DNA content as humans can be plated on one CD?

(2) Estimate the information content in a CD and compare this to the information in the score needed to generate the music of a typical piano piece.

(3) The reaction rate in chemistry can be expressed as rate $= \nu_0 \exp(-\Delta G/k_B T)$, where $\nu_0$ is a characteristic vibration frequency. A length scale of molecular interaction is of the order 1 Å.

 (a) First estimate $k_B T$ in kJ/mol.

 (b) Now let the typical energy for a fluctuation be $\sim k_B T$ and assume that the reacting molecule has an effective mass of 10 Da. Estimate the typical frequency $\nu_0$ of vibration (assume a harmonic potential).

 (c) What is the reaction rate for $\Delta G = 5$ kcal/mol, 10 kcal/mol, 15 kcal/mol, 20 kcal/mol and 1 eV per molecule?

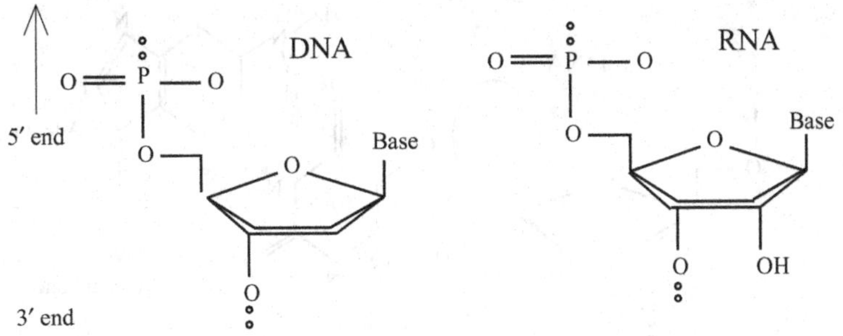

Figure 3.2. One unit of the DNA (left) and RNA (right) sugar–phosphate backbone. The RNA has an extra OH group. Single-stranded RNA is more flexible than single-stranded DNA.

(4) Assume that we can make 1 g of macromolecules. How many different proteins and DNA pieces, respectively, could we sample if each had to contain the information equivalent to a 300 amino acid protein?

### *DNA, the base pairs*

A single strand of DNA consists of a sugar–phosphate backbone (thus there is one negative charge per monomer), and one of four different bases (A, T, G, C) attached to each sugar. By way of nomenclature, adenine (A) and guanine (G) are called purines; thymine (T) and cytosine (C) are pyrimidines. There is also a pyrimidine called uracil (U), which replaces T when DNA is translated into RNA (Fig. 3.2). The polymer has a direction: there is a 5′ end and a 3′ end of the strand. Two complementary DNA strands come together (with anti-parallel orientation) to form the double helix: A pairs with T, G pairs with C.

In the double-helix structure, the sugar–phosphate backbone is on the outside with the bases on the inside, stacked like the steps of a ladder. The bases pair through hydrogen bonds, two hydrogen bonds for A–T, three for G–C, so the latter pairing is stronger (Fig. 3.3). In addition, there are stacking interactions between adjacent bases on the same strand. The binding energy for a base pair is 1–2.5 kcal/mol (depending on the base pair), whereas the energy associated in initiating a "bubble" in a fully stacked DNA double helix is about 5 kcal/mol.

The persistence length of double-stranded DNA is about $l_p \sim 50$ nm, whereas single-stranded DNA is very flexible, with a persistence length of $\sim 1$ nm. For the normal physiological form of DNA (B-DNA) the helix diameter is 18 Å and there is one helix turn every 34 Å or 10.4 base pairs. There is a major groove and a minor groove. But remember that this is the *average* structure; local twist angles and conformations depend on the sequence. The basis for the recognition of specific sequences by DNA binding proteins is through exposing the bases within

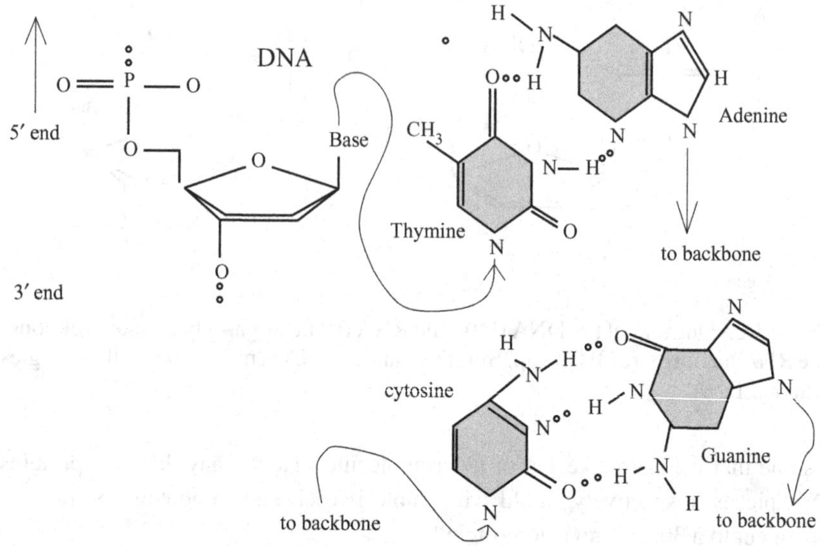

Figure 3.3. Base pairing for DNA with, respectively, two and three hydrogen bonds. The bases are attached to the sugar–phosphate backbone. Notice also that adenine and guanine are the same units as used in ATP and GTP, but in that case they act together with a tri-phosphate group instead of the one-phosphate group used in the DNA/RNA backbone.

the grooves: the edges of the base pairs form the floor of the grooves. Each of the four pairs (AT, TA, GC or CG) exposes a unique spatial pattern of hydrogen-bond acceptors and donors, as well as a single positive charge; these patterns are recognized by DNA-binding proteins. In all cases there are one H-bond acceptor and two H-bond donors on the major groove.

It is remarkable that the $\alpha$-helix structures of proteins typically fit well in the major groove of DNA, thus achieving ideal exposure of amino acid side chains toward the DNA base pairs in the major groove. A DNA binding protein mainly recognizes the unique pattern of H-bond donors and acceptors associated with a short sequence (typically 6–9 base pairs). More subtle effects involving local sequence dependent conformations and rigidity may also play a role.

Finally we would like to emphasize that DNA has other structural features apart from the double helix. In particular, two double strands of DNA can change partners, as in the so-called Holliday junction shown (after Holliday, 1964) in Fig. 3.4. This structure is associated with recombination events in living cells. The formation of the junction requires the breaking and rejoining of DNA strands between two double-stranded pieces of DNA. The junction is facilitated by helper proteins (in prokaryotes by RecA, which binds to single-stranded (ss) DNA, and the proteins RuvA, RuvB and RuvC that bind to the junction and help its migration (Ariyoshi *et al.*, 2000).

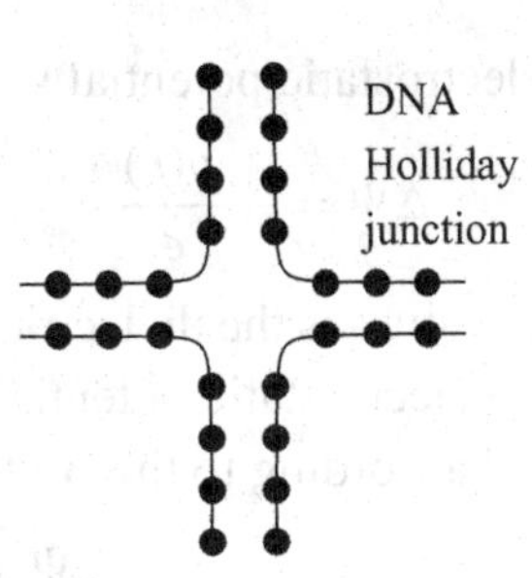


Figure 3.4. The DNA Holliday junction. It is possible for the DNA structure to allow two DNA strands to exchange partners. Sliding of the Holliday junction allows recombination of large sections of DNA with the only cost being the creation and elimination of the crossing between the strands. The initiation of the junction is by DNA repair mechanisms. The propagation of the junction is facilitated by certain proteins.

### *DNA in water and salt*

Water plays a prime role in protein–DNA interactions. DNA is always surrounded by a shell of well-ordered water molecules. In many cases changes in bases within the protein–DNA interaction region that do not directly hydrogen bond the protein still have a big impact on the overall binding properties. The attractive interaction between proteins and DNA is in part electrostatic (DNA is negative and proteins are positive), but also contains a large entropic part. For example, Takeda *et al.* (1992) report that the total $\Delta G = \Delta H - T\Delta S$ is negative (attractive) whereas $\Delta H$ is positive (repulsive) for the protein Cro binding DNA.

An important aspect of DNA is that it is charged. One length scale that characterizes electrostatic interactions is the Bjerrum length:

$$l_{\mathrm{b}} = \frac{e^2}{4\pi\epsilon k_{\mathrm{B}}T} \tag{3.1}$$

where $l_{\mathrm{b}} = 0.7$ nm in water, $\epsilon$ is the dielectric constant of water, and we are using, for simplicity, Gaussian units. The significance of $l_{\mathrm{b}}$ is that it is the separation distance at which the interaction energy between two unit charges is equal to $k_{\mathrm{B}}T$ ($\sim$0.617 kcal/mol at room temperature). Thus at distances $r > l_{\mathrm{b}}$ this interaction is weak, while at distances $r < l_{\mathrm{b}}$ it is substantial. The potential energy is then $k_{\mathrm{B}}T(l_{\mathrm{b}}/r)$ per charge unit.

If water contains salts, the electrostatic interactions will be screened and this introduces a second length scale into the problem. Call $\rho = ezn$ the ion charge density, where $z$ is the charge valence. Around a central charge $q$ there will be a higher density of counter ions and a lower density of ions with the same sign of charge, $n < n(\infty)$, compared with the charge density $n(\infty)$ far from the screened central charge. The negative ions will thus have a lower density $n < n(\infty)$ closer to the DNA than far away. The electrostatic potential $\Psi$ fulfils the Maxwell equation

$$\Delta\Psi = -\frac{\rho(r)}{\epsilon} \tag{3.2}$$

where $\rho = ezn$ is the charge density, $\epsilon$ is the dielectric constant of the medium, and the electric field in terms of the electrostatic potential is $\vec{E} = -\nabla\Psi$. On the other hand the charges are distributed according to the Boltzmann factor:

$$\rho = \rho(\infty)\exp\left(-\frac{ez\Psi}{k_B T}\right) \tag{3.3}$$

thus giving the self-consistent potential

$$\epsilon\Delta\Psi = -ezn(\infty)\exp\left(-\frac{ez\Psi}{k_B T}\right) \tag{3.4}$$

When linearized for $ez\Psi << T$ this becomes

$$\epsilon\Delta\Psi = -ezn(\infty)\left(1 - \frac{ez\Psi}{k_B T}\right) \tag{3.5}$$

or, for a point charge, it can be reformulated into spherical coordinates

$$\epsilon\frac{1}{r^2}\frac{\mathrm{d}}{\mathrm{d}r}\left(r^2\frac{\mathrm{d}\Psi}{\mathrm{d}r}\right) = -ezn(\infty)\left(1 - \frac{ez\Psi}{k_B T}\right) \tag{3.6}$$

which can be solved by the ansatz $\Psi = \text{const} \cdot \exp{(-\kappa r)}/r$. We thus obtain a potential that decays exponentially

$$\Psi = \Psi_0 \exp(-\kappa r) \tag{3.7}$$

with $\Psi_0 = q/(4\pi\epsilon r)$ for a point charge $q$. The characteristic scale $1/\kappa$ is called the Debye length

$$\kappa^{-1} = \sqrt{\frac{\epsilon k_B T}{e^2 z^2 n(\infty)}} = \left(\sqrt{4\pi l_b z^2 n(\infty)}\right)^{-1} \tag{3.8}$$

At cellular concentrations of salt the Debye length is about 1 nm. Under these conditions the electrostatic self-interaction of DNA is responsible for about half the persistence length of DNA.

*Question*

(1) Use a generalized version of Eq. (3.5) in Eq. (3.8) for several types of ion, and prove that

$$\kappa = \sqrt{4\pi l_b \sum_i z_i^2 n_i(\infty)} \tag{3.9}$$

where the summation runs over all ion species. Normally there will be several ion species because the overall charge is neutral.

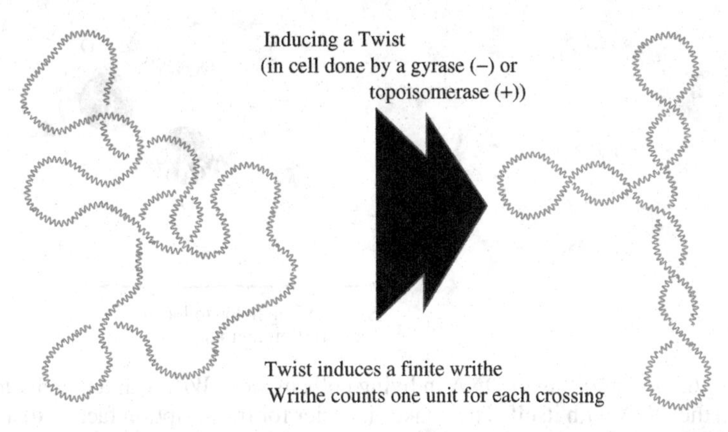


Figure 3.5. Introduction of a twist in a relaxed circular double-stranded molecule of DNA (on the left) induces DNA that supercoils around itself on a large length scale. The DNA double strand is shown as a single line; the right-hand part of the figure shows the induced *Writhe*. In the cell, the twist is introduced by a gyrase enzyme.

### *DNA supercoiling*

Double-stranded (ds) DNA has a persistence length of about 50 nm. This sets the scale for typical curvatures of DNA in the cell. In vivo, however, DNA also exhibits other larger-scale deformations, the so-called supercoils (Fig. 3.5). These are imposed by certain proteins (gyrases) that rotate the DNA around itself (similar to the twist in a telephone cord). Supercoils are thought to have a role for DNA condensation (Osterberg *et al.*, 1984). Supercoiling is characterized by two types of twisting:

- *Twist*, the number of turns of the DNA helix;
- *Writhe*, the number of times the helix crosses itself on a planar projection.

That is, *Writhe* is the number of super-helical turns that are present. The large-scale coiling of DNA consists of both *Twist* and *Writhe*; the linking number is their sum $\mathcal{L} = Twist + Writhe$, and is measured relative to the relaxed DNA, $\Delta\mathcal{L} = \mathcal{L} - \mathcal{L}(\text{relaxed})$. For circular DNA the linking number can be changed only by breaking the double-stranded DNA. Let us consider an example. A 5200 base pair DNA with a preferred twist of 10.4 base pairs will have a relaxed twist number $Twist = 5200/10.4 = 500$.

There is a characteristic length scale $K = 1050$ bp for the loops in supercoiled DNA; the elastic energy of a molecule of length $N$ bp, with an excess linking number $\Delta\mathcal{L}$, is of the form:

$$\Delta G = k_B T (K/N)(\Delta\mathcal{L})^2 \tag{3.10}$$

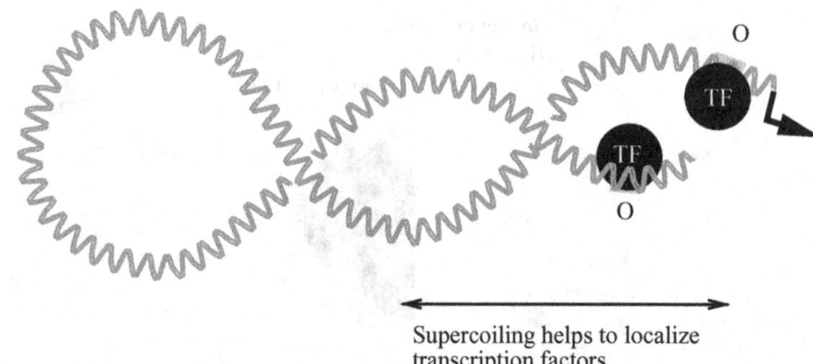


Figure 3.6. Supercoiling of DNA in living cells induces *Writhe* and thereby tends to align the DNA with itself. This makes it easier for transcription factors to act on regulating promoters that are separated by long distances along the DNA. The TF refer to transcription factors, and the regions on DNA marked O refer to operators, both regulating the promoter indicated by an arrow. All these terms are explained further in Chapter 7.

The main biological effect related to supercoils is that the *Twist* can be changed by special motor proteins (called gyrases) that temporarily cut the DNA and twist it around its main axis. Most gyrases induce negative *Twist*, but reverse gyrases also exist (observed in bacteria that live at high temperatures). When *Twist* is changed, the *Writhe* will change too, and the DNA will coil around itself on a large scale, helping to confine the DNA in the cell. Negative supercoil is common in eubacteria, including *E. coli*. Theory for supercoils can be found in LeBret (1978) and Marko & Siggia (1994, 1995).

Supercoiling has implications for the larger-scale organization of DNA within living cells. In particular, as it tends to wind the DNA around itself it also tends to make operator sites that are separated by several thousand bases along the DNA much closer to each other than they would be without supercoil. Thus it will facilitate the ability to regulate gene transcription activity at a distance; see Fig. 3.6.

### *Questions*

(1) Consider a small piece of DNA that is fixed to stick horizontally out of a wall. The perpendicular displacement of the free end of this rod $x$ is related to its radius of curvature by $x = L^2/2R$. At temperature $T$, what is the average mean square displacement $\langle x^2 \rangle$ of the end of the DNA?
(2) Consider a piece of DNA of length $L = 100$ nm and bending modulus $B = 50T$ nm. Assume that the DNA makes one large circular curve. What is the distribution of angles between tangent vectors at the two ends of the DNA piece?
(3) DNA molecules in the human cell are wrapped around disk-shaped protein complexes called histones, with one turn consisting of 146 base pairs. What is the bending energy per turn?

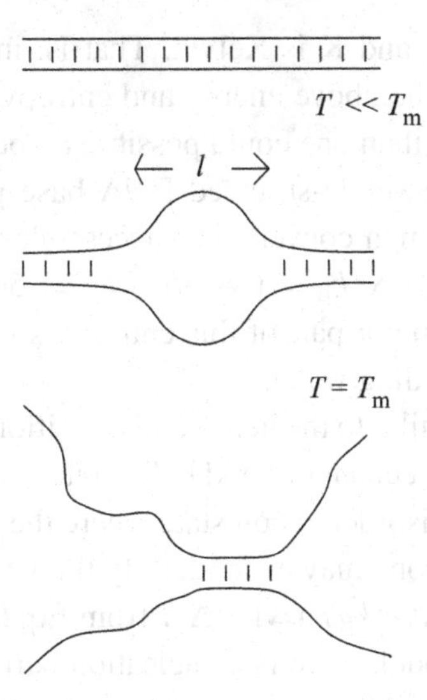


Figure 3.7. DNA melting. As the temperature is raised past the average melting point $T_m$, the double-stranded DNA first melts by forming bubbles, then by forming longer stretches of single-stranded DNA.

(4) Assume that a gyrase twists the 5200 base pair circular DNA with the preferred *Twist* = 500 from a state where *Writhe* = 0 to a state where initially *Writhe* = 0 and *Twist* = 475. What is the preferred supercoil (*Writhe*)? What is the statistical distribution in the supercoil number?

(5) Calculate the probability that one half of the DNA in Question (4) is in an unwound state, where there is no supercoil (local *Writhe* = 0, and local *Twist* relaxed). What is the probability for such an event?

## DNA melting

The double-stranded helix in DNA can melt in a process that to some degree resembles the helix–coil transition (see Fig. 3.7). Thus we interpret the helix state "h" as the double-stranded state, and the coil state "c" as the single-stranded unbounded state. For DNA the loop initiation factor is of order $\sigma \sim 10^{-4\pm1}$. The enthalpy and entropy differences between double-stranded and single-stranded states depend hugely on temperature and salt concentrations. At 25 °C and 1 M $Na^+$ they are (Borer *et al.* (1974))

$$\Delta H \approx (-6\,\text{to}\,-15)\,\text{kcal/(bp}\cdot\text{mol)} \tag{3.11}$$

$$\Delta S \approx (-0.013\,\text{to}\,-0.035)\,\text{kcal/(bp}\cdot\text{mol}\cdot\text{K)} \tag{3.12}$$

$$\Delta G = \Delta H - T\Delta S \approx (-1.6\,\text{to}\,-4.8)\,\text{kcal/(bp}\cdot\text{mol)} \tag{3.13}$$

where bp means base pair, and K is Kelvin. That is, in this simple model each base pair contributes with the above energy and entropy. Notice that the entropic contribution is much larger than one could possibly associate with conformational degrees of freedom for the single-stranded DNA base pair; i.e. for $T = 300$ K, then $TS \sim$3–9 kcal/mol, which converted to accessible number of states $n_{\text{acc}}$ for one pair of bases $\ln(n_{\text{acc}}) = S/k_B = (3\text{–}9)/(k_B T) = 5\text{–}15$, would mean $n_{\text{acc}} >$ 1000 different states. The major part of this entropy is in fact associated with the interaction with the surrounding water.

DNA melting is quite similar to the helix–coil transition mentioned in Chapter 2; that is, it is a transition between an ordered helix state where the two DNA strands form a double helix and a disordered coil state where the two strands are separated over long distances. In fact one may even identify the $s$ parameter from the helix–coil transition with $\exp(-\Delta G/k_B T)$ with $\Delta G$ from Eq. (3.13). Also in analogy to the helix–coil transition model, there is a nucleation barrier for changing between the double-stranded DNA and the two single-stranded DNA states. However, the analogy stops there. The simple helix–coil picture fails in two regards: first, the different base pairs have different binding affinity; and second, when there is a limited coil region, two single strands have to meet again, and this "bubble" costs some additional loop entropy.

To take the second of these points first, there is an additional entropy cost $\Delta S$ of bringing the two strands together given by

$$e^{\Delta S/k_B} \propto \frac{\text{conformation space of ends together}}{\text{conformation space of dangling end}} = \frac{1}{V(l)} \tag{3.14}$$

where $V(l)$ is the typical volume that otherwise would be spanned by the end-to-end distance of two single-stranded DNA pieces of length $l$ that start from the same point. As $V(l) = l_p(l/l_p)^\gamma$, with $\gamma \approx 1.5$–$1.8$, the entropy cost associated to a bound strand of length $l$ is

$$\frac{\Delta S}{k_B} \sim \ln(l_p) - \gamma \cdot \ln(l/l_p) \quad \text{with} \quad \gamma = 1.5 \rightarrow 1.8 \tag{3.15}$$

In fact loop free energies have been measured, including the loop initiation cost. For loop length $l > 15$ bp it can be fitted by (Rouzina & Bloomfield, 2001):

$$\Delta G = G(\text{loop}) - G(\text{closed}) = 3\ \text{kcal/mol} + 1.8\ k_B T \cdot \ln(l) \tag{3.16}$$

where $l$ is measured in base pairs (bp) and the constant takes into account the initiation threshold. $G(\text{closed})$ is the free energy associated to the double-stranded DNA without any melted parts.

If we ignored the loop closure cost, melting of homogeneous DNA would be equivalent to the helix–coil model. However, the loop closure imposes an entropy cost of $\propto \ln(l)$ for any loop, and because

$$\ln(l_1 + l_2) < \ln(l_1) + \ln(l_2) \tag{3.17}$$

this entropy favors loops to merge. Thus the long-range interaction imposed by $\ln(l)$ will tend to make fewer and longer loop segments than the independent loop formation assumed in the helix–coil model. That is, the melted regions and the bound regions separate more than in the helix–coil case, and thus DNA melting will be more cooperative than the simple helix–coil transition.

The size of the effect induced by the logarithmic loop closure depends on the prefactor $\gamma$ to the ln. When $\gamma$ becomes larger than 2, the DNA melting transition is first order, in contrast to the helix–coil transition that was not a phase transition. To see that a large $\gamma$ makes the transition first order, we now investigate the relative weight of one bubble versus two bubbles at the transition point $T = T_m$ for DNA melting.

In Fig. 3.8 we examine entropy of respectively the one-bubble system and the two-bubble system (see, for example, Tang & Chate, 2001). For each bubble the entropy is given a contribution from all possible positions of either of its two ends,

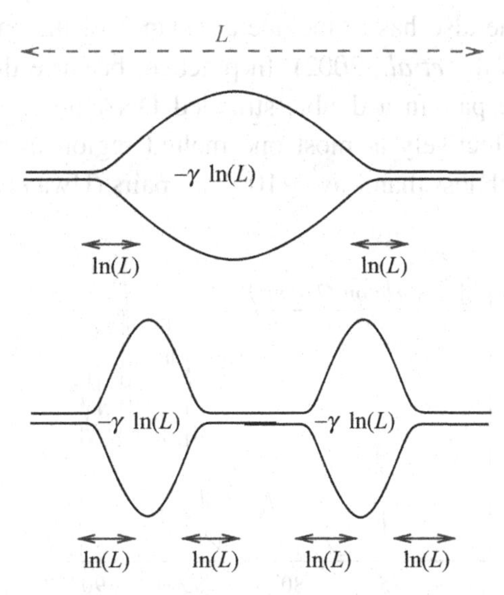


Figure 3.8. Competition between one large bubble and two smaller ones. By separating into two bubbles we gain translational entropy of the additional two ends (net gain $2\ln(L)$), but lose entropy because one should merge ends of two bubbles instead of one. The loss equals $\gamma(\ln(L) + \ln(L) - \ln(L)) \sim -\gamma\ln(L)$ when we ignore that lengths of the bubbles are several factors smaller than the total length $L$ of the double-stranded DNA. At the transition point bubbles are of a length of the same order of magnitude as $L$.

and a penalty for closing the loop when forming the bubble:

$$\frac{1}{k_B} S(\text{one bubble}) = +2\ln(L) - \gamma \ln(L) \tag{3.18}$$

We ignore the nucleation barrier because this becomes relatively less important when $L \to \infty$. Also, for simplicity, we count all lengths as the total length $L$ of the double-stranded DNA. In principle, of course, the bubble is smaller, and the positions of the ends constrain each other. However, at the melting point everything will be of order $L$, and we only make a mistake that does not depend on $L$. The entropy of making two bubbles is given by

$$\frac{1}{k_B} S(\text{two bubbles}) = +2\ln(L) - \gamma \ln(L) + 2\ln(L) - \gamma \ln(L) \tag{3.19}$$

Thus the entropy gain of going from two bubbles to one bubble is

$$\Delta S = S(\text{one bubble}) - S(\text{two bubbles}) = k_B(\gamma - 2)\ln(L) \tag{3.20}$$

which is $< 0$ for $\gamma = 1.8$ (the Flory expectation value). Thus in principle for large $L$ two and therefore also multiple bubbles are favored (see also Fig. 3.9). However, because $\gamma$ is close to 2, the one-bubble approximation is fairly good for small $L$. In fact the $\gamma = 1.8$ deduced from simple self-avoidance in principle may instead be $\sim 2.1$, because one also has to include avoidance of the part of the DNA that is not in the loop (Kafri *et al.*, 2002). In practice, because the cost of initiating a single-melted base pair in a double-stranded DNA piece is of the order 3–5 kcal/mol, there is effectively at most one melted region in even heterogeneous DNA strands of length less than, say, $\sim 10^3$ base pairs (Hwa *et al.*, 2003). Thus for

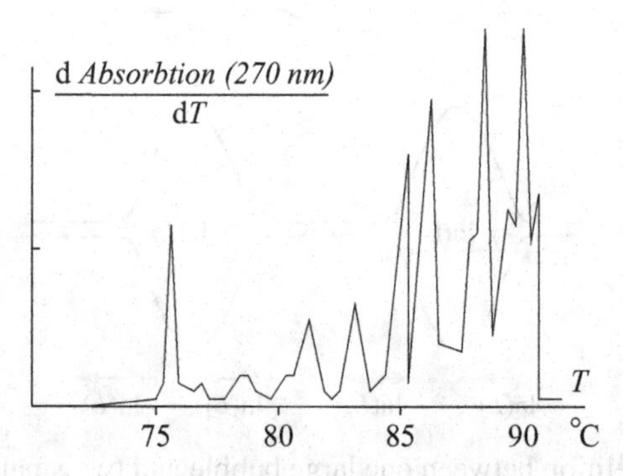


Figure 3.9. DNA melting of a 4662 bp heterogeneous sequence of DNA. Multiple melting steps are observed; the first at 76 °C corresponds to the melting of a 245 bp bubble in a particularly AT-rich domain. The last steps, on the other hand, correspond to a CG-rich region. The typical width of the melting steps is about 0.5 K (Blake, 1999).

DNA it is worthwhile to investigate the single-bubble approximation, which, when the melting happens from one end, is equivalent to the **zipper model** (Hwa *et al.*, 2003).

Let us now consider the zipper model (see Schellman, 1955). This is a simplification where we allow the DNA to melt from only one point, say one end. Thus no bubbles are allowed. Let the length of the double-stranded DNA be $N$. Then the possible states of the system can be labeled by the number of paired base pairs $n$ counted from the clamped position to the last bases that are paired. Thus the bases at position $i = n + 1, n + 2, \ldots, N$ are free (corresponding to the coil state). The energy of this state is

$$E(n) = \sum_{i=1}^{i=n} \epsilon_i = -\epsilon \cdot n \tag{3.21}$$

where the last part of the equation implies that we ignore variation in base pairing energies. The entropy of state $n$ is:

$$S(n)/k_{\mathrm{B}} = (N - n)\ln(g) \tag{3.22}$$

where $g$ is the number of possible conformations for one base pair when they are not bound to each other. For simplicity we ignore excluded volume effects of melted base pairs, which in principle would decrease this entropy slightly. The statistical weight associated to this state labeled $n$ is thus

$$Z(n) = \mathrm{e}^{(N-n)\ln(g)+n\epsilon/k_{\mathrm{B}}T} = \mathrm{e}^{N\ln(g)} \left(\mathrm{e}^{\epsilon/T-\ln(g)}\right)^n = g^N z^n \tag{3.23}$$

with $z = \exp(\epsilon/T - \ln(g))$. The total partition sum is

$$Z = \sum_{n=1}^{N} Z(n) = g^N \sum_{1}^{N} z^n = g^N \frac{z^{N+1} - 1}{z - 1} \tag{3.24}$$

The total partition sum allows us to calculate the total free energy

$$F = -k_{\mathrm{B}}T\ln(Z) \tag{3.25}$$

as well as the energy

$$E = k_{\mathrm{B}}T^2 \frac{\mathrm{d}\ln(Z)}{\mathrm{d}T} \tag{3.26}$$

and the heat capacity

$$C = \frac{\mathrm{d}E}{\mathrm{d}T} \tag{3.27}$$

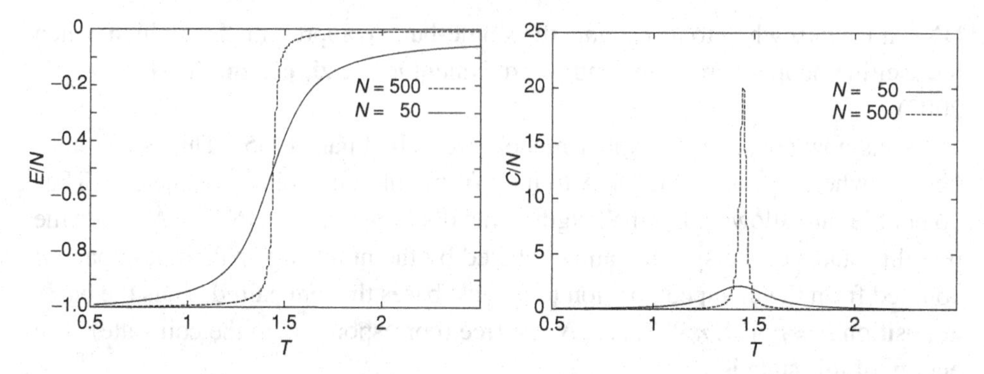


Figure 3.10. The zipper model with energy per base pair $\epsilon = 1$ and entropy per base pair $k_B \ln(2)$. The two curves in each plot refer to different system sizes, and the increase in sharpness of the phase transition illustrates that the phase transition in the zipper model is first order. The transition occurs at temperature $k_B T_M = \epsilon/\ln(2)$.

This behavior is shown in Fig. 3.10. Notice that in the zipper model, the transition gets (infinitely) sharper as $N \to \infty$. In fact the transition is a first-order phase transition.

In practice, heterogeneity of the bindings in DNA alters the sharp transition expected of homogeneous DNA, because different regions melt at different temperatures. Further, different DNA strands have different melting curves, and the melting temperature can differ by up to 10 K between different DNA strands. In general the melting temperature $T_m$ is defined as the point where the DNA is half melted. $T_m$ is typically between 70 and 80 °C.

As we increase the temperature, DNA melting appears to occur in steps. This would also be found in a zipper model with variable energy from base pair to base pair. The steps correspond to a walk in a one-dimensional potential landscape where one can be trapped in various valleys along the way; for a discussion see Hwa *et al.* (2003). DNA stacking energies and entropies can be found in SantaLucia *et al.* (1996).

## *Questions*

(1) Calculate the heat capacity in the zipper model, and find the scaling of heat capacity at the transition point with system size $N$.

(2) Consider the zipper model with $\epsilon = -1$ and $g = 10$. Calculate the transition temperature $T_m$, and calculate the average number of free base pairs at $T - 0.01$ for a system of size $N = 50$, and a system with $N = 500$.

(3) Consider a zipper model with variable energies, $\epsilon(i) = -1 \pm 1$ and write a computer model to calculate the specific curve for one realization of such a zipper. Compare two

different realizations, and calculate numerically the average heat capacity over a large ensemble of zippers. Use, for example, the system size $N = 500$ with $g = 2$.

(4) When double-stranded DNA melts completely, the two strands separate from each other. If the two strands are confined in a volume $V$ of $1\,\mu m^3$, calculate the additional entropy gain associated with the translational motion relative to each other. How does the melting temperature change with $V$?

(5) Again consider the example of melting a double stranded DNA in a volume $V$. Assume that the individual base pair energies all are $\epsilon = -0.7$ kcal/mol and the polymer degree of freedom has $g = 2$ states per melted base pair. First ignore strand separation and estimate the melting temperature for an $N = 500$ double-stranded piece of DNA. What would be the melting temperature within a volume $V = 1\,\mu m^3$, in a volume $V = 1\ m^3$ and in a volume $V = 10^9\ m^3$?

(6) Consider a segment of double-stranded DNA where there are two subsequent regions of five base pairs that can not pair. The two regions are separated by two bases that could pair with two others on the opposing strand with free energy $\Delta G < 0$. Assume that the open segment can be assigned free energies similar to those of DNA bubbles. What values of $\Delta G$ would make the formation of two bubbles possible?

## RNA

RNA is an information-carrying molecule like DNA, but like proteins it is capable of performing catalytic functions, as well as ligate (put together by covalent bond) and sometimes also cleave (break a covalent bond) other polymers. RNA is made of four types of base, namely A, G, C and U. Thus T from DNA is replaced by an U in RNA. This has the effect that in RNA the G can base pair with both C and with U, thereby opening for more sliding between RNA strands than would be the case for DNA. This helps single-stranded RNA from being captured in metastable states upon folding. Also the RNA backbone has an additional OH group, which makes it more flexible, and allows the RNA to form elaborate three-dimensional structures.

RNA folding is more complicated than DNA because it forms hairpins and loops (actually DNA does also, if left single stranded, but that has presumably no function). Therefore RNA structure is far from one-dimensional (see Fig. 3.11). However, it is more tractable than proteins, because most of the energy in RNA sits in the secondary structure: i.e. it is essentially given by the number of contact energies, one for each paired base pair, and some entropies for the loops. This allows accurate calculation of equilibrium properties. However, the fact that each base pair adds a few $k_B T$ in energy also implies that after forming a few tenths of base pair contacts, the system may get sufficiently frozen to prevent global equilibrium to be reached. Thus one could expect that large RNA molecules never reach their ground state. Miraculously, it seems that biologically relevant RNAs indeed have a native state that coincides with their ground state.

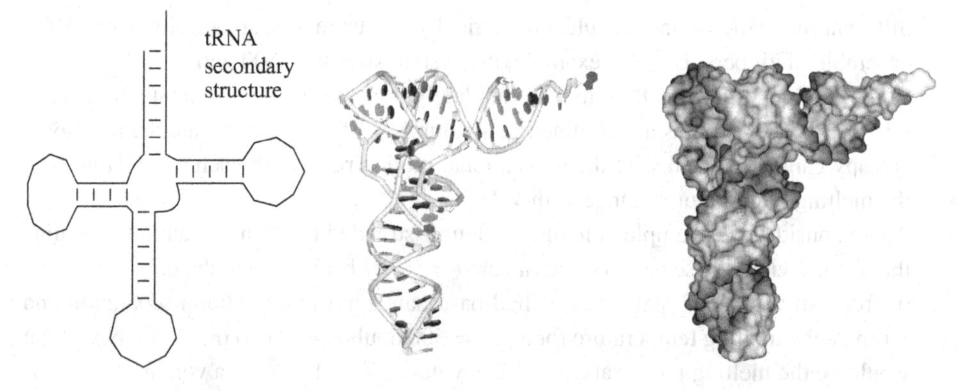


Figure 3.11. Secondary structure of an RNA molecule, here a transfer RNA (tRNA) responsible for the genetic code. Note that this "cover leaf" structure folds up to the final L-shaped tertiary structure characteristic of tRNAs, with a recognition site in the lower end and an amino acid bound to the upper right end. The right-hand panel shows the water-accessible surface of the folded tRNA.

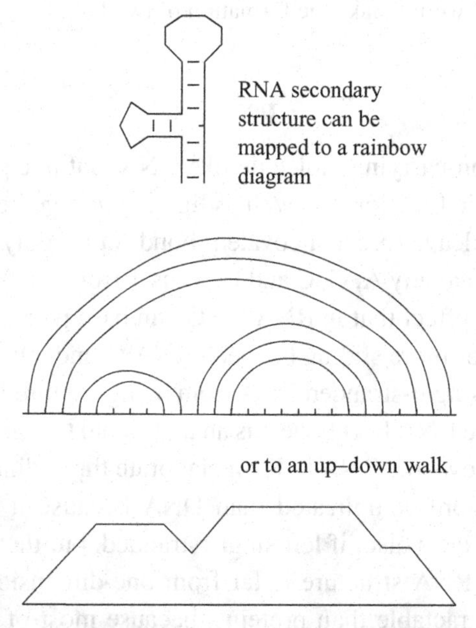


Figure 3.12. Mapping of RNA secondary structure to an arch or rainbow diagram (top), and a random walk (below) with a step up each time a bond is initiated, and a step down each time a bond is terminated (mountain diagram). This mapping is possible when pseudo-knots are ignored.

Secondary RNA folding energies can be calculated if we ignore pseudo-knots. Presenting each base pairing as a connecting line, the partition sum can be treated by considering all non-crossing arches (see Fig. 3.12). Reading from left to right along the RNA string, and defining a step up each time a new base pairing is formed,

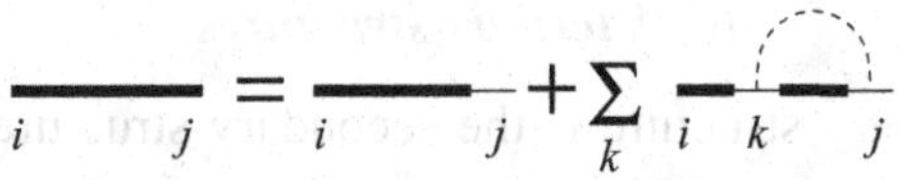

$$\underset{i\qquad j}{\overline{\qquad}} = \underset{i\qquad j}{\overline{\qquad}} + \sum_k \underset{i\quad k\quad j}{\overline{\qquad}}$$

Figure 3.13. Recursion relation for determining partition sum for an RNA sequence. The recursion relation resembles the one for the Hartree equation.

and a step down with each closing, this is also equivalent to all one-dimensional interfaces going from left to right. Each time the zero level is reached, one ends a hairpin.

The partition sum is calculated for a given RNA string of length $N$ by an iterative procedure involving $\sim N^3$ steps. The reason this can be done is because the binding options for a secondary structure can be limited to a "rainbow" type path of non-crossing arches shown in Fig. 3.12. This means that if base $i$ binds to base $j > i$ then a base $k$ outside the interval $[i, j]$ cannot bind to any base in $[i, j]$. Thereby one can recursively calculate the partition sum $[Z(i, j)]$ for all secondary structures of sequence $[i, j]$. See also Fig. 3.13.

In practice one can de-convolute the partition sum between each pair $i = 1, \ldots, N$ and $j = 1, \ldots, N$ into either the option where base $i$ is free, $i$ binds to $j$, or the option where $i$ binds to one of the other $k = i + 1, \ldots, j - 1$ bases

$$Z(i, j) = Z(i + 1, j) + Z(i + 1, j - 1)w(i, j) \tag{3.28}$$

$$+ \sum_{k=i+1}^{j-1} Z(i + 1, k - 1) \cdot Z(k + 1, j)w(i, k) \tag{3.29}$$

Here $Z(i, j)$ is partition function from $i$ to residue $j$, with $Z(i, i) = Z(i, i + 1) = 1$ and with stacking energies entering through $w(i, j) = \exp(-\Delta G(i, j)/k_B T)$; these can be found in Freir *et al.* (1986). Loop entropies are typically listed in a linear approximation (to avoid complications from log()s in the recursion relation). Typical values for interaction energies are $\Delta G_i \sim -2$ to $-4$ kcal/mol (at 20 °C), whereas loop entropies are about 2–3 kcal/mol for loops smaller than length 8.

Finally we reiterate that our approximation to limit RNA secondary structures to be of the "rainbow" type means that we have ignored secondary structures with pseudo-knots. A pseudo-knot is the interaction of distant loops, and they therefore break the rainbow diagram approximation. One may view such pseudo-knots as the next level of interactions, leading from the secondary structure to the final folded three-dimensional tertiary structure of the RNA.

### *RNA tertiary structures*

The basis for the tertiary structure is the secondary structure supplemented with interactions between the secondary structure elements. Some general statements can be made about the tertiary RNA structure: it typically consists of a number of helix segments 5–15 base pairs in length interrupted by bulges or loops. Both helices and nucleotides in the loops stack into fairly stiff structures, the helices with persistence length of 75 nm. Nearby stacks interact, and form an overall stiff structure. Also it is noteworthy that the free energy of the loops is mostly associated with stacking, and not entropy.

The timescales for the steps in RNA folding have been measured (see Sclavi *et al.*, 1998) and are of order:

- zipping of helical regions is very fast, closing of hairpin loops is 10 μs;
- establishment of first secondary elements takes 0.01–0.1 s, whereas final secondary folding is of the order of seconds;
- tertiary folding may take minutes or more.

Prediction of tertiary structure formation in RNA is still an open problem.

RNA is negatively charged ($-2e$ per base pair), and its tertiary folding properties especially depend on the presence of positive ions in the medium. The more positive these ions are, the better the folding. In particular ions with $+2e$ (e.g. $Ca^{2+}$ and $Mg^{2+}$) or $+3e$ (e.g. $(Co(NH_3)_6)^{3+}$) increase folding stability. On the other hand, such ions also slow down the folding process as they add roughness to the folding landscape, i.e. they increase the waiting time in intermediate states. Especially remarkable is the very cooperative effect of $Mg^{2+}$ on folding stability. Changing $Mg^{2+}$ concentration the folding–unfolding of RNA switches suddenly at $K = 2$ mM. In a living cell the total $Mg^{2+}$ is about 20 mM, but the free concentration may be close to 2 mM. In fact a free $Mg^{2+}$ concentration of close to 2 mM may be optimal in the sense that it facilitates folding but contributes anyway to stability of the final native state for functional RNA molecules (like tRNA, or ribosomal RNA). Another aspect of in vivo RNA structures is regulation of folding through proteins. In *E. coli* the very abundant protein CspA is known to prevent the RNA folding (Newkirk *et al.*, 1994), and thereby it has a role for ribosomal access to translating the mRNA.

### *Questions*

(1) Ribosomes are large, and contain sub-units of RNA with mass of about $1.8 \times 10^6$ Da. What is the total number of bases in these sub-units? What would be the number of amino acids needed to make similar sized sub-units? How many base pairs of DNA are needed to code for sub-units, and how many would be needed to code for similar weight sub-units if they were made of proteins?

(2) A large (numerical) exercise is to consider all possible folds of a homopolymer RNA strand, and use the Hartree approximation to calculate the free energy as a function of temperature, say for a strand of length $N = 50$. One may subsequently compare this to a strand of length $N = 200$ to see if the melting transition is first order.

## Manipulating molecular information

### *DNA computing*

A key to the success of modern biology has been our ability to measure, manipulate and thereby determine the information in DNA. Two key aspects of modern DNA technology are PCR and electrophoresis. In PCR (polymerase chain reaction) one can amplify the number of a specific piece of DNA by many orders of magnitude. In electrophoresis one can determine the length of the DNA, and thereby separate a sample of DNA strands according to length. The latter is in fact a key element in sequencing of DNA. To introduce these basic DNA manipulation techniques we will describe how they have been utilized in a proposal to use the DNA to perform calculations. Then in two sections we will describe the PCR method and the physics behind gel-electrophoresis.

Consider the following problem. We have seven cities, each being connected to at least one of the other cities by a directed link. The question is whether one can devise a travel path through all cities starting in city 7 and ending in city 3, such that each city is visited exactly once. Adleman (1994) solved this problem by assigning each city a DNA code, using say a sequence of 20 bases; see Fig. 3.14.

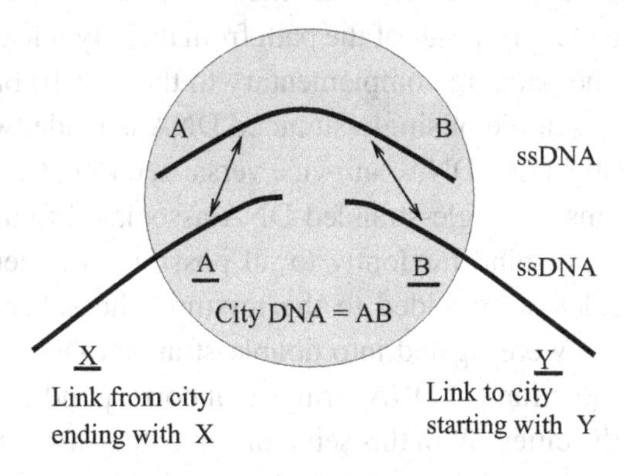


Figure 3.14. Mapping of directed network to single-stranded (ss) DNA pieces. For each node (shaded circle above) one makes an ss DNA sequence AB, consisting of sequence A followed by sequence B. Each incoming link to AB is constructed from an ss DNA strand of which the second half consists of the complementary ss DNA to A (A with underline). Similarly each outgoing link to AB is constructed from an ss DNA strand of which the first half consists of the complementary ss DNA to B.

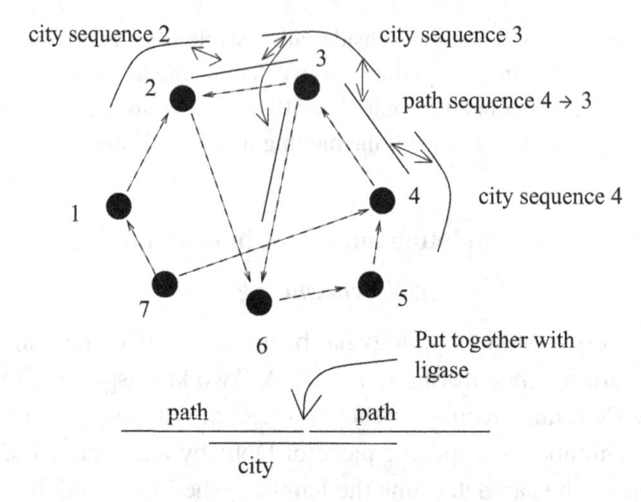


Figure 3.15. DNA computing for a complete problem, where the dots are cities and the arrows are allowed connections between cities. For each city we devise a 20 base pair DNA sequence; for each link we design a base sequence that is complementary to the cities it connects. By making single-stranded DNA associated to all cities and all allowed links, and mixing them together in a single test tube, the single-stranded DNA will bind randomly to their complementary partners. Thus one makes an ensemble of DNA strings that correspond to all allowed travel itineraries between the cities (all paths in the directed network).

Each allowed link between two cities is similarly assigned a 20 base DNA code constructed as shown. This code for the directed links is made to complement half the sequence of each of the involved cities. That means the first 10 bases are complementary to the last 10 bases of the path from the city it leaves from. Further, the last 10 bases of the path are complementary to the first 10 bases of the city of arrival. Thereby an ensemble of single-stranded DNA is made, where "city" DNA partially complements "link" DNA, and vice versa; see Fig. 3.15.

By mixing solutions of single-stranded DNA associated to all cities and to all allowed paths, they will bind randomly to all possible complementary partners. When **ligase** molecules were added to the mixture, the subsequently matching single-stranded DNAs were ligated into double-stranded DNA. One then obtains an ensemble of double-stranded DNA strings that correspond to all possible travel itineraries between the cities. With this setup one may then ask whether there is one path that connects all cities, visits each city once and furthermore starts in city 7 and ends in city 3. To test this (1) select DNA pieces that are exactly $7 \times 20 = 140$ base pairs long by gel-electrophoresis, and (2) test whether single-stranded versions of this DNA have complements to all 7 cities. The first step is done by screening the ensemble of strings for the ones that are exactly 140 base pairs long. This is done by gel-electrophoresis. The second step tests the actual content of the DNA, and could involve attachment of each of the known specific "city" DNA to magnetic beads. In a sequence of seven such tests, each time first denature the DNA, and then select the part of the single-stranded DNA that binds to the city DNA in question. Only single-stranded DNA that binds to magnetic beads with city DNA is retained, the rest is washed away.

If some DNA remains after all these tests one knows that there exists a path that connects all cities, and visits each city only once. However, each of the above screens reduces the numbers of DNA, and thus one needs to amplify the remaining DNA along the way. This is done by PCR, a method that in itself allows for an additional screen of amplifying certain DNA sequences. For example, if we want paths that start only in city 7 and end in city 3, we may use primers corresponding to these two cities in the PCR amplification. In the next section we will describe PCR and the use of primers.

### *Questions*

(1) How many DNA molecules are in 1 ml of solution with a concentration of 1 $\mu$M double-stranded DNA?

(2) Consider a network of $N$ cities, each with two outgoing links and two incoming links. Consider Adleman's algorithm and calculate how big $N$ can be if we want the solution of strands in Question (1) to sample all paths up to length $N$.

(3) What length of DNA strand is needed if we want to differentiate between, say, $N = 1000$ different nodes? First solve this problem where the complementarity is perfect. Then discuss how this number must be increased if any base pair can mismatch with a probability $p \sim 0.1$.

## *Polymerase chain reaction*

**The polymerase chain reaction** is a method of replicating DNA molecules, using a DNA polymerase enzyme that is resistant to high temperatures. To do this, a short piece of DNA is required as a primer to signal where to begin making the complementary copy. DNA polymerase will start adding bases (A, C, T or G) to the primer to create a complementary copy of the template DNA strand. This makes a piece of double-stranded DNA. Heating the system makes this melt into two single-stranded pieces. Cooling again makes new primers bind to the ends of the two strands, thus opening for new polymerase action. Thus we have effectively doubled the DNA between the defining primers. The primers have to be such that for a given piece of double-stranded DNA there is one primer in one end complementary to one strand, and another primer in the other end, complementary to the other strand (see Fig. 3.16).

The main ingredient in this system is the DNA polymerase (DNAP) molecule, and the discovery that opened up the temperature cycling was the isolation of DNAP from a bacterium that normally lives under extremely high temperatures. In this way a DNAP was obtained that did not melt when the temperature was high enough

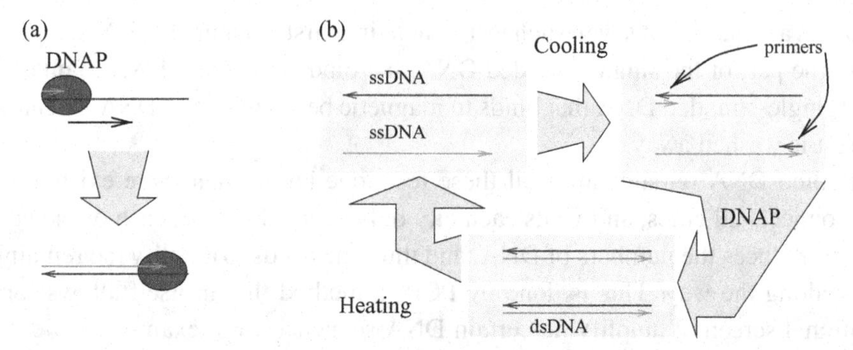


Figure 3.16. (a) DNA polymerase (DNAP) binds to double-stranded (ds) DNA, and extends this along the single-stranded (ss) DNA by subsequently binding bases to it. (b) By shifting between low and high temperature one can respectively melt ds DNA, and replicate ss DNA to obtain ds DNA. In this way one can in principle amplify a single ds DNA molecule into a macroscopic amount of DNA. The amplification relies on (1) an ample supply of base pairs and energy in form of ATP/GTP, (2) a large number of primers for replication from both ends of the ds DNA, and (3) the availability of DNAP that is resistant to high temperatures.

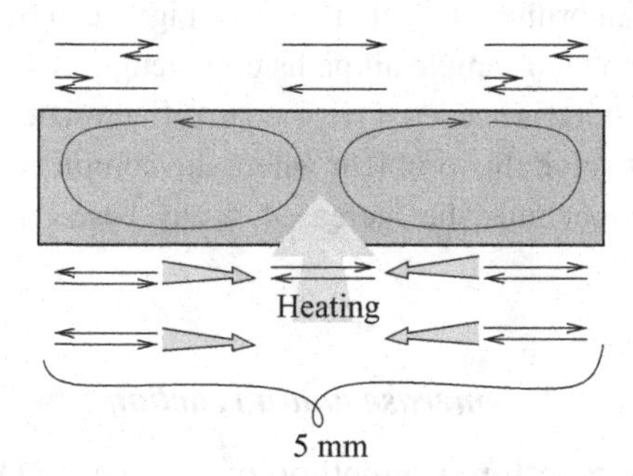


Figure 3.17. PCR in a thermal convection setup constructed by Braun *et al.* (2003). The setup is reminiscent of what one may have found in thermal vents in the primordial Earth (Braun & Libchaber, 2004). The DNA melts as the center temperature reaches 95 °C, whereas the DNAP replicates the ss DNA in the colder periphery (at about 60 °C). The cycle time is about 30 s. The heating is done by a laser.

to melt the DNA. In principle the replication could go on ad infinitum, provided that there is enough raw material present. However, the DNAP makes errors, and they get exponentially amplified. After 30–40 cycles, the PCR becomes too imprecise for practical purposes.

A particularly interesting implementation of the PCR idea is shown in Fig. 3.17; the temperature cycling is done automatically by heating a small convection cell in the middle. This simple setup automatically cycles the temperature between a hot middle and a colder periphery where DNAP replicates the separated DNA strands.

### *Reptation and gel-electrophoresis*

The physical problem here is to separate DNA fragments ranging in size from 1 to $10^9$ bp! Because DNA is (negatively) charged (it is an acid; in water, the phosphate groups are ionized, which confers one negative charge per nucleotide), it migrates in an electric field. This is the basis for electrophoresis. However, if the medium is just water, different sized fragments all migrate at the same speed, so no separation occurs. The reason is as follows.

Calling **u** the velocity of the DNA molecule and **F** the applied force, we have

$$\mathbf{u} = \mu \mathbf{F} \tag{3.30}$$

where $\mu$ is a mobility (units of time/mass). Let us consider a chain of $N$ segments, each of size $a$ and charge $q$; in a field $E$ the force is

$$F = NqE \tag{3.31}$$

while the mobility can be estimated from the Stokes formula

$$F_{\text{drag}} \propto \eta N a u \tag{3.32}$$

which gives the drag force on an object of size $Na$ moving at velocity $u$ through a fluid of viscosity $\eta$. Therefore the mobility of the chain is

$$\mu \propto \frac{1}{\eta a N} \tag{3.33}$$

We then obtain for the migration velocity

$$u \propto \frac{NqE}{\eta a N} \tag{3.34}$$

independent of the size of the chain $N$!

As in the literature, we introduce the "electrophoretic mobility" $\mu_e = u/E$ (this has different units from a usual mobility!). The result is then that the electrophoretic mobility is independent of $N$:

$$\mu_e = \frac{u}{E} \propto \frac{q}{\eta a} \tag{3.35}$$

and consequently no separation occurs.

In summary, if we do electrophoresis in water the applied force is $\propto N$, but the drag is also $\propto N$, so that the mobility is independent of $N$.

To remedy this situation one can think of either changing the $N$ dependence of the charge, or changing the $N$ dependence of the friction. It turns out that the second alternative is simpler in practice, and it can be realized by running the electrophoresis in a gel.

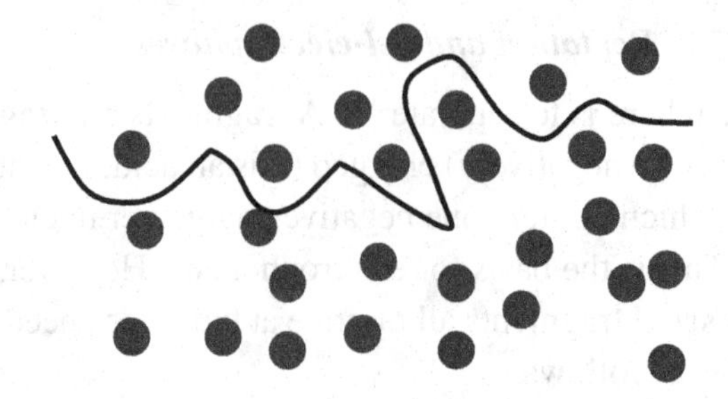

Figure 3.18. Polymer reptating through a gel.

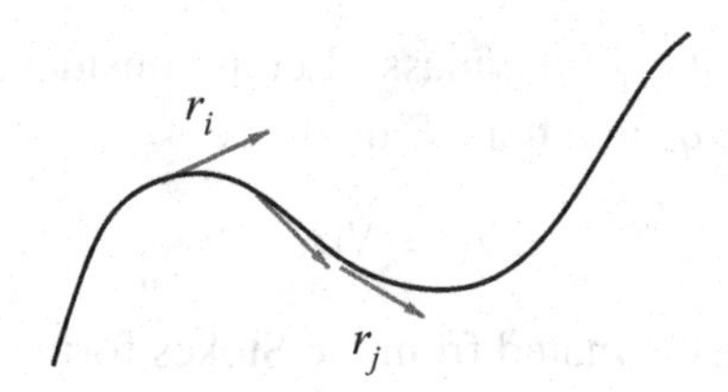


Figure 3.19. Polymer with unit tangent vectors indicated.

A gel can be visualized as a series of obstacles, or alternatively a series of interconnected pores. Consider a polymer chain in the gel, see Fig. 3.18: if the chain is much longer than the pore size (the distance between cross-links in the gel) then its lateral motion is restricted, and the chain moves in a snake-like fashion (i.e. the "tail" following along the path chalked out by the "head"), which is called "reptation". This drastically alters the mobility.

We now discuss reptation of a polymer chain, following the original article by De Gennes (1971). Consider a chain of $N$ links (of size $a$) $\mathbf{r_1}, \mathbf{r_2}, \ldots, \mathbf{r_N}$, see Fig. 3.19. Let $a$ be the persistence length, so the directions of the links are statistically independent: $\langle r_i r_j \rangle = a^2 \delta_{ij}$. In Fig. 3.20 we schematize the motion of the chain in the gel as consisting of the migration of "defects".

Call $b$ the chain length stored in a defect. Consider a monomer along the chain, e.g. point B in Fig. 3.20. When a defect passes by, this monomer moves by an amount $b$. Call $D$ the diffusion coefficient of the defects along the chain; this quantity is characteristic of the microscopic "jumping" process, independent of $N$. If $\rho(n)$ is the number of defects per unit length of the chain, we can write an expression for the current of defects $j(n)$ at monomer position $n$ along the chain:

$$j(n) = -D\frac{1}{a}\frac{\mathrm{d}\rho}{\mathrm{d}n} + \mu_{\mathrm{d}}\rho\varphi(n) = D\left(-\frac{1}{a}\frac{\mathrm{d}\rho}{\mathrm{d}n} + \frac{1}{k_{\mathrm{B}}T}\rho\varphi(n)\right) \tag{3.36}$$

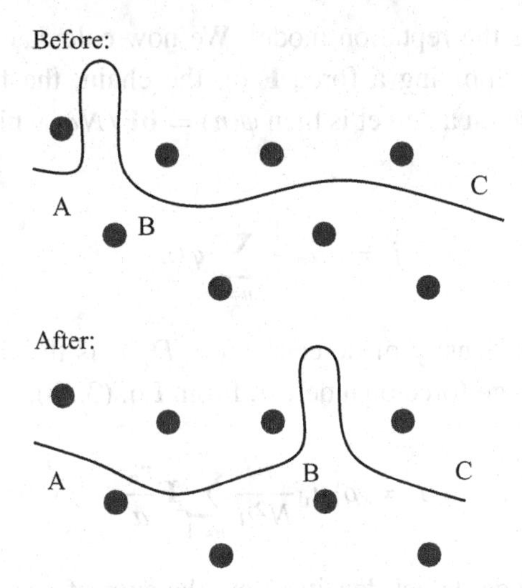


Figure 3.20. Reptation by migration of "defects", first located between A and B. If the length stored in the defect is $b$, then when the defect moves, the monomer at position B moves by an amount $b$. In practice, the typical defect size $b$ is related to the pore size, and thus it is a property of the gel.

The second term on the right-hand side is the drift imposed by an external force, i.e. $\varphi(n)$ is the force on one defect. We used the Einstein relation $D = k_B T \mu_d$ to relate the mobility $\mu_d$ of the defects to their diffusion coefficient $D$ (see the Appendix for this relation). To obtain the mobility of the chain, we have to introduce an external force and calculate the resulting velocity of the center of mass (CM) of the chain (for this reason we allow for an external force $\varphi(n)$ in Eq. (3.36)). We call $\mathbf{f}(n)$ the force applied on a "monomer", and $\varphi(n)$ the resulting force on a "defect". Thus we now have to relate $\varphi$ and $f$. If one defect moves by $a\delta n$ along the chain, then $\delta n$ monomers are displaced, each by $\mathbf{b}$ (see Fig. 3.20), where $\mathbf{b}$ is a vector of magnitude $b$, tangential to the chain. By thus equating the work done seen from the side of the monomers with the work done calculated by the defect:

$$\delta n \mathbf{f}(n) \times \mathbf{b} = \text{work done} = a\delta n \varphi(n) \tag{3.37}$$

$$\Rightarrow \varphi(n) = \frac{1}{a}\mathbf{b} \times \mathbf{f}(n) \tag{3.38}$$

Because the only way a monomer can move is if a defect crosses that point in the chain, the velocity of the $n$th monomer is related to the current of defects at position $n$:

$$\frac{\mathrm{d}\mathbf{r_n}}{\mathrm{d}t} = \mathbf{b}\, j(n) \tag{3.39}$$

The above equation is the reptation model. We now calculate the chain mobility in this model. When applying a force $\mathbf{F}$ on the chain, the force per monomer $\mathbf{f}(n) = \mathbf{F}/N$. The force on a defect is then $\varphi(n) = \mathbf{bF}/Na$, which implies a steady state current of defects

$$j = \rho\mu_{\mathrm{d}}\frac{1}{N}\sum_{n=1}^{N}\varphi(n) \tag{3.40}$$

Here $\rho$ is the average density of defects, $\mu_{\mathrm{d}} = D/T$ is the defect mobility, and $\sum\varphi(n)/N$ is the average force on a defect. From Eq. (3.38):

$$j = \langle\rho\rangle\mu_{\mathrm{d}}\frac{b}{N^2a}\sum_{n=1}^{N}\mathbf{F}\frac{\mathbf{r_n}}{a} \tag{3.41}$$

where $\langle\rho\rangle$ is the average defect density. Now the sum of vectors along the chain is the end-to-end vector:

$$\sum_{n}\mathbf{r_n} = \mathbf{h} \tag{3.42}$$

Taking $\mathbf{F}$ in the $x$-direction:

$$j = \frac{\rho\mu_{\mathrm{d}}bF}{(Na)^2}h_x \tag{3.43}$$

The velocity of the CM is

$$\mathbf{v}_{\mathrm{CM}} = \frac{1}{N}\sum_{i=1}^{N}\frac{\mathrm{d}\mathbf{r}_i}{\mathrm{d}t} \tag{3.44}$$

and using Eq. (3.39) we have

$$\mathbf{v}_{\mathrm{CM}} = \frac{b}{Na}j\sum_{i=1}^{N}\mathbf{r}_i = \frac{bj}{Na}\mathbf{h} \tag{3.45}$$

This is the velocity of the CM for a given conformation of the chain specified by its EED $\mathbf{h}$. Thus all conformations with the same $\mathbf{h}$ have the same $\mathbf{v}_{\mathrm{CM}}$. To get the statistical average of $\mathbf{v}_{\mathrm{CM}}$ we have to average over all values of $\mathbf{h}$; since $\langle h_x\mathbf{h}\rangle = \langle h_x^2\rangle\mathbf{x}$ we obtain:

$$\langle v_{\mathrm{CM}}\rangle = \frac{b^2\rho\mu_{\mathrm{d}}F}{(Na)^3}\langle h_x^2\rangle \tag{3.46}$$

and the mobility of the chain, $\langle v_{\mathrm{CM}} \rangle / F$, is:

$$\mu = \frac{b^2 \rho \mu_{\mathrm{d}}}{(Na)^3} \langle h_x^2 \rangle \tag{3.47}$$

If $F$ is sufficiently small (we come back to this point later), then:

$$\langle h_x^2 \rangle = \frac{1}{3} N a^2 \tag{3.48}$$

(i.e. the chain is not stretched) and

$$\mu = \frac{b^2 \rho \mu_{\mathrm{d}}}{3N^2 a} \tag{3.49}$$

i.e. $\mu \propto N^{-2}$ or $D \propto N^{-2}$ for the diffusion coefficient, which is very different from the mobility in the solvent without gel, which is $\propto N^{-1}$. This is the basis for gel-electrophoresis separation, as we shall see next. However, we have to remember that we have calculated the mobility for $F \to 0$; this yields the right diffusion coefficient but for finite applied external force $F$ we have to see how $\langle h_x^2 \rangle$ is modified.

Before proceeding we note that the above analysis can be reproduced by a simple scaling argument. In the gel we can think of the chain behaving as if it was trapped in a tube. The friction for motion *along the tube* is proportional to the tube length $L$, which is proportional to $N$, so if we apply a force $F_{\mathrm{p}}$ tangential to the tube the velocity of the chain along the tube will be

$$v_{\mathrm{p}} = \mu_{\mathrm{tube}} F_{\mathrm{p}} \tag{3.50}$$

with $\mu_{\mathrm{tube}} = \mu / N$, where $\mu$ is independent of $N$. Therefore the diffusion coefficient for diffusion *along the tube* is

$$D_{\mathrm{tube}} = \mu_{\mathrm{tube}} k_{\mathrm{B}} T = \frac{\mu_1 k_{\mathrm{B}} T}{N} \propto N^{-1} \tag{3.51}$$

The renewal time is the time it takes to diffuse one tube length along the tube:

$$\tau_{\mathrm{r}} = \frac{L^2}{D_{\mathrm{tube}}} \propto \frac{N^2}{N^{-1}} = N^3 \tag{3.52}$$

Thus $\tau_{\mathrm{r}}$ is the time after which the chain is trapped in a new, independent tube. To estimate the translational diffusion coefficient of the chain, consider that after a time $\tau_{\mathrm{r}}$ the CM of the chain has moved by only a radius of gyration $R_0 = aN^{1/2}$, therefore the diffusion coefficient for reptation is

$$D_{\mathrm{rep}} = \frac{(\text{characteristic length})^2}{\text{characteristic time}} = \frac{R_0^2}{\tau_{\mathrm{r}}} \propto \frac{N}{N^3} = N^{-2} \tag{3.53}$$

which is our previous result as mobility and $D$ are related.

*Electrophoresis in a gel*

Consider a gel of pore size $a$; we model the chain as $N$ segments $\mathbf{r}_1, \mathbf{r}_2, \ldots, \mathbf{r}_N$ of size $a$. We take $q$ to be the charge per segment. In a field $\mathbf{E}$, the total force *along the tube* is

$$F_{\mathrm{p}} = \sum_{i=1}^{N} q\mathbf{E}\frac{\mathbf{r}_i}{a} = \frac{q}{a}\mathbf{E}\mathbf{h} \tag{3.54}$$

where $\mathbf{h} = \sum \mathbf{r}_i$ is the EED of the tube. The corresponding velocity along the tube is

$$v_{\mathrm{p}} = \mu_{\mathrm{tube}}\frac{qE}{a}h_x = \mu\frac{qE}{a^2N}h_x \tag{3.55}$$

where again $\mu_{\mathrm{tube}} = \mu/N$. In Eq. (3.55) we have taken $\mathbf{E}$ in the $x$-direction.

The velocity of the CM is

$$\mathbf{v}_{\mathrm{CM}} = \frac{1}{N}\sum_{i=1}^{N}\frac{\mathrm{d}\mathbf{r}_i}{\mathrm{d}t} \tag{3.56}$$

$$\Rightarrow \langle v_{\mathrm{CM}}\rangle = \frac{1}{N}\left\langle\sum_{i=1}^{N} v_{\mathrm{p}}\frac{\mathbf{r}_i}{a}\right\rangle = \frac{1}{Na}\langle v_{\mathrm{p}}\mathbf{h}\rangle \tag{3.57}$$

$$\Rightarrow \mathbf{v}_{\mathrm{CM}} = \frac{\mu qE}{a^3N^2}\langle h_x\mathbf{h}\rangle = \frac{\mu qE}{a^3N^2}\langle h_x^2\rangle\mathbf{x} \tag{3.58}$$

so the electrophoretic mobility is (remember that we have different units from usual mobility)

$$\mu_{\mathrm{e}} = \frac{\langle v_{\mathrm{CM}}\rangle}{E} = \frac{\mu q}{a^3N^2}\langle h_x^2\rangle \tag{3.59}$$

We now distinguish two extreme cases.

(1) The polymer is not oriented by the field (weak fields and/or short polymers): then $3\langle h_x^2\rangle = Na^2$, and $\langle h_x^2\rangle \propto N \Rightarrow \mu_{\mathrm{e}} \propto 1/N$ and there is separation of the polymers by length.
(2) The polymer is oriented (stretched): then $\langle h_x^2\rangle \propto (Na)^2$ and $\mu_{\mathrm{e}}$ is independent of $N$. In this case there is no separation!

The key quantity that determines whether we are in weak (1) or in strong (2) fields is the degree of alignment for individual segments determined by the ratio of energy associated with electric alignment to the thermal energy:

$$\chi = \frac{qaE}{2k_{\mathrm{B}}T} \tag{3.60}$$

This quantity, or rather $\chi^2$, should be compared to $1/N$ (see Question (3) below). This limitation implies that separation of DNA molecules longer than about 50 000 bp under steady fields is not possible. The limitation on separation can be overcome by varying the electric field, thereby playing with the time it takes to reorient the migrating DNA molecules (Schwartz & Cantor, 1984). In fact, pulsed field electrophoresis was one of the major experimental discoveries that made it possible to separate DNA strands to well over $10^7$ base pairs.

## *Questions*

(1) Consider a repeating polymer in an electric field. Let $h_x$ be the projection of polymer segment $x$ along the field $E$: $h_x = a\sum_{i=1}^{N}\cos\theta_i$ and thus $h_x^2 = a^2\sum_{i,j}\cos\theta_i\cos\theta_j$. Find $\langle h_x^2\rangle$ in terms of $N$, $a$, $\langle\cos(\theta)\rangle$ and $\langle\cos(\theta)^2\rangle$.

(2) The head follows the tail, thus the alignment of $\theta$ is defined by the energy distribution of the head segment: $p\cos(\theta) \propto \exp(1/2(Eqa\cos(\theta)/k_\mathrm{B}T)) = \exp(\chi\cos(\theta))$, where the charge $q$ of the segment is assumed to be uniform along the rod. Calculate $\langle\cos(\theta)\rangle$ and $\langle\cos^2(\theta)\rangle$ in terms of $\chi$ (the dimensionless field).

(3) Show by expanding to second order in $\chi$ that the mobility $\mu_\mathrm{e} = (\mu_{1q}/3a)((1/N) + (1/3\chi^2))$.

(4) Consider DNA with persistence length 50 nm and assume that it has a charge $q = 200e$ per persistence length (about two charge units per base pair). Use an electric field $E = 1000$ V/m and calculate the length above which DNA cannot be separated by constant field electrophoresis.

## References


Adleman, L. (1994). Molecular computation of solutions to combinatorial problems. *Science* **266**, 1021.

Ariyoshi, M., Nishino, T., Iwasakidagger, H., Shinagawadagger, H. & Morikawa, K. (2000). Crystal structure of the Holliday junction DNA in complex with a single RuvA tetramer. *Proc. Natl Acad. Sci. USA* **97**, 8257–8262.

Blake, R. D. (1999). Statistical mechanical simulation of polymeric DNA melting with MELTSIM. *Bioinformatics* **15**, 370.

Borer, P. N., Dengler, B., Tinoco Jr, I. & Uhlenbeck, O. C. (1974). Stability of ribonucleicacid double stranded helices. *J. Mol. Biol.* **86**, 843.

Braun, D., Goddard, N. L. & Libchaber, A. (2003). Exponential DNA replication by laminar convection. *Phys. Rev. Lett.* **91**, 158103.

De Gennes, P. G. (1971). Concept de reptation pour une chaîne polymérique. *J. Chem. Phys.* **55**, 572.

Freir, S. M. *et al.* (1986). Improved free-energy parameters for predictions of RNA duplex stability. *Proc. Natl Acad. Sci USA* **83**, 9373–9377.

Holliday, R. (1964). A mechanism for gene conversion in fungi. *Genet. Res.* **5**, 282–304.

Hwa, T., Marinari, E., Sneppen, K. & Tang, Lei-han (2003). Localization of denaturation bubbles in random DNA sequences. *Proc. Natl Acad. Sci. USA* **100**, 4411–4416.

Kafri, Y., Mukamel, D. & Peliti, L. (2002). Melting and unzipping of DNA. *Eur. Phys. J.* **B27**, 135–146.

LeBret, M. (1978). Theory for supercoils can be found in *Biopolymers* **17**, 1939; (1984) **23**, 1835.

Marko, J. & Siggia, E. (1994). Fluctuations and supercoiling of DNA. *Science* **265**, 506. (1995). Statistical mechanics of supercoiled DNA. *Phys. Rev. E* **52**, 2912.

Newkirk, K., Feng, W., Jiang, W., Tejero, R., Emerson, S. D., Iouye, M. & Montelione, G. T. (1994). Solution of NMR structure of the major cold shock protein (CspA) from *Escherichia Coli*: identification of a binding epitope for DNA. *Proc. Natl Acad. Sci. USA* **91**, 5114.

Osterberg, R., Persson, D. & Bjursell, G. (1984). The condensation of DNA by chromium(III) ions. *J. Biomol. Struct. Dyn.* **2**(2), 285–290.

Rouzina, I. & Bloomfield, V. A. (2001). *Biophys. J.* **80**, 882–893.

SantaLucia, Jr. J. *et al.* (1996). *Biochemistry* **35**, 3555.

Schellman, J. A. (1955). The stability of hydrogen bonded peptide structures in aqueous solution. *Compt. Rend. Trav. Lab. Carlsberg. Ser. Chim.* **29**, 230–259.

Schwartz, D. C. & Cantor, C. R. (1984). Separation of yeast chromosome-sized DNAs by pulsed field gradient gel electrophoresis. *Cell* **37**, 67.

Sclavi, B., Sullivan, M., Chance, M. R., Brenowitz, M. & Woodson, S. A. (1998). RNA folding at millisecond intervals by synchrotron hydroxyl radical footprinting. *Science* **279**, 1940.

Takeda, Y., Ross, P. D. & Mudd, C. P. (1992). Thermodynamics of Cro protein – DNA interactions. *Proc. Natl Acad. Sci. USA* **89**, 8180.

Tang, L. H. & Chate, H. (2001). *Phys. Rev. Lett.* **86**, 830.

## Further reading

Allemand, J. F., Bensimon, D., Lavery, R. & Croquette, V. (1998). Stretched and overwound DNA forms a Pauling-like structure with exposed bases. *Proc. Natl Acad. Sci. USA* **95**, 14 152.

Bartel, D. P. & Unrau, P. J. (1999). Constructing an RNA world. *Trends Biochem. Sci.* **24**, M9–12.

Bates, A. D. & Maxwell, A. (1993). *DNA Topology*. Oxford: IRL Press.

Braun, D. & Libchaber, A. (2004). Thermal force approach to molecular evolution. *Phys. Biol.* **1**, 1–8.

Bundschuh, R. & Hwa, T. (2002). Statistical mechanics of secondary structures formed by random RNA sequences. *Phys. Rev. E* **65**, 031903.

Bustamante, C., Smith, S., Liphardt, J., Smith, D. (2000). Single-molecule studies of DNA mechanics. *Curr. Opin. Struct. Biol.* **10**, 279.

Charlebois, R. L. & Stjean, A. (1995). Supercoiling and map stability in the bacterial chromosome. *J. Mol. Evol.* **41**, 15–23.

Chu, G. (1991). Bag model for DNA migration during pulsed-field electrophoresis. *Proc. Natl Acad. Sci. USA* **88**, 11 071.

Clark, S. M., Lai, E., Birren, B. W. & Hood, L. (1988). A novel instrument for separating large DNA molecules with pulsed homogeneous electric fields. *Science* **241**, 1203.

Crick, F. H. C. (1963). On the genetic code. *Science* **139**, 461–464. (1968). The origin of the genetic code. *J. Mol. Biol.* **38**, 367.

Crick, F. H. C. *et al.* (1961). General nature of the genetic code for proteins. *Nature* **192**, 1227–1232.

Cule, D. & Hwa, T. (1997). Denaturation of heterogeneous DNA. *Phys. Rev. Lett.* **79**, 2375.

Dickerson, R. E., Drew, H. R., Conner, B. N., Wing, R. M., Fratini, A. V. & Kopka, M. L. (1982). The anatomy of A-, B-, and Z-DNA. *Science* **216**, 475–485.

Doi, M. & Edwards, S. F. (1986). *The Theory of Polymer Dynamics*. Oxford: Clarendon Press.

Essevaz-Roulet, B., Bockelmann, U. & Heslot, F. (1997). Mechanical separation of the complementary strands of DNA. *Proc. Natl Acad. Sci. USA* **94**, 11 935.

Faulhammer, D., Cukras, A. R., Lipton, R. J. & Landweber, L. F. (2000). Molecular computation: RNA solutions to chess problems. *Proc. Natl Acad. Sci. USA* **97**, 1385.

Fonatane, W. & Schuster, P. (1998). Continuity in evolution: on the nature of transitions. *Science* **280**, 1451.

Gilbert, W. (1986). The RNA world. *Nature* **319**, 618.

Higgs, P. G. (1996). Overlaps between RNA secondary structures. *Phys. Rev. Lett.* **76**, 704.

Hirao, I. & Ellington, A. D. (1995). Re-creating the RNA world. *Curr. Biol.* **5**, 1017.

Irvine, D., Tuerk, C. & Gold, L. (1991). Selexion. Systematic evolution of ligands by exponential enrichment with integrated optimization by non-linear analysis. *J. Mol. Biol.* **222**, 739.

Joyce, G. F. (1989). Amplification, mutation, and selection of catalytic RNA. *Gene* **82**, 83.

Joyce, G. F., Schwartz, A. W., Miller, S. L. & Orgel, L. E. (1987). The case for an ancestral genetic system involving simple analogs of the nucleotides. *Proc. Natl Acad. Sci. USA* **84**, 4398.

Kindt, J., Tzlil, S., Ben-Shaul, A. & Gelbart, W. M. (2001). DNA packaging and ejection forces in bacteriophage. *Proc. Natl Acad. Sci. USA* **98**, 13 671.

Lipton, R. J. (1995). DNA solution of hard computational problems. *Science*, **268**, 542.

Lubensky, D. K. & Nelson, D. R. (2002). Single molecule statistics and the polynucleotide unzipping transition. *Phys. Rev.* E **65**, 031917.

Morgan, S. R. & Higgs, P. G. (1996). Evidence for kinetic effects in the folding of large RNA molecules. *J. Chem. Phys.* **105**, 7152–7157.

Orgel, L. E. (1968). Evolution of the genetic apparatus. *J. Mol. Biol.* **38**, 381.

Poland, D. & Scherage, H. A. (editors) (1970). *Theory of Helix–Coil Transition in Biopolymers*. New York: Academic Press.

Record, M. T. & Zimm, B. H. (1972). Kinetics of helix coil transition in DNA. *Biopolymers* **11**, 1435.

Schultes, E. A. & Bartel, D. P. (2000). One sequence, two ribozymes: implications for the emergence of new ribozyme folds. *Science* **289**, 448–452.

Sinden, R. R. (1987). Supercoiled DNA: biological significance. *J. Chem. Educ.* **64**, 294–301.

Southern, E. M., Anand, R., Brown, W. R. A. & Fletcher, D. S. (1987). A model for the separation of large DNA molecules by crossed field gel electrophoresis. *Nucleic Acids Res.* **15**, 5925.

Stadler, B. M., Stadler, P. F., Wagner, G. P. & Fontana, W. (2001). The topology of the possible: formal spaces underlying patterns of evolutionary change. *J. Theor. Biol.* **213**, 241.

Strick, T., Allemand, J. F., Bensimon, D., Bensimon, A. & Croquette, V. (1996). The elasticity of a single supercoiled DNA molecule. *Science* **271**, 1835–1837.

Szostak, J. W. (1988). Structure, and activity of ribozymes. In *Redesigning the Molecules of Life*, ed. S. A. Benner. Heidelberg: Springer-Verlag, pp. 87–114.

Thirumalai, D. & Woodson, S. A. (1996). Kinetics of folding of proteins and RNA. *Acc. Chem. Res.* **29**, 433–439.

Unrau, P. J. & Bartel, D. P. (1998). RNA-cat-analyzed nucleotide synthesis. *Nature* **395**, 260.

Wartel, R. M. & Benight, A. S. (1985). Thermal denaturation of DNA molecules: a comparison of theory with experiment. *Phys. Rep.* **126**, 67.
Watson, A. & Crick, F. (1953). A structure for Deoxyribose Nucleic Acid. *Nature*, April 2.
Woese, C. (1967). *The Genetic Code*. New York: Harper and Row.
Xie, Z., Srividya, N., Sosnick, T. R., Pan, T. & Scherer, N. F. (2004). Single-molecule studies highlight conformational heterogeneity in the early folding steps of a large ribozyme. *Proc. Natl Acad. Sci. USA* **101**, 534–539.
Zamore, P. D., Tuschl, T., Sharp, P. A. & Bartel, D. P. (2000). RNAi: dsRNA directs the ATP-dependent cleavage of mRNA at 21 to 23 nucleotide intervals. *Cell* **101**, 25.
RNA structure: www.mb-jena.de/RNA.html
RNA structure: bioinfo.math.rpi.edu/zukerm/

# 4

# Protein structure

Giovanni Zocchi & Kim Sneppen

## General properties

Proteins are polymers (made of amino acids) that fold into a rigid globule with a well-defined structure. A typical protein consists of ~300 amino acids; the size of the folded globule is maybe ~5 nm, and the protein can fold into this native state within a fraction of a second. Proteins are the actuators, the molecular machines that perform all tasks necessary to the living cell. This includes pumping ions in and out, moving vesicles along the actin network, translating DNA, and, in general, catalyzing all metabolic chemical reactions. Proteins are designed to carry out regulatory functions (bind and dissociate) within a closed volume (cell, cell compartment) of order $V \sim 1\ \mu\text{m}^3$, which may typically contain 10–1000 copies of the regulatory protein. The requirement is then that at concentrations of order 1 molecule/$(\mu\text{m})^3 \approx 10^{-9}$ M a binding reaction between the two proteins:

$$\text{A} + \text{B} \leftrightarrow \text{AB} \tag{4.1}$$

is neither totally shifted to the left nor to the right, i.e.

$$[\text{A}] \approx [\text{B}] \approx [\text{AB}] \sim 10^{-9}\ \text{M} \tag{4.2}$$

which requires a dissociation constant $K$ of order

$$K = \frac{[\text{A}][\text{B}]}{[\text{AB}]} \sim 10^{-9}\ \text{M} \tag{4.3}$$

or a corresponding binding free energy

$$\Delta G = k_\text{B}T\ \ln(K) \sim -10 \text{ to } -15 \text{ kcal/mol} \tag{4.4}$$

a number that is somewhat universal for all protein interactions. In fact one also finds $\Delta G \sim -10\,\text{kcal/mol}$ as a typical value for the difference in free energy between the folded and unfolded (single-domain) protein. Presumably this is not a coincidence;

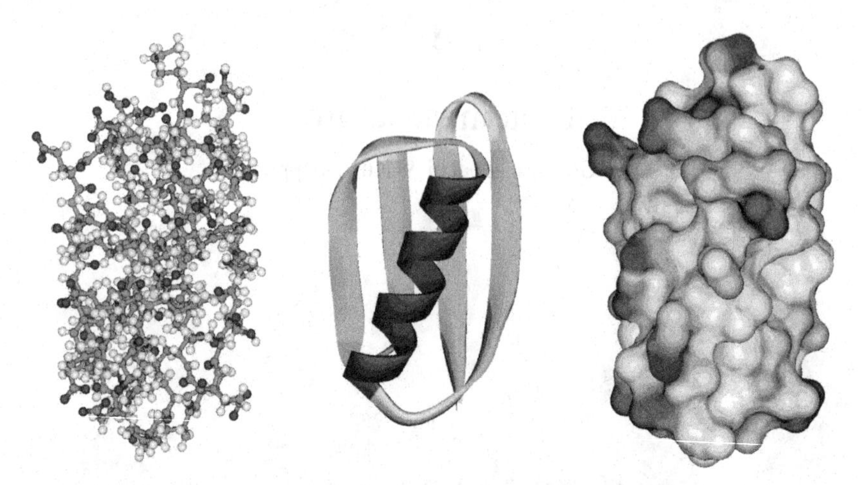

Figure 4.1. Typical protein structure: 1pgb (protein G) consisting of one β-sheet with four strands and one α-helix. On the left are shown all atoms including hydrogens (the white balls). In the middle the secondary structures are emphasized, and the right-hand panel shows the surface of the protein that is accessible to a water molecule.

if protein stability was much larger, it would be difficult to modify proteins with the then much weaker protein–protein interactions.

The relationship between cell volume and macromolecular interactions goes both ways. Thus, conversely, if 10 kcal/mol is the typical protein–protein or protein–DNA interaction, this limits the cell volume/compartment to be of order $\sim$1–10 μm$^3$. Otherwise one needs a much higher copy number of protein species in each cell, and changes in chemical composition would be a much more tedious process.

A protein is a heteropolymer built of a sequence of amino acids. There are a wide variety of amino acids; see Table 4.1. The sequence of these amino acids defines the primary structure of the protein. Typically, a protein can spontaneously fold into the well-defined, rigid three-dimensional structure in which it functions (see Fig. 4.1). The function is highly specific. It may for example include ordered motion (allosteric proteins), catalytic properties (which can also include conformational changes), or DNA binding and regulation of transcriptional activity.

## *Questions*

(1) Consider two specific proteins in a cell with 30% of weight being other protein species. Assume that proteins have an average size of 300 amino acid. Estimate what the average non-specific interaction between proteins is allowed to be if this should not dominate a specific interaction of $-15$ kcal/mol between the two proteins.

Table 4.1. *The amino acids*

| Amino acid | Hydrophobicity (kcal/mol) | Stabilization of α-helix (kcal/mol) | Volume ($A^3$) | Number of non-H atoms | "MJ" hydrophobicity | Frequency of use (%) |
|---|---|---|---|---|---|---|
| Cys (C) | −1.0 | −0.23 | 108 | 2 | −2.7 | 1.7 |
| Ser (S) | 0.3 | −0.35 | 86 | 2 | −0.1 | 6.9 |
| Thr (T) | −0.4 | −0.11 | 113 | 3 | −0.6 | 5.8 |
| Pro (P) | −1.4 | 3 | 121 | 3 | −0.5 | 5.1 |
| Ala (A) | −0.5 | −0.77 | 86 | 1 | −1.2 | 8.3 |
| Gly (G) | 0 | 0 | 58 | 0 | −0.5 | 7.2 |
| Asn (N) | 0.2 | −0.07 | 116 | 4 | −0.2 | 4.4 |
| Asp (D) | 2.5 | −0.15 | 108 | 4 | 0.3 | 5.3 |
| Glu (E) | 2.5 | −0.27 | 129 | 5 | 0.4 | 6.1 |
| Gln (Q) | 0.2 | −0.33 | 142 | 5 | 0.1 | 4.0 |
| His (H) | −0.5 | −0.06 | 150 | 6 | −1.1 | 2.2 |
| Arg (R) | 3.0 | −0.68 | 197 | 7 | −0.2 | 5.7 |
| Lys (K) | 3.0 | −0.65 | 166 | 5 | 0.7 | 5.7 |
| Met (M) | −1.3 | −0.50 | 161 | 4 | −3.3 | 2.4 |
| Ile (I) | −1.8 | −0.23 | 165 | 4 | −3.9 | 5.2 |
| Leu (L) | −1.8 | −0.62 | 165 | 4 | −4.4 | 9.0 |
| Val (V) | −1.5 | −0.14 | 137 | 3 | −3.1 | 6.6 |
| Phe (F) | −2.5 | −0.41 | 187 | 7 | −4.4 | 3.9 |
| Tyr (Y) | −2.3 | −0.17 | 191 | 8 | −2.3 | 3.2 |
| Trp (W) | −3.4 | −0.45 | 225 | 10 | −3.0 | 1.3 |

The hydrophobic values of the amino acids are given by Nozaki & Tanford (1971). Helix propensity is given by O'Neil & DeGrado (1990), with negative values meaning more α-helix propensity than Gly. Volume is measured as an increase in the volume of water by adding amino acids (Zamyatnin, 1984). The column marked "MJ" is from Li *et al.* (1997) decomposing the 20 × 20 Mirazawa Jernigan matrix for amino acid interactions into 20 hydrophobicity values. The right-hand column is overall frequency of amino acids in proteins (Creighton, 1992). Amino acids are grouped according to the chemistry of the side group: STPAG–small hydrophilic, MILV–small hydrophobic and FYW–aromatic. C is in a special group because it makes the covalent sulfur cross-links and thus stabilizes the protein beyond any other type of interaction. From an evolutionary viewpoint, it is believed that some amino acids originated before others, with G, A, V, D, S and E being the early ones and H, M, Y and W being the latest.

(2) Consider rescaling the *E. coli* cell to 1 $m^3$. If the number of different proteins was maintained at 5000, what would be the number of proteins of each type? What would be expected from the typical binding energies between the proteins?
(3) Consider rescaling the *E. coli* cell to 1 $m^3$. If the number of copies of each protein was maintained to be of order 1000, what would be the number of different proteins? What

would be expected from the typical binding energies between the proteins? What can one say about the characteristic cell generation time for such sized organisms with a single copy of DNA?

## The amino acids

As already mentioned, a protein has a one-dimensional backbone consisting of amino acids. In water at high temperature this polymer is in a random swollen state. At room temperature it is folded into a compact state that essentially is a solid. In other words, the relative positions of the monomers are fixed, and the native protein may thus, in Schrödinger's words, be called a non-periodic crystal. This is the "native" functional state of the protein, which is held together by the (non-covalent) interactions between the amino acids in water.

The chemical structure of the amino acids is shown in Fig. 4.2: there is an amino group and a carboxyl group, separated by a C atom that is denoted $C_\alpha$. The 20 different amino acids that occur in proteins differ by the side chain R. The simplest side chain is just an H atom, corresponding to the smallest amino acid, glycine. The side chains can be grouped as follows:

- hydrophobic, e.g. $-CH_3$ (alanine), and
- hydrophilic, which are either charged, e.g. $-CH_2-COO^-$ (aspartic acid) or polar, e.g. $-CH_2-OH$ (serine).

Polar (having a permanent dipole moment) or charged molecules are hydrophilic ("likes water") because of the dipole–dipole or charge–dipole interactions, respectively, with the polar water molecules. A hydrophilic molecule is also typically one that forms hydrogen bonds with water. Conversely, a non-polar molecule that cannot form hydrogen bonds is hydrophobic ("dislikes water"). Hydrophobic molecules dissolved in water tend to cluster (i.e. they attract) in order to minimize their surface of contact with the water. This effective attractive interaction is called the

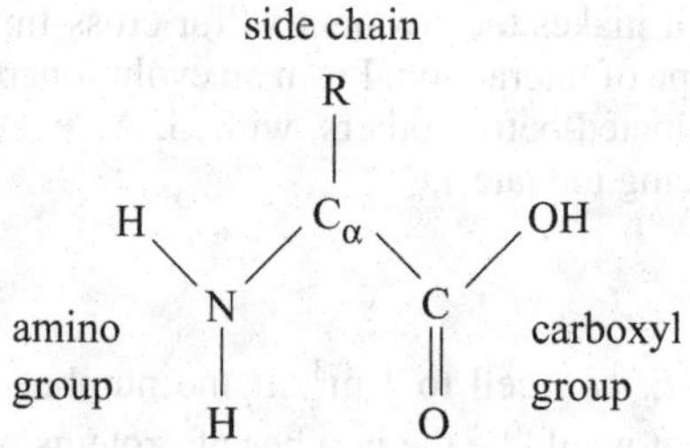

side chain
R
H
$C_\alpha$
OH
N
C
amino group
carboxyl group
H
O

Figure 4.2. A single amino acid, consisting of a CCN backbone and a side group R that can take one of 20 different forms. The simplest are where R = H (glycine) and R = $CH_3$ (alanine).

hydrophobic interaction. Later we describe this effect more quantitatively, and we see that a protein in water folds into a compact structure with the hydrophobic amino acids mostly on the inside. Hydrophobicity is believed to be the main force that drives the protein into a collapsed globular state.

Now we go through a brief description of the amino acids (see also Table 4.1). The simplest amino acid is glycine (G); its side chain consists of only one H atom. The presence of glycine confers particular flexibility to the polypeptide chain because of the minimal steric constraint imposed by its small size.

The hydrophobic amino acids are (in order of increasing number of carbons in the side chain, and thus increasing hydrophobicity): alanine (A), valine (V), leucine (L), isoleusine (I) and phenylalanine (F). In the folded protein the hydrophobic amino acids are mostly found in the interior, because they are repelled by the water. F is special because its side chain is a benzene ring, and this large planar structure enforces considerable steric constraints on the structure inside the protein.

Tyrosine (Y) and tryptophan (W) also have large planar side groups. These large amino acids tend to be exposed on the surface of a protein. Thus they are often associated with the function of the protein.

The polar (hydrophilic) amino acids are serine (S) and threonine (T), and also aspargine (N) and glutamine (Q). The first two are small and polar, owing to an OH group in the side chain.

The charged amino acids are essentially always on the surface of globular proteins, and are thus presumably important for defining/limiting interactions between a protein and other proteins and/or DNA. The charged amino acids are subdivided into, respectively, the positively charged arginine (R), lysine (K) and histidine (H), and the negatively charged aspartate (D) and glutamate (E). Charged amino acids of opposite sign can bind to each other through a so-called salt bridge. In some cases salt bridges can be essential for maintaining stability of a protein. The side chains of these amino acids are ionized (charged) to an extent that depends on the pH. At low pH the basic (negative) side chains are ionized, whereas at high pH the acidic (positive) side chains are ionized. Thus the overall charge of the protein depends on the pH. The pH value for which the protein is on average neutral is called the isolectric point.

Proline (P) is a special structural amino acid: as a result of its cyclic structure it allows for specific sharp turns of the polypeptide chain. Cysteine (C) is the only amino acid that can cross-link the polypeptide chain, by forming covalent disulfide bonds:

$$-\mathrm{CH_2}-\mathrm{SH}+\mathrm{HS}-\mathrm{CH_2}- \;\rightarrow\; -\mathrm{CH_2}-\mathrm{S}-\mathrm{S}-\mathrm{CH_2}- \qquad (4.5)$$

The folded structure of large proteins is typically stabilized by several such bonds. Finally there is the amino acid methionine (M), which also contains sulfur. However, it cannot form disulfide bridges. Methonine and cysteine are the only amino acids that contain more than H, C, O and N atoms.

The interaction properties of the 20 amino acids in the protein structure have been investigated by Miyazawa & Jernigan (1985, 1996). Their method is indirect, and consists of translating relative occurrences of contacts between amino acids in a sample of real proteins into effective two-body interaction strengths. That is, one deduces the interaction strength between amino acids of type $i$ and type $j$ from the overall frequencies of them being neighbors in known protein structures:

$$P_{ij} = \kappa P_i P_j \exp(-M_{ij}/k_B T) \tag{4.6}$$

where $P_{ij}$ is the probability of $i$ and $j$ being neighbors (deduced from the observed frequencies), $M_{ij}$ is an effective interaction potential matrix that parametrizes nearest-neighbor interactions, and $P_i$, $P_j$ are the frequencies of occurrence of the individual amino acids. Note that $\kappa$ is an overall normalization constant showing that a constant can be added to all $M_{ij}$ without changing the validity of the fit. This overall normalization in part reflects the entropy loss for bringing the polymer from a random swollen state to a specific compact state.

Li *et al.* (1997) demonstrated that the matrix $M_{ij}$ can be re-parametrized as

$$M_{ij} = C_0 + C_1(h_i + h_j) + C_2 h_i h_j \tag{4.7}$$

where the $h$ values are tabulated as "MJ" hydrophobicity in Table 4.1, and where the quadratic term is about a factor 6 smaller than the linear term. We notice that these $h$ values correlate strongly with the directly measured hydrophobicity in Table 4.1 (second column). Thus averaging over all protein structures somehow reflects the typical strength of interactions between amino acids. This is surprising! There is no reason to believe that different amino acids in different proteins are in thermal equilibrium with each other.

The average number of nearest neighbors for an amino acid in a protein is $\sim$3 (excluding the two along the polypeptide chain). The number of nearest neighbors for an amino acid completely inside a protein is $6 \pm 2$ (again excluding the two along the chain). The number of nearest neighbors is nearly independent of amino acid type (size), but varies a lot from structure to structure. The typical distance between the center of mass of amino acids in contact is $\sim$6–7 Å.

We finally stress that, in addition to the hydrophobic interactions, the amino acids have different propensities for forming (for example) $\alpha$-helices (Lyu *et al.*, 1990; O'Neil & DeGrado, 1990; Padmanabhan *et al.*, 1990). In Table 4.1 we list energies associated with stabilizing $\alpha$-helices. The different helix propensities presumably reflect side chain entropic effects (Creamer & Rose, 1994). Effective values of the helix–coil parameters for amino acids in proteins can also be found in Andersen & Tong (1997).

## *Questions*

(1) Li *et al.* (1997) found to a good approximation that the "MJ" matrix $M_{ij} \approx \langle M_{ij} \rangle + \lambda_1 V_{1i} V_{1j} + \lambda_2 V_{2i} V_{2j}$ in terms of the two eigenvalues with largest absolute values was $\lambda_1 = -22$, $\lambda_2 = 19$, and the $V$s were the corresponding eigenvectors. Assume a linear relationship $\vec{V}_2 = \vec{\alpha} + \gamma \vec{V}_1$ (Li *et al.* found all elements of $\alpha = -0.3$ and $\gamma = -0.9$) to recover $M_{ij} = C_0 + C_1(h_i + h_j) + C_2 h_i h_j$ and identify $h_i$ in terms of the first eigenvector. Find the constants $C_1$, $C_2$.

(2) Plot hydrophobicity vs. MJ hydrophobicity. Equilibrium physics is also equivalent to the expectation derived from a minimal information approach, considering $I = -\sum p_i \ln(p_i)$ minimized over possible states $i$, with constraints of conservation of average energy. Discuss the Miyazawa–Jernigen results in this perspective.

(3) Which is the amino acid that best forms α-helixes? Which is the worst?

(4) Test whether there is correlation between helix propensities and hydrophobicity, and between helix propensity and size of amino acids.

(5) Discuss why C and W are used so rarely.

(6) What is the average radius of a compact globular 100 amino acid protein?

(7) Use the average size of amino acids to estimate the distance between centers of neighboring amino acids in a compact protein.

## Protein primary structure

Amino acids can polymerize to form polypeptide chains. The carboxyl group of one amino acid condenses with the amino group of the next, forming the peptide (C–N) bond (Fig. 4.3). Thus a polypeptide chain has an amino terminus and a carboxy

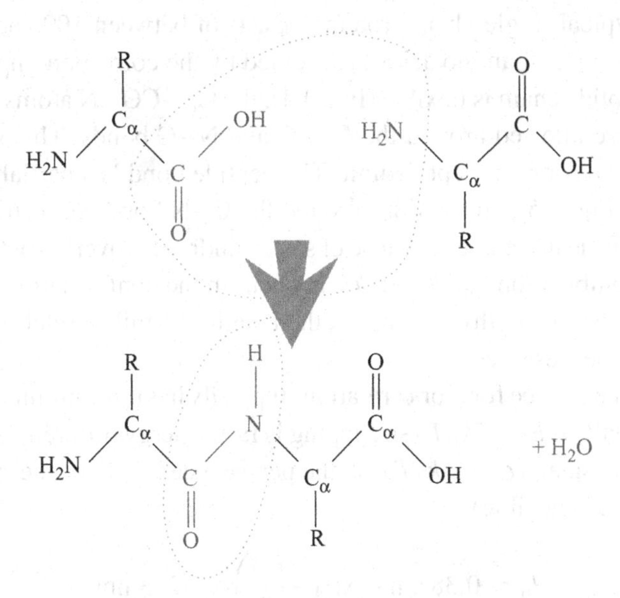


Figure 4.3. Formation of peptide bond between two amino acids. The C–C–N is the backbone element and R denotes the side chain of the amino acid.

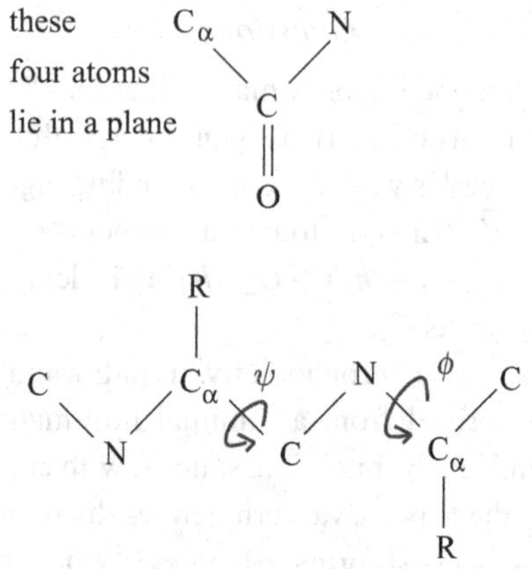

Figure 4.4. Peptide chain with the two angles $\phi$ and $\psi$ defined.

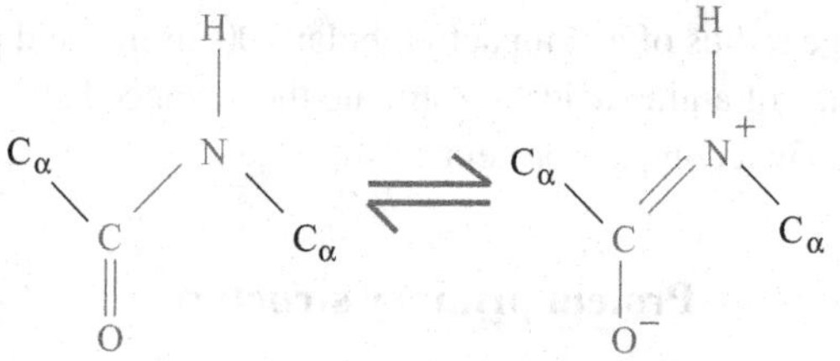

Figure 4.5. The peptide bond has a partial double-bond character, so rotations around the C–N bond are restricted, and the C–CO–N atoms lie in a plane.

terminus. A typical single chain protein consists of between 100 and 1000 amino acids. The sequence of amino acids is specified by the corresponding gene.

The polypeptide chain is flexible (Fig. 4.4): the $C_\alpha$–CO–N atoms lie in a plane, but rotations are allowed around the $C_\alpha$–C and N–C bonds. This gives two degrees of freedom for each peptide unit. The peptide bond has partial double-bond character (see Fig. 4.5), so rotations around the C–N bond are restricted, and the C–CO–N atoms lie in a plane. Because of steric hindrance (overlap between atoms), only certain combinations of $\phi$ and $\psi$ can occur in the conformation of a real protein. Given the bond lengths and angles, the length of a fully stretched polypeptide chain is 3.8 Å per residue.

The energy landscape for $\phi$ or $\psi$ rotations typically has three minima, see Fig. 4.6; barriers are small ($\Delta E < 5k_B T$) so flipping is fast. Energy differences between the minima are also small ($\epsilon \sim 2\, k_B T$), so the persistence length of the random coil is small (the chain is flexible):

$$l_p \sim 0.38\,\text{nm} \cdot \exp\left(\frac{\epsilon}{k_B T}\right) \sim 2\text{–}3\,\text{nm} \tag{4.8}$$

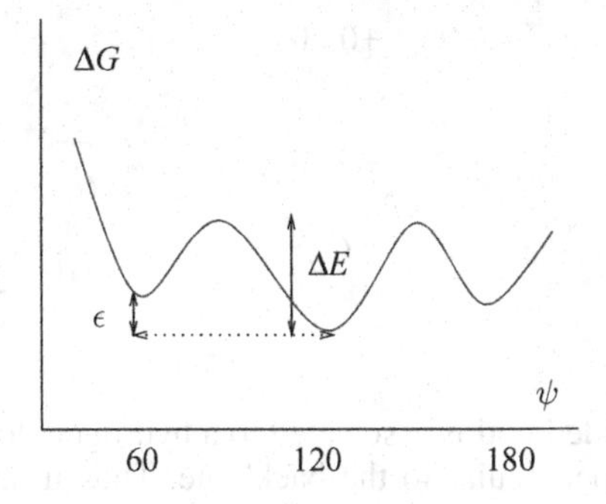


Figure 4.6. Schematic energy landscape for $\psi$ rotations.

where the 0.38 nm is the length per amino acid residue. This pesistence length corresponds to ~5 amino acids (Brandt & Flory, 1965; Miller & Goebel, 1968), which is close to the diameter of a typical single domain protein. The estimate is very uncertain, and recent direct measurements (Lapidus *et al.*, 2000) of the rate of contact formation for ends of small peptide chains rather suggest $l_p \sim 0.8$ nm.

The overall lesson is that each amino acid adds substantial entropy associated to possible orientations. Taking into account $\phi$ and $\psi$ there are in total about 6 conformational minima per amino acid. Thus a protein with 100 residues will have a total number of local energy minima $\sim 6^{100} \approx 10^{77}$.

## *Questions*

(1) There are 20 kinds of amino acid. What is the total number of different sequences of length 100 that theoretically could be produced?
(2) What is the expected distribution of the number of the amino acid Ala in proteins of length 100 amino acids?
(3) Assume that proteins consist of either α-helixes or β-sheets, and that each such secondary structure consists of 10 amino acids. How many different secondary structure permutations could be made of a protein that is 100 amino acid long?

## Secondary and tertiary structure

A polypeptide chain in water that wants to fold into a compact structure has the following problem. In each peptide unit, the O linked to the C and the H linked to the N can form hydrogen bonds, being hydrogen bond acceptors and donors, respectively (see Fig. 4.7). In a swollen state of the chain, these hydrogen bonds would be satisfied with water molecules. This makes it unfavorable to fold to a compact structure because this excludes the water. The protein remedies this situation by choosing compact structures where different residues along the chain can hydrogen bond to each other. Because of the limited directionality of hydrogen

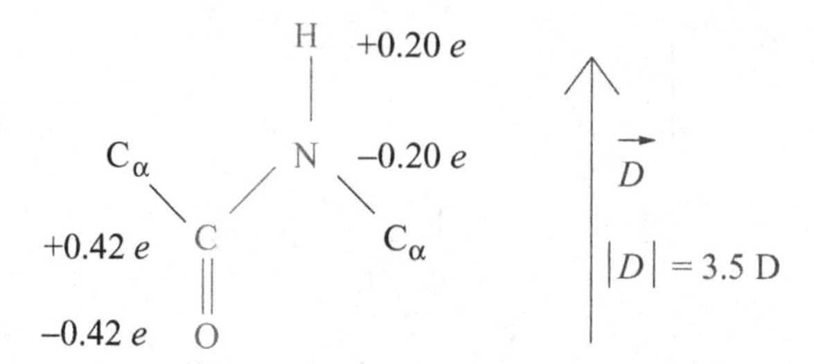


Figure 4.7. Each peptide bond is associated to a hydrogen donor and a hydrogen acceptor, oriented perpendicular to the backbone. Thus it has a dipole moment $D = 3.5$ Debye (D), where for comparison an $e+$ and $e-$ charge separated by 0.1 nm have a dipole moment of 4.8 D. In Chapter 2 we saw how these acceptor–donor correlations may force homopolymers to fold into fairly specific structures.

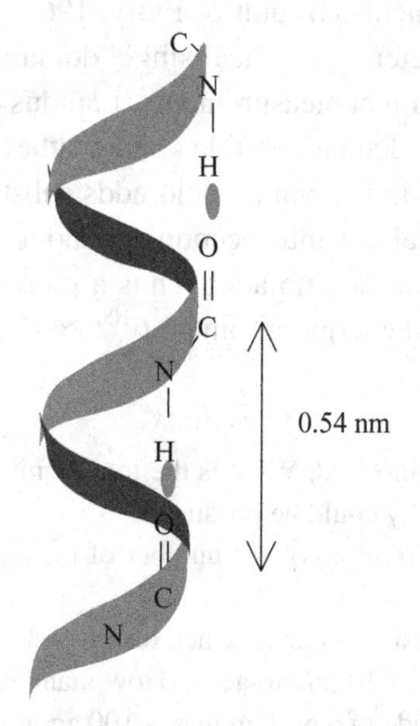


Figure 4.8. The α-helix, held together by hydrogen bonds.

bonds, this severely constrains the possible structures. There are two such regular conformations of polypeptides: the α-helix and the β-sheet. They are the regular elements of the secondary structure. These structures were in fact already predicted in the model of homopolymer folding with peptide like monomers at the end of Chapter 2.

In the α-helix (which is right-handed), each NH hydrogen of the backbone forms a hydrogen bond with the carbonyl oxygen of the backbone four residues down the chain (Fig. 4.8). Thus there are two hydrogen bonds per residue. The pitch of the helix is 5.4 Å, corresponding to 3.6 residues per turn. The amino acid

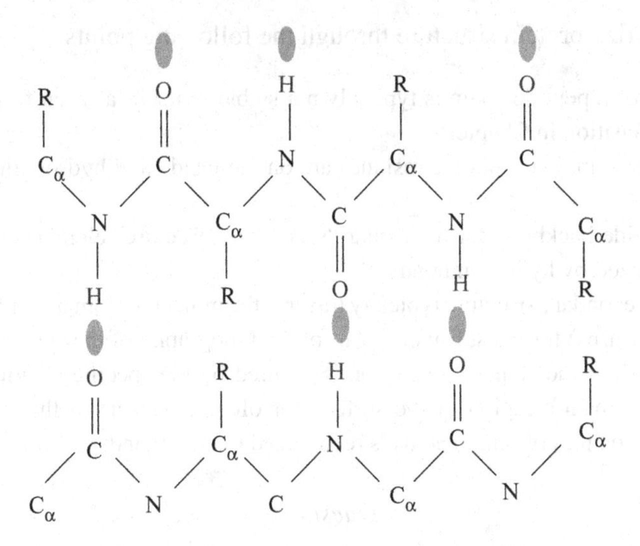


Figure 4.9. The β-sheet, held together by hydrogen bonds.

side chains stick outwards, conferring a specific chemical identity to the helix. For example, a distinctive pattern is an α-helix, which is globally amphiphilic, because the sequence is such that the polar residues all end up on one side of the helix and the hydrophobic residues end up on the other side. Then as a consequence of hydrophobic interactions, such helices can pack together in a specific tertiary structure.

The second regular element of secondary structure that satisfies all hydrogen bond requirements is the β-sheet (Fig. 4.9). A β-sheet is formed by parallel (or anti-parallel) β-strands (which are the extended conformation of the polypeptide chain), forming hydrogen bonds with each other. Again, each O of the backbone hydrogen bonds to an NH group of the backbone on the adjacent strand. The amino acid side chains protrude alternately above and below the plane of the β-sheet.

In the folded protein, these elements of secondary structure pack together to form the tertiary (three-dimensional) structure. The general architecture is that elements of the secondary structure, α-helices or β-strands, traverse the molecule from one side to the other, and are connected by loops at the surface of the protein (as an example see Fig. 4.1). From the global structure one can see that there is an important difference between the α-helix and the β-sheet: the former is a local structure (with contacts between amino acids separated by only four residues) along the polypeptide chain, but the latter is non-local in the sense that interacting amino acids may be very separated along the backbone chain. As a consequence it is easier to predict the α-helices than β-sheets from a sequence.

We summarize protein structure through the following points.

- The α-helix of a peptide chain is typically not stable in itself, as seen from data on the helix–coil transition in Chapter 2.
- In a folded protein, hydrophobic residues are on the inside and hydrophilic ones on the outside.
- The polypeptide backbone forms regular "secondary structure" elements (α-helices, β-sheets) stabilized by hydrogen bonds.
- Elements of secondary structure typically traverse the molecule from one side to the other; loop regions (turns) form a substantial part of the topography of the surface.
- The specificity of molecular recognition is ensured by the specific chemical and topographical pattern in a region of the surface (or often a pocket) of the molecule. This particular disposition of amino acids is recognized by the ligand.

*Question*

(1) Consider two dipoles like the one in Fig. 4.7, and place them along a line with their centers at a distance of 0.54 nm apart. Let one dipole be directed along the axis, and calculate the energy as a function of their relative angle of orientation.

## Forces that fold proteins

The stability of proteins is the result of residue–residue and residue–water interactions. All interactions are finally electrostatic in origin, but it is convenient to classify them as follows:

- van der Waals interaction (always present between any groups);
- electrostatic interaction between charged groups (salt bridges);
- hydrogen bonds;
- hydrophobic forces; and
- disulfide bridges (the only covalent bonds between non-contiguous residues, separated by a large energy gap from all other interactions).

One must remember that what matters for the stability of the folded state is the difference between the residue–residue and residue–water interactions. This difference is small for the van der Waals interaction. Given that the number of disulfide bonds is small, the main forces that shape the folded state are hydrophobicity and the requirement that all possible hydrogen bonds are formed. Indeed, in a folded structure the hydrophobic residues are typically found in the inside, where they are shielded from the water, whereas the hydrophilic ones are on the surface, in contact with the water. The ordering of the folded structure, however, is a consequence of the requirement to form hydrogen bonds. This induces secondary structure, which is stabilized by a global optimization of fitting them into a global compact three-dimensional tertiary structure while putting hydrophobic residues on the inside.

From this it is clear that the hydrophobic interaction occupies, in the life sciences, a position of importance comparable with any of the four fundamental interactions in the physical sciences. Yet our knowledge of it is very limited. We do not, for example, really know its range.

As a result of the above interactions, each amino acid–amino acid contact contributes a free energy of order

$$\Delta G_{\text{aa}} \sim (-5 \text{ to } +1) \text{ kcal/mol} \tag{4.9}$$

which consists of large contributions from $H$ and $T\Delta S$, which are of similar magnitude, but of opposite sign. Thus any estimate on effective interaction strength will easily make a large total error. Upon folding the contact energies have to counteract a polymer folding entropy (effective number of possible orientations for each amino acid, including its side chain, $g \approx 8$):

$$\Delta G_{\text{polymer}} \approx k_{\text{B}}T\Delta S \sim k_{\text{B}}T \ln(g) \sim 2\, k_{\text{B}}T \tag{4.10}$$

per residue. We note that estimates of the effective potentials vary considerably; in particular they vary between the scientific community engaged in protein design and researchers in molecular dynamics. We here focus on the two main forces in protein folding: the hydrogen bonds and the hydrophobic interactions.

**Hydrogen bonds** are responsible for secondary structure formation in proteins, RNA and DNA, and they also provide the mechanism for reading the genetic code. The hydrogen bond potential is very specific, with a steep angular dependence (Fig. 4.10). Gordon *et al.* (1999) suggest the following functional form:

$$E_{\text{HB}} = D_0 \cdot \left(5(R_0/R)^{12} - 6(R_0/R)^{10}\right) \cos^4(\theta) \tag{4.11}$$

where the angle $\theta$ counts the deviation from the ideal alignment between donor and acceptor, and $D_0 \approx 8$ kcal/mol. Thus hydrogen bonds are directional (see Fig. 4.11, and compare the shape of the above potential to the spherically symmetric van der Waals potential), and quite strong. As mentioned before, these features force the inside of the protein to be in either a β-sheet or α-helix structure.

The other main stabilizing force is the **hydrophobic interaction**. Hydrophobicity arises because a compound is not able to form hydrogen bonds to the surrounding water; as a consequence its solubility is low. The hydrophobic force is, in essence, an effect of the absence of hydrogen bonds between the hydrophobic substance and the surrounding water.

Gordon *et al.* (1999) report a free energy difference associated with exposing hydrophobic groups to water of 4.8 kcal/(mol nm$^2$), which is comparable to the typical strength of hydrogen bonds given above.

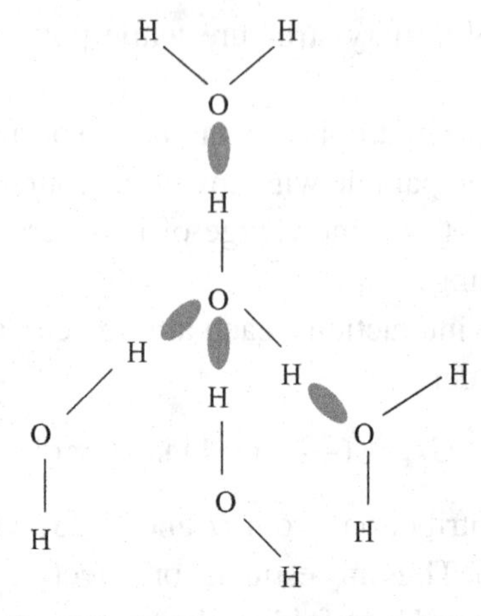


Figure 4.10. The structure of water, where oxygen binds with hydrogen bonds to hydrogens of neighboring water molecules. The energy needed completely to break a hydrogen bond is $\sim 8\ k_B T \sim 5$ kcal/mol (Suresh & Naik, 2000). Breaking it in water essentially means to reshuffle it with some water hydrogen bonds and the associated energy is only about $\sim 0.5\ k_B T$ to $2\ k_B T$ (Silverstein *et al.*, 2000). Hydrogen bonds are cooperative in water; breaking one weakens the strength of neighboring hydrogen bonds (Heggie *et al.*, 1996). Thus water organizes itself in cooperative clusters, which are predicted to be of size 400 (Luck, 1998). The lifetime of a hydrogen bond is about 1–20 ps (Keutsch & Saykally, 2001); the time of staying unbroken (time to move from one to another partner) is only 0.1 ps.

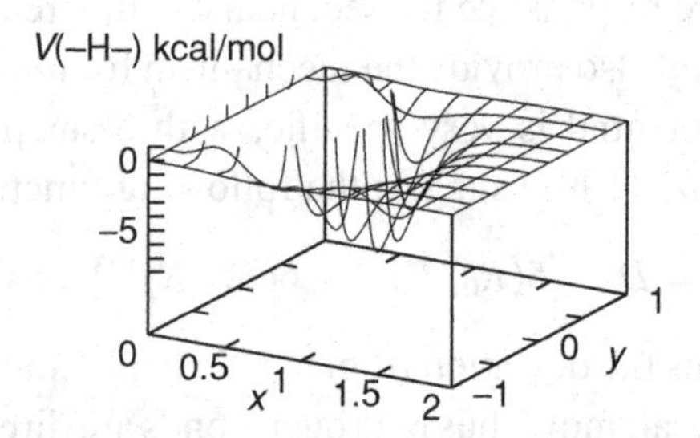


Figure 4.11. Hydrogen bonding potential.

The free energy difference $\Delta G$ associated with transferring a hydrophobic substance ("oil") into water can be measured from the solubility constant $K$:

$$\Delta G = G(\text{oil in water}) - G(\text{oil in oil}) = -k_B T \ln K \tag{4.12}$$

where $K$ is the equilibrium constant for the "reaction" of dissolving oil in water, i.e.

$$K = \frac{\text{mol fraction of oil in water}}{\text{mol fraction of oil in oil}} \tag{4.13}$$

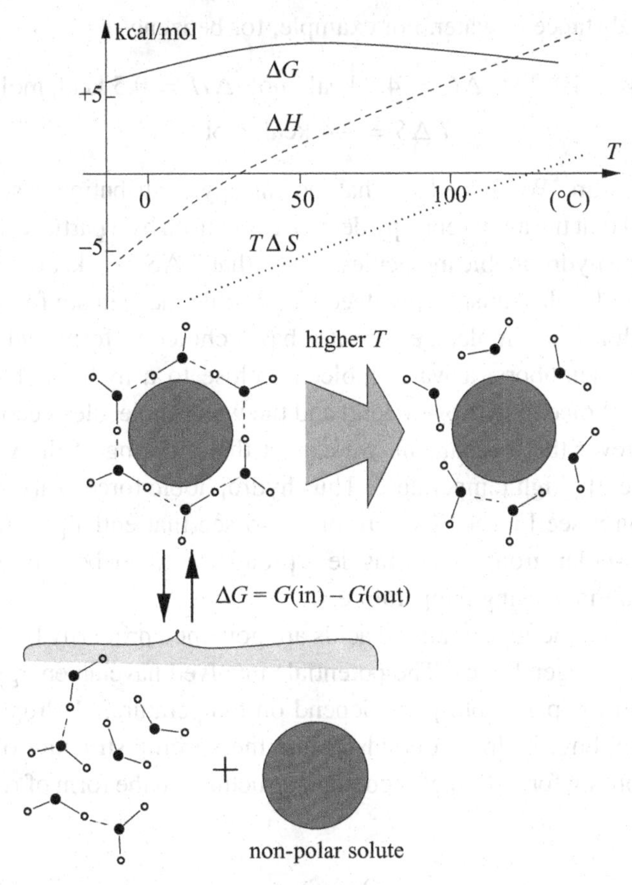


Figure 4.12. Variation of thermodynamic quantities with temperatures for a hydrophobic molecule, for example pentane, in water (relative to not being in water) as reported by Privalov & Gill (1988). That $\Delta H$ and $T\Delta S$ follow each other results from the approximately constant gap in heat capacity between pentane in water and pentane in pentane: $C_p = dH/dT = TdS/dT$. In the lower part of the figure we illustrate the thermodynamics in terms of the model of Southall *et al.* (2002). At low temperatures the non-polar solute imposes a highly ordered layer of water molecules connected by hydrogen bonds (dashed lines). At higher temperatures the hydration layer effectively disappears.

(Note that the mol fraction of oil in oil is equal to 1.) Hydrophobic substances have low solubility in water ($K \ll 1$), which means $\Delta G$ is large and positive.

On the other hand, $\Delta H$ is typically small (positive or negative) at room temperature. The enthalpy change $\Delta H$ can be measured by calorimetry. Because $\Delta G = \Delta H - T\Delta S$, for these substances $\Delta S$ is large and negative at room temperature. Thus there is a large entropy decrease of the system when dissolving a hydrophobic substance in water. For example, for benzene

$$K = 4 \times 10^{-4}\ \text{M},\ \Delta G = 4.5\ \text{kcal/mol},\ \Delta H = 0.5\ \text{kcal/mol} \rightarrow \quad (4.14)$$

$$T\Delta S = -4\ \text{kcal/mol} \quad (4.15)$$

at room temperature. We indeed see that the entropy contribution is substantial.

It is believed that the above entropy decrease is caused by a partial ordering of the water around the hydrophobic molecules. Given that $T\Delta S \sim 4$ kcal/mol $\sim 7\ k_BT$, several water molecules must be involved. A possible mechanism for this ordering is that although a water molecule in water has a choice to form hydrogen bonds with any of its neighbors, a water molecule close to a hydrophobic surface is restricted in its choice to hydrogen bond and the water molecules become partially oriented. Whatever the mechanism, we expect this ordering of the water to melt away at sufficiently high temperature. Thus hydrophobic forces vary substantially with temperature; see Fig. 4.12 where one also see that enthalpies (energies) of transfer of non-polar groups from inside a protein to water become increasingly unfavorable with increasing temperature.

In summary, interactions of amino acids are governed primarily by the presence or absence of hydrogen bonds. The potentials involved have an energy (enthalpy) part and an entropy part; both parts depend on temperature. Hydrophobic forces drive protein folding; hydrogen bonds define the specific structure of the folded protein by enforcing formation of secondary structures in the form of $\alpha$-helices and $\beta$-sheets.

## *Question*

(1) Assume that the boiling of water is associated only with the breaking of all hydrogen bonds. Use a hydrogen bond energy of 5 kcal/mol to estimate the latent heat for boiling 1 kg of water.

## References

Andersen, N. H. & Tong, H. (1997). Empirical parameterization of a model for predicting peptide helix/coil equilibrium populations. *Protein Sci.* **6**, 1920–1936.

Brant, D. A. & Flory, P. J. (1965). The configuration of random polypeptide chains. I. Experimental results. *J. Am. Chem. Soc.* **87**, 2788–2791.

Creamer, T. P. & Rose, G. D. (1994). Alpha-helix-forming propensities in peptides and proteins. *Proteins* **19**, 85–97.

Creighton, T. E. (1992). *Protein Folding*. New York: W. H. Freeman and Company.

Gordon, D. B., Marshall, S. A. & Mayo, S. L. (1999). Energy functions for protein design. *Curr. Opinion Struct. Biol.* **9**, 509–513.

Heggie, M. I., Lathan, C. D., Maynard, S. C. P. & Jones, R. (1996). Cooperative polarisation in ice I and the unusual strength of the hydrogen bond. *Chem. Phys. Lett.* **249**, 485–490.

Keutsch, F. N. & Saykally, R. J. (2001). Water clusters: untangling the mysteries of the liquid, one molecule at a time. *Proc. Natl Acad. Sci. USA* **98**, 10 533–10 540.
Lapidus, L., Eaton, W. & Hofrichter, J. (2000). Measuring the rate of intramolecular contact formation in polypeptides. *Proc. Natl Acad. Sci. USA* **97**(13), 7220–7225.
Li, H., Tang, C. and Wingreen, N. S. (1997). Nature of driving force for protein folding: a result from analyzing the statistical potential. *Phys. Rev. Lett.* **79**, 765.
Luck, W. A. P. (1998). The importance of cooperativity for the properties of liquid water. *J. Mol. Struct.* **448**, 131–142.
Lyu, P. C., Liff, M. I., Marky, L. A. & Kallenbach, N. R. (1990). Side chain contribution to the stability of alpha helical structure in peptides. *Science* **250**, 669–673.
Miller, W. G. & Goebel, C. V. (1968). Dimensions of protein random coils. *Biochemistry* **7**, 3925.
Miyazawa, S. & Jernigan, R. L. (1985). Estimation of effective inter-residue contact energies from protein crystal structures: quasi-chemical approximation. *Macromolecules* **18**, 534.
(1996). Residue–residue potentials with a favorable contact pair term and an unfavorable high packing density term, for simulation and threading. *J. Mol. Biol.* **256**, 623.
Nozaki, Y. & Tanford, C. (1971). The solubility of amino acids and two glycine peptides in aqueous ethanol and dioxane solutions. *J. Biol. Chem.* **246**, 2211.
O'Neil, K. T. & DeGrado, W. F. (1990). Thermodynamic parameters for helix formation for the twenty commonly-occurring amino acids. *Science* **250**, 646–651.
Padmanabhan, S., Marqusee, S., Ridgeway, T., Laue, T. M. & Baldwin, R. L. (1990). Relative helix-forming tendencies of nonpolar amino acids. *Nature* **344**, 268–270.
Privalov, P. L. & Gill, S. J. (1988). Stability of protein structure and hydrophobic interaction. *Adv. Protein Chem.* **39**, 191.
Roseman, M. A. (1988). Hydrophilicity of polar amino acid side chains is markedly reduced by flanking peptide bonds. *J. Mol. Biol.* **200**, 513.
Silverstein, K. A. T., Haymet, A. D. J. & Dill, K. A. (2000). The strength of hydrogen bonds in liquid water and around nonpolar solutes. *J. Am. Chem. Soc.* **122**, 8037–8041.
Southall, N. T., Dill, K. A. & Haymet, A. D. J. (2002). A view of the hydrophobic effect. *J. Phys. Chem. B* **106**, 521–533.
Suresh, J. & Naik, V. M. (2000). Hydrogen bond thermodynamic properties of water from dielectric constant data. *J. Chem. Phys.* **113**, 9727.
Zamyatnin, A. A. (1984). Amino acid, peptide, and protein volume in solution. *Ann. Rev. Biophys. Bioeng.* **13**, 145–165.

## Further reading

Anfinsen, C. (1973). Principles that govern the folding of protein chains. *Science* **181**, 223.
Chothia, C. (1974). Hydrophobic bonding and accessible surface area in proteins. *Nature* **248**, 338–339.
(1976). The nature of the accessible and buried surfaces in proteins. *J. Mol. Biol.* **105**, 1–14.
Cowan, R. (1988). A simple formula for solvent-accessible areas of amino acid side-chains. *Biochem. Biophys. Res. Commun.* **156**, 792–795.

Fernandez, A., Kardos, J. & Goto, Y. (2003). Protein folding: could hydrophobic collapse be coupled with hydrogen-bond formation? *FEBS Lett.* **536**, 187–192.

Fisinger, S., Serrano, L. & Lacroix, E. (2001). Computational estimation of specific side chain interaction energies in α helices. *Protein Sci.* **10**, 809–818.

Israelachvili, J. (1992). *Intermolecular and Surface Forces*. London: Academic Press Limited.

Padmanabhan, S., York, E. J., Stewart, J. M. & Baldwin, R. L. (1996). Helix propensities of basic amino acids increase with the length of the side-chain. *J. Mol. Biol.* **257**, 726–734.

Pauling, L., Corey, R. B. & Branson, H. R. (1951). The structure of proteins: two hydrogen-bonded helical configurations of the polypeptide chain. *Proc. Natl Acad. Sci. USA* **37**, 205–211; 235–240; 729–740.

Vaisman, I. I., Brown, F. K. & Tropsha, A. (1994). Distance dependence of water structure around model solutes. *J. Phys. Chem.* **98**, 5559.

Yang, A. S. & Honig, B. (1995). Free energy determinants of secondary structure formation: I. Alpha-helices. *J. Mol. Biol.* **252**, 351–365.

# 5

# Protein folding

Kim Sneppen & Giovanni Zocchi

For a typical protein, the global stability at room temperature is of the order 10 kcal/mol $\sim$20 $k_B T$ for a $\sim$100 residue domain. Thus proteins are stabilized with a free energy of a fraction of $k_B T$ per amino acid. Nonetheless, proteins have well-defined structures that are stabilized by cooperative interactions between their many small parts. Thus there is essentially no "partial melting" of the protein; it stays folded and specifically functional to do whatever job it is supposed to do. When a large enough fluctuation does indeed melt the protein, the structure can (mostly) reassemble.

The miracle of protein structure and function in the noisy Brownian environment of ongoing $k_B T$ perturbations may be envisioned as a car engine exposed to perturbations that could easily tear any part of it at any time. Imagine that a car would melt at maybe 70 °C, but would stay together and remain functional nearly all the time. However, should it disassemble, it would reassemble by itself and resume functioning. Proteins may be regarded as "intelligent" and robust nano-technology made of programmable matter.

Proteins can fold into their native state in a remarkably short time; one infers from this that the enormous phase space available to the polypeptide chain must be sampled not randomly, but following one or more "folding pathways". Thus the amino acid sequence contains not only the plan for the final structure, but it also specifies how to get there (see Fig. 5.1).

When a protein folds, elements of the secondary structure ($\alpha$-helices and $\beta$-sheets) are formed and, subsequently or simultaneously, the protein collapses into a compact three-dimensional structure where the helices, sheets and loops stabilize each other and produce a definite tertiary structure. The associated timescales are 10–100 ps for individual amino acid movements; a minimum of $\sim$1000 such moves to fold a reasonably sized protein would give an absolute lower limit for the folding time of 10–100 ns. However, the fastest proteins take about 10 μs to fold, some proteins take minutes to fold, and some (about 5–10% of proteins in an *E. coli*)

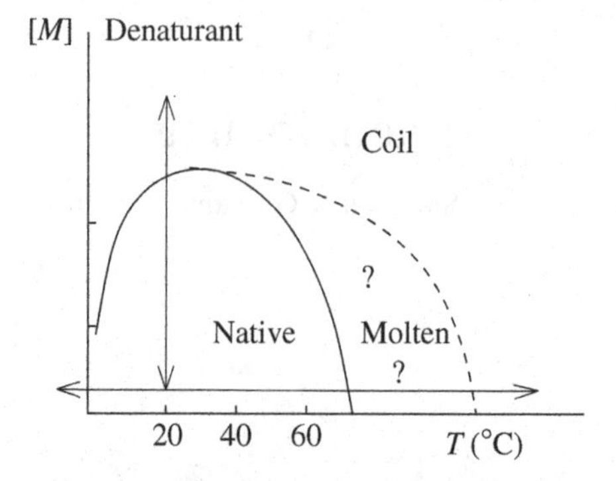


Figure 5.1. Sketch of the protein phase diagram. With no or little denaturant unfolding is an all-or-none transition to a denatured state that presumably still has sizable contact energies and maybe remnants of secondary structure. Thus the heat-denatured protein may be in a molten state not far from that of a collapsed homopolymer. The cold denatured protein tends to be highly swollen after denaturation.

The horizontal and vertical, respectively, arrows refer to the two main modes of experimental analysis of protein folding: **calorimetry** (pioneered by Privalov and collaborators) and **m-value analysis** (pioneered by Fersht, Jackson and collaborators).

cannot fold by themselves and need help from other proteins (chaperones) to fold. Concerning the secondary structure, the timescale for α-helix formation is estimated to about 0.1–1 μs; for β-sheets, the timescale is 1–100 μs.

In this chapter, we first discuss the thermodynamics of protein stability, including measures of cooperativity, mutation analysis, and cold unfolding. Then we describe a series of simple statistical mechanics models, each of which captures some aspect of folding. The models are only qualitative, but are necessary because it is hard to build an understanding from the thermodynamics alone.

## Calorimetry: denaturation by heating

As is general in condensed matter systems, the principal way of exploring the thermodynamic properties is to measure the specific heat. The heat capacity of proteins as a function of temperature can be measured through (delicate) calorimetry experiments. The heat capacity teaches us about possible fluctuations in the energy distribution of the protein

$$C_{\mathrm{p}} = \frac{\mathrm{d}\langle H \rangle}{\mathrm{d}T} = \frac{\langle H^2 \rangle - \langle H \rangle^2}{k_{\mathrm{B}} T^2} \tag{5.1}$$

where $H$ is the enthalpy (playing the role of energy for constant pressure) and where the subscript "p" refers to heat capacity at constant pressure (for derivation see the Appendix). We use thermodynamic quantities for constant pressure

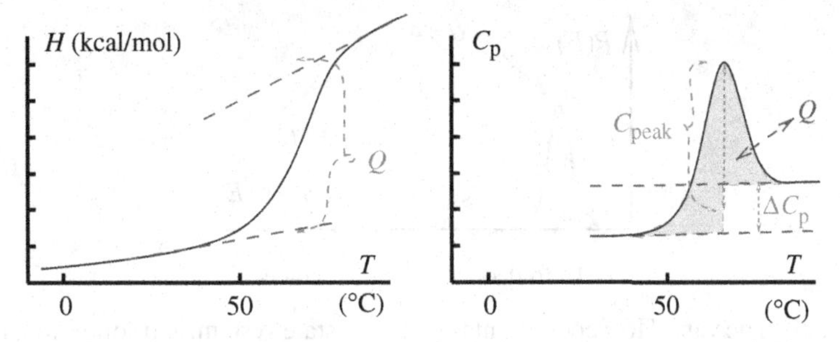


Figure 5.2. Schematic melting curve for a typical protein. On the left we show energy (enthalpy) per protein; on the right we show the heat capacity. Latent heat $Q$ is indicated by the shaded area. Notice the gap in heat capacity $\Delta C_p$ between the folded and unfolded states, reflecting the exposure of hydrophobic residues upon unfolding. The peak heat capacity is counted from the midpoint of the transition, as indicated. Throughout the text we focus on the part of the heat capacity that is associated with the difference between a folded and unfolded protein. Thus $C_p$ in the text corresponds to the curve defined by the shaded area in the right-hand figure.

conditions because all experiments on proteins are done at constant pressure and not at constant volume. We also replace the usual free energy $F = E - TS$ used at constant volume with the $G = H - TS = E - PV - TS$ that should be used for probability calculations at constant pressure (see Appendix).

Figure 5.2 shows the heat capacity as a function of temperature $T$ for a typical single-domain protein. The peak in the heat capacity signals a phase transition; a condition where the protein fluctuates violently between two extreme structures: one with low energy and one with high energy. The phase transition is from the native folded state (f) to a melted or unfolded state (u) with higher energy (and much more entropy). When increasing the temperature from below the phase transition to above it, the protein absorbs an energy

$$\Delta H = \int_{T \ll T_m}^{T \gg T_m} C_p \mathrm{d}T = H_u - H_f = Q \tag{5.2}$$

where the integral includes the whole transition region around the melting point $T_m$ where the heat capacity $C_p$ is maximal ($= C_{peak}$). $Q$ is the latent heat of the transition. In practice, for real proteins, the evaluation of $\Delta H$ is complicated because both $H_u$ and $H_f$ depends of $T$, and they do so with different slopes.

In principle the sharpness of a phase transition can be quantified by the coefficient $\alpha$

$$\alpha = \frac{(H_u - H_f)^2}{\langle H^2 \rangle - \langle H \rangle^2} \tag{5.3}$$

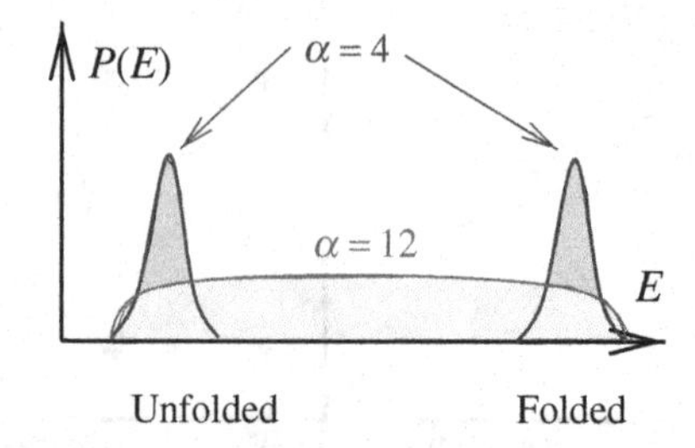


Figure 5.3. The van't Hoff coefficient for the two-state system, and softer transition of protein folding. The curves show the probability distributions at transition state, and the coefficient $\alpha$ measures the extent to which the system is in one of two distinct states at this point.

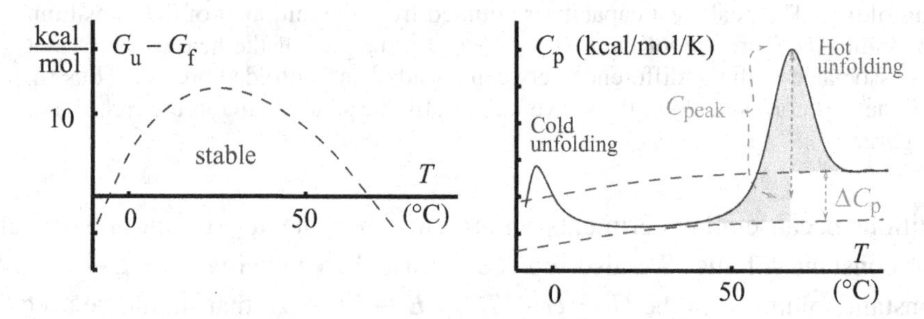


Figure 5.4. Left: curve of $\Delta G = G(\text{unfolded}) - G(\text{folded}) = G_\text{u} - G_\text{f}$ for a typical globular protein as function of temperature. Right: illustration of calorimetry for the same protein. Notice that there are two phase transitions.

When $\alpha$ is small the transition is sharp. The sharpest possible transition is a two-state system, where the system does not populate any intermediates at the transition point; see Fig. 5.3. In that case, at the midpoint of the transition

$$\alpha = 4 \tag{5.4}$$

This corresponds to the reciprocal variance for a distribution with two distinct peaks ($P(H = H_\text{f}) = 0.5$, $P(H = H_\text{u}) = 0.5$) (see Dommersnes *et al.*, 2000). In the opposite limit where all states between $H_\text{u} = H_\text{unfolded}$ and $H_\text{f} = H_\text{folded}$ are equi-probable then $\alpha = 12$ (see the Questions on p. 101).

In real proteins we resort to calorimetry, and thus obtain curves as seen in Fig. 5.4. The peak of the $C_\text{p}$ curve thus determines the denominator in Eq. (5.3). The numerator can in principle be determined by integrating the excess heat capacity (shaded area in Fig. 5.4), $\Delta H = \Delta H_\text{cal} = Q$. This latter quantity is in the protein literature denoted calorimetric enthalpy. Thus

$$\alpha = \frac{\Delta H_\text{cal}^2}{k_\text{B} T_\text{m}^2 C_\text{peak}} \tag{5.5}$$

Privalov and collaborators also use the so-called van't Hoff enthalpy defined from the van't Hoff equations (Privalov, 1979; Privalov & Potekhin, 1986)

$$\Delta H_{\mathrm{vH}} = k_{\mathrm{B}} T^2 \frac{\mathrm{d}\ln(K)}{\mathrm{d}T} \tag{5.6}$$

where, for a two-state system, the reaction constant $K(T) = \exp(-(G_{\mathrm{u}} - G_{\mathrm{f}})/k_{\mathrm{B}}T) = N_{\mathrm{u}}/N_{\mathrm{f}}$ is the ratio of unfolded to folded proteins at temperature $T$. The $\Delta H(T_{\mathrm{m}})$ deduced from this equation is called the van't Hoff enthalpy $\Delta H_{\mathrm{vH}}$. The ratio $\Delta H_{\mathrm{vH}}/\Delta H_{\mathrm{cal}} = 1$ for a two-state system. If there is a populated transition state, then $\Delta H_{\mathrm{vH}}/\Delta H_{\mathrm{cal}} < 1$.

Privalov and collaborators quantify cooperativity through the dimensionless measure (Privalov, 1979; Privalov & Potekhin, 1986)

$$\alpha_{\mathrm{Priv}} = \frac{\Delta H_{\mathrm{cal}} \Delta H_{\mathrm{vH}}}{k_{\mathrm{B}} T_{\mathrm{m}}^2 C_{\mathrm{peak}}} \tag{5.7}$$

where $\alpha_{\mathrm{Priv}} = 4$ implies a perfect two-state folding. Here $\Delta H_{\mathrm{cal}}$ is the area under the peak in the heat capacity curve (left-hand part of Fig. 5.4), and thus $\Delta H_{\mathrm{cal}}/(T C_{\mathrm{peak}})$ its width in units of temperature. Thus all calorimetric uncertainty is condensed into an estimate of this width. As a consequence $\alpha_{\mathrm{Priv}}$ is less sensitive to problems in determining $\Delta H_{\mathrm{cap}}$ than the one defined through Eq. (5.3). For typical small proteins Privalov reports a value

$$\alpha_{\mathrm{Priv}} = 4.2 \tag{5.8}$$

which is remarkably close to the two-state system, considering the rather small free energy difference $\Delta G \sim 10$ kcal/mol between the folded and unfolded states.

Coming back to the specific heat curve in Fig. 5.4, another remarkable feature is that there is a difference between the heat capacity of the folded and unfolded states ($\Delta C_{\mathrm{p}} \simeq 0.3$ cal/(g · K), similar for all proteins). This is not a small effect: for comparison, recall that the specific heat of water is 1 cal/(g · K). This large difference is peculiar to proteins; it is caused by the interactions between the water and the hydrophobic residues that are exposed upon unfolding.

A substantial gap in the specific heat, $\Delta C_{\mathrm{p}} > 0$, implies the possibility of a second (cold) unfolding transition (Privalov, 1990). That is, $\Delta C_{\mathrm{p}} > 0$ implies that the free energy difference $\Delta G = \Delta(H - TS) = G_{\mathrm{unfolded}} - G_{\mathrm{folded}}$ (i.e. the "stability" of the protein) has a non-monotonous behavior when plotted against $T$. In the Questions (p. 101) we go through this argument and prove that a constant temperature independent gap in heat capacity implies

$$\Delta G \propto (T - T_0) \; - \; T \ln(T/T_0) \tag{5.9}$$

where the $T$ term comes from the enthalpy increase with temperature, and the $-T\ln(T)$ is the entropic contribution. If $\Delta G$ has two zeroes, then there are two transitions (Fig. 5.4), and correspondingly two peaks in the specific heat curve. This is a very peculiar situation: the protein melts not only upon heating, but also upon cooling!

The gap in $C_p$ means that the unfolded state effectively has more degrees of freedom than the folded state (the Dulong–Petit law[1]). One may interpret this in terms of additional energy levels of the water that appear through the interaction with exposed hydrophobic residues. Also it means that the difference between folded and unfolded enthalpy, $\Delta H = H_u - H_f$, of the protein increases by about 2 kcal/mol per degree Celsius. Therefore the absorbed heat at melting tends to be larger for high melting temperatures than for low melting temperatures.

Conversely, water binds to the unfolded amino acid chain, which at low enough temperatures causes $\Delta H$ to become sufficiently negative to drag the protein apart, i.e. to melt. Later in this chapter we will present a minimalistic model for this cold unfolding process. As with the warm unfolding, it is associated with a gap in the heat capacity, but as the melting occurs upon lowering the temperature, a negative $\Delta H$ (folded $\rightarrow$ melted) reflects an energy gain associated with water binding to the unfolded chain.

Cold unfolding is observed experimentally; however, because this transition usually occurs below 0 °C, one has to take the necessary precautions so that the solvent (e.g. water + glycerol) does not freeze. Indeed, the specific heat of a protein, when observed over a sufficiently wide temperature range, displays this effect. Proteins are typically most stable around room temperature; see the left-hand panel of Fig. 5.4.

A guess on the overall phase diagram for proteins was presented in Fig. 4.1. The molten phase at high temperature is controversial, the question being whether the observed expansion of the protein disrupts specific bindings only, or whether it additionally disrupts non-specific bindings sufficiently to allow the polymer to expand into a dilute gas-like phase. For the collapsed but homogeneous state, the gap in heat capacity is half that of the folded state, suggesting that half the hydrophobic part is still exposed to the solvent.

In summary, warm unfolding is driven by the entropy increase of the chain, and cold unfolding is driven by the enthalpy decrease of the water. Protein folding is often a two-state transition where all intermediates are highly unstable. In fact the two-state transition is not only a thermodynamic phenomenon, but presumably also signals mechanistic brittleness. In the pulling of proteins examined by Makarov *et al.* (2002) the protein titin, dominated by $\beta$-sheets, is seen to be able to sustain

[1] The Dulong–Petit law states that there is an energy of $k_B T/2$ for each degree of freedom that moves and an additional $k_B T/2$ for each degree of freedom that vibrates. It is proven by showing that any energy of the form $E_q = \frac{1}{2}kq^2$ gives an average energy $\langle E \rangle = \int dq\, E_q \exp(-E(q)/k_B T)/\int dq \exp(-E(q)/k_B T) = k_B T/2$.

an external stress of up to 200 pN. Similarly, Schlierf *et al.* (2004) and Carrion-Vazquez *et al.* (2003) report that ubiquitin is able to sustain up to 200 pN of stress, when pulled along the peptide chain. Thus this protein is indeed able to sustain large forces while cooperatively keeping everything in order.

### *Questions*

(1) Show that $\Delta H = H(\text{unfolded}) - H(\text{folded})$ varies with temperature $T$ as $\Delta H(T) = \Delta C_\text{p}(T - T_0) + \Delta H(T_0)$ if we assume that $\Delta C_\text{p} = C_\text{p}(\text{unfolded}) - C_\text{p}(\text{folded})$ is independent of $T$. Use the relation $C_\text{p}\text{d}T = \text{d}H$.

(2) Show that, in the same approximation, $\Delta S(T) = \Delta C_\text{p} \ln(T/T_0) + \Delta S(T_0)$. Use the relation[2] $C_\text{p} = T\partial S/\partial T$.

(3) Show that $\Delta G \sim \Delta H - T\Delta S \sim T - T\ln(T/T_0)$ is non-monotonous with $T$ (see Fig. 5.4).

(4) Expand $\Delta G(T) = G(\text{unfolded}) - G(\text{folded})$ around the upper melting point $T_0$ where $\Delta G(T_0) = 0$. Show that the protein is stable just below this temperature, but also that it loses stability when the temperature becomes too low. What thermodynamic quantity drives unfolding at very low temperature?

(5) With an overall stability of $\Delta G = 10\,k_\text{B}T$, how great a deformation could a protein that can sustain 200 pN survive before denaturing?

## Folding–unfolding rates: *m*-values

Protein folding and unfolding **rates** can be studied by using stopped-flow measurements. Upon rapid mixing of unfolded protein in water (i.e. by diluting the concentration of the denaturant that was used to unfold the protein; this can be done within 2–5 ms), one can observe the formation of native structure, i.e. folding, by spectroscopic methods, for example changes in fluorescence or CD (circular dichroism) signal. The typical result is to observe an exponential relaxation towards the new equilibrium state; the characteristic time constant gives the folding time. One can generalize this approach to measure how the protein folds in the presence of different concentrations of denaturant; then one can study how the folding rate $k_\text{f}(D)$ depends on denaturant concentration $[D]$. Similarly, one can start from the folded state and study how the unfolding rate $k_\text{u}(D)$ depends on denaturant concentration.

In the following discussion we use "N" for the native state, "TR" for the transition state, and "U" for the unfolded protein; this notation is summarized in Fig. 5.5. As the denaturant one may typically use urea, which is believed to mostly act as a good

[2] This is in analogy with constant volume conditions, where $F = E - TS$ implies that $S = E/T - F/T = E/T - k_\text{B}\ln(Z)$. Differentiating $\text{d}S/\text{d}T = E/T^2 - C/T - k_\text{B}\text{d}\ln(Z)/\text{d}T = E/T^2 - C/T - E/T^2 = C/T$, from which one obtains $C = T\text{d}S/\text{d}T$. The relation $S = E/T - \ln(Z)$ also follows directly from $S = \sum p_i \ln(p_i)$, using $p_i = \text{e}^{-E_i/k_\text{B}T}/Z$.

$$U \underset{k_u}{\overset{k_f}{\rightleftharpoons}} N$$

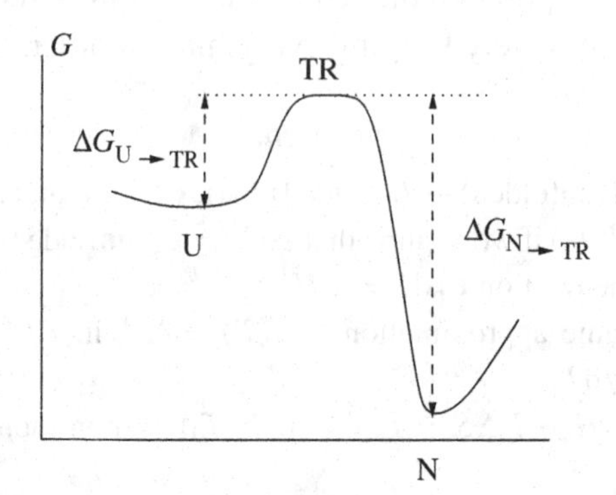


Figure 5.5. Schematic view of protein folding. U is the denatured state, TR is the transition state, and N is the native state. The $x$-axis represents a reaction coordinate, and could be the fraction of native contacts formed (for example). Notice that such pictures easily could be misleading, as the real $x$-axis is multidimensional: for a given number of native contacts there could in principle be many ways to fold the protein. In the $\phi$ value analysis one assumes that there is only one pathway, as illustrated here.

solvent for hydrophobic residues, thus destabilizing the hydrophobic core of the protein.

In the experiments one observes a linear relationship between the log of the unfolding rates and the denaturant concentration $[D]$:

$$\ln(k_u) = \ln\left(k_u^{H_2O}\right) + m_u[D] \tag{5.10}$$

where $k_u^{H_2O}$ stands for the unfolding rate in the absence of denaturant. In terms of free energy barriers (see Fig. 5.5):

$$\Delta G_{N\to TR} = \Delta G_{N\to TR}^{H_2O} - m_u[D]k_B T \tag{5.11}$$

This assumes that the rate of barrier hopping can be expressed as: rate $= r\exp(-\Delta G/k_B T)$ (see Kramer's formula in the Appendix), where $\Delta G$ is the barrier height and $r$ is a microscopic attempt rate. For small single-domain proteins the transition is generally in two states, and a similar relation holds for the folding rate

$$\Delta G_{U\to TR} = \Delta G_{U\to TR}^{H_2O} + m_f[D]k_B T \tag{5.12}$$

Experimentally we measure the relaxation of a mixture of folded and unfolded proteins, after quenching to a new denaturant concentration (see Fig. 5.6). The relaxation rate is given by $k_f(D) + k_u(D)$ versus $D$ (see the Question below), and

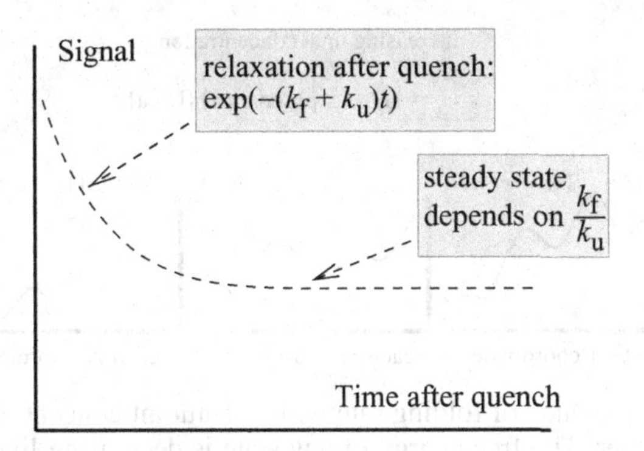


Figure 5.6. Typical time course of a signal based on a mixture of folded and unfolded proteins, after a quenching from (for example) folding conditions to denaturant-induced unfolding conditions.

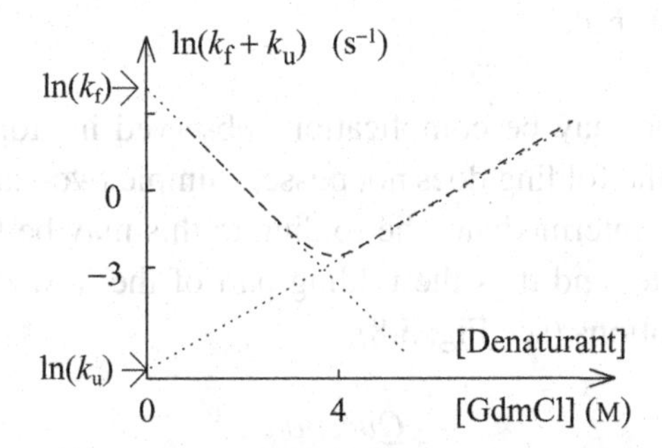


Figure 5.7. Chevron plot: folding and unfolding rates (respectively) after a fast quench of unfolded (native) protein into low (high) concentration of denaturant. The linear behavior of the rates allows extrapolation to zero denaturant conditions, where actual rates are so high/low that they cannot be measured directly. The slope of the unfolding curve is called the *m*-value, and it correlates to the buried surface area of amino acids in the protein.

thus we express $\ln(k_f(D) + k_u(D))$ versus $D$ as

$$\ln(k_{obs}) = \ln(k_f(D) + k_u(D)) = \ln\left(k_f^{H_2O} e^{-m_f[D]} + k_u^{H_2O} e^{m_u[D]}\right) \tag{5.13}$$

One now obtains a characteristic V-shaped curve, which is usually called a chevron plot; see Fig. 5.7. The straight-line extrapolations shown with dotted lines can be used to extrapolate $k_u$, $k_f$ to zero and high denaturant concentration. Thus one can infer folding times that are faster than the inherent mixing time limit of 2–5 ms.

The ratio of forward and backward rates must give the equilibrium constant for any denaturant concentration

$$\frac{k_f}{k_u} = K(D) = \frac{[N]}{[U]} \tag{5.14}$$

which can be measured by equilibrium experiments.

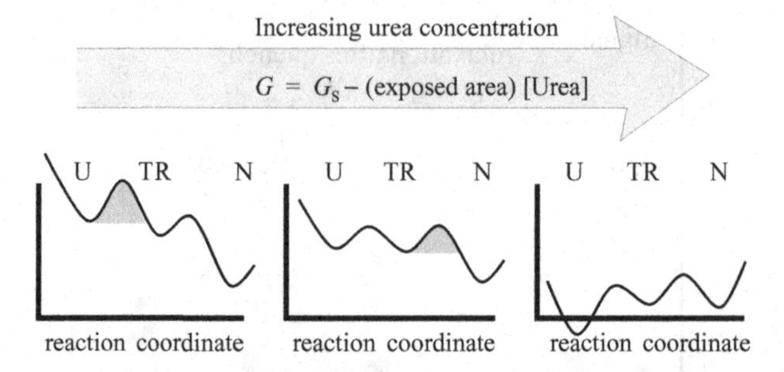


Figure 5.8. Dependence of folding rate with denaturant concentration [Urea] for a three-state folder. The free energy of any state is decreasing linearly with area exposed to urea. Unfolded protein therefore depends more on [Urea], than transition state (TR), and so on. As a consequence, folding of a three-state folder is sensitive to a different barrier at low [Urea], than it is at high [Urea]. This makes a characteristic roll-over of the Chevron plot for low [Urea]. For three-state folders it is no longer V-shaped.

Occasionally there may be complications observed in stopped-flow measurements, especially if the folding does not possess simple two-state behavior. Thus if there is a metastable intermediate, the folding to this may be faster than the folding to the native state, and thus the folding part of the V will level off for small denaturant concentrations (see Fig. 5.8).

*Question*

(1) Consider a two-state folder in a stop-flow experiment. Changing urea concentration [Urea], the rate of change of the unfolded protein fraction $\mathrm{d}U/\mathrm{d}t = -k_\mathrm{u}U + k_\mathrm{f}N$, whereas the rate of change of the folded protein fraction $\mathrm{d}N/\mathrm{d}t = k_\mathrm{u}U - k_\mathrm{f}N$. The experiment measures the overall signal asssociated to the state of the system as a function of time. Show that the decay towards equilibrium is governed by the rate $k_\mathrm{f} + k_\mathrm{u}$, whereas the equilibrium state is given by $k_\mathrm{f}/k_\mathrm{u}$.

## Single point mutations: $\phi$-value analysis

Given the folding and unfolding rates estimated from extrapolating to $D = 0$ in Fig. 5.7, one determines the free energy barriers according to

$$\Delta G(\mathrm{TR} \to \mathrm{U}) = k_\mathrm{B}T \ln(k_\mathrm{f}^{\mathrm{H_2O}}/r_0) \quad \text{and} \quad \Delta G(\mathrm{TR} \to \mathrm{N}) = k_\mathrm{B}T \ln(k_\mathrm{u}^{\mathrm{H_2O}}/r_0) \tag{5.15}$$

Here $r_0$ is a microscopic rate that we do not have to specify because we need only the ratio $k_\mathrm{f}/k_\mathrm{u}$. Please notice that it is the folding rate $k_\mathrm{f}$ that determines the barrier from the unfolded state to the transition state, and it is the unfolding rate that determines the barrier from the native state to the transition state. The sign of the above equation is given by the fact that a high barrier between U and TR implies

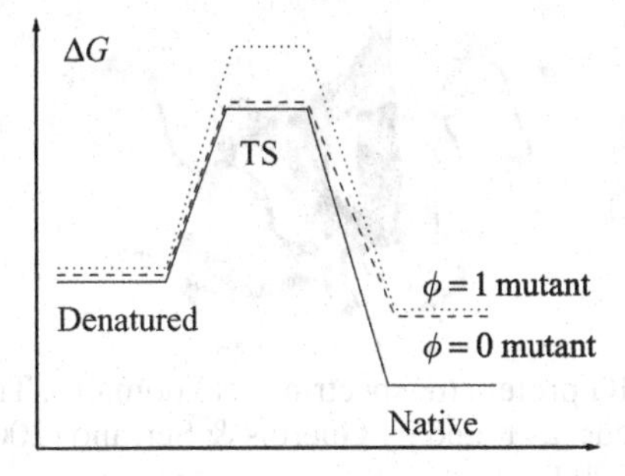


Figure 5.9. Representation of the physical significance of $\phi$ values measured from barrier changes relative to native energy change, as suggested by Fersht. Energy barriers are deduced from the so-called Chevron plot, i.e. the folding and unfolding rates extrapolated at zero denaturant concentration. In real cases one typically sees $\phi$ values between 0 and 0.6.

$\Delta G(\mathrm{TR} \to \mathrm{U}) = G(\mathrm{U}) - G(\mathrm{TR}) \ll 0$, and also implies that the folding rate $k_\mathrm{f}$ is very low ($\ll \tau_0$). Because free energy barriers should add up, the stability of the folded state is then (see Fig. 5.5)

$$\Delta G(\mathrm{N} \to \mathrm{U}) = \Delta G(\mathrm{TR} \to \mathrm{U}) - \Delta G(\mathrm{TR} \to \mathrm{N}) = k_\mathrm{B}T \ \ln\left(k_\mathrm{f}^{\mathrm{H_2O}}/k_\mathrm{u}^{\mathrm{H_2O}}\right) \tag{5.16}$$

The key idea in the Fersht–Jackson school is to study $\phi$ values (Jackson & Fersht, 1991). These are defined for single-point mutations of each amino acid residue especially to an alanine (a "neutral" amino acid that does disrupt or add new interactions). One performs the above study for both the wild type and the mutant. Then the $\phi$ value is defined (see Fig. 5.9) as

$$\phi = \frac{\Delta\Delta G(\mathrm{TR} \to \mathrm{U})}{\Delta\Delta G(\mathrm{N} \to \mathrm{U})} \tag{5.17}$$

where $\Delta\Delta G(\mathrm{TR} \to \mathrm{U}) = \Delta G(\mathrm{TR} \to \mathrm{U};\ \text{mutant}) - \Delta G(\mathrm{TR} \to \mathrm{U};\ \text{wild type})$ is the change in barrier height between the wild type and the mutant, etc. One can now ask whether the mutation modifies the stability (= free energy difference with respect to the unfolded state) of the native state only or also of the transition state[3]. When $\phi = 0$, the native state is modified but the transition state is not; the point mutation corresponds to an amino acid in a region of the protein that does not participate in the folding nucleus. When $\phi = 1$, the native contacts around the studied amino acids are fully formed in the transition state. Typically $\phi$ values are between 0 and about 0.6. Notice that $\phi$ values can be defined only if the stability of

[3] Implicitly we thereby ignore the possible effect of mutations on the denatured state, meaning that we assume absence of persistent contacts in the unfolded protein state.

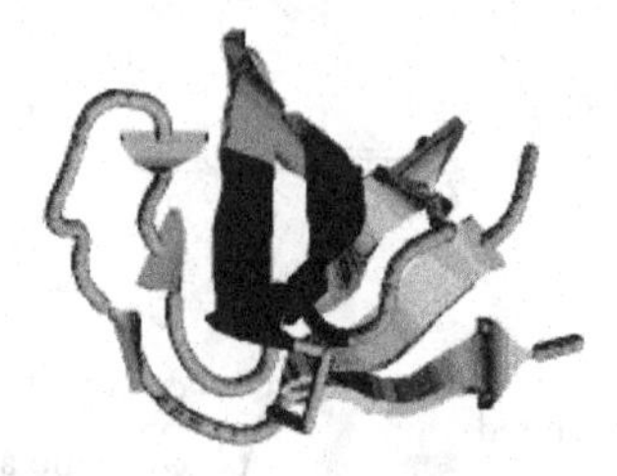

Figure 5.10. The 1SHG protein ($\alpha$-spectrin, SH3 domain). The black region indicates the folding nucleus, as found by Guerois & Serrano (2000). Crystallography by Musacchio *et al.* (1992).

the native state depends on the given amino acid. Furthermore, the whole concept of $\phi$ values depends on a well-defined unique transition state for comparison. If there were alternate pathways, then amino acids that were important in only one of the pathways typically would not be detected in the overall $\phi$ value.

For an explicit example of a protein where a folding nucleus was determined see Fig. 5.10.

## Models

Numerous models for protein folding have been considered, including very detailed models based on single-atom resolution, schematic models based on beads on a lattice, and models of the protein–solvent interaction that basically aim at capturing the thermodynamics.

The very detailed models quickly run into problems because the force fields are not known to sufficient accuracy to give a stability of $\Delta G \sim -10$ kcal/mol by adding up a huge number of interactions that are several kcal/mol each.

The schematic lattice models can be envisioned as in Fig. 5.11 (drawn in two dimensions). The circles are amino acids, assumed to be of equal size, but differing in interactions. Figure 5.11 illustrates the simplest lattice model, where amino acids are assumed to be either hydrophilic or hydrophobic, and all other properties are ignored. The energy of a configuration is set equal to the number of hydrophobic amino acids that are on the surface of the protein; in Fig. 5.11 that is zero. In the figure, if one of the ends opens up, the hydrophobic core is exposed to water, which raises the energy.

The specificity of the above potentials is quite weak, and although models can typically predict sequences that have only one ground state, the energy gap between the ground state and the rest is too small to give a two-state folding transition. This presumably reflects either wrong potentials, or more likely the fact that the residue–residue potential is not spherically symmetrical and in addition also depends on nearby amino acids. For example, hydrogen bonds are very directional and impose

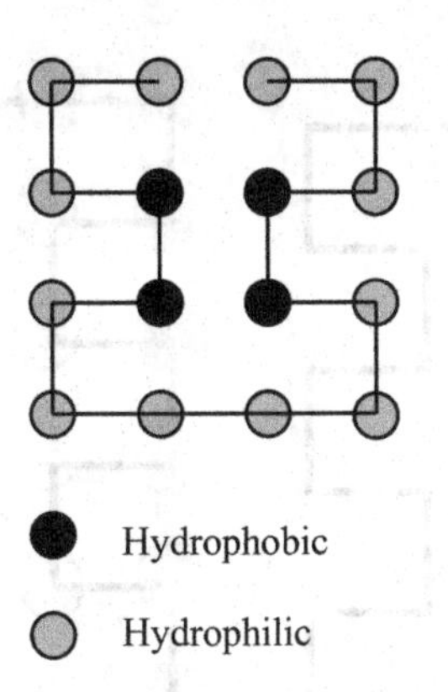


Figure 5.11. The hydrophobic–polar (HP) model on a lattice, introduced by Dill (1985) and later analyzed (see review by Dill *et al.* (1995)). If one wants to represent a protein as hydrophobic–polar units only, the corresponding typical amino acids would be ($S$ and $V$) or ($T$ and $L$).

severe constraints on the structure of the core of the protein. Nevertheless the lattice polymer model is so simple that it opens for discussion some basic concepts related to protein folding. This includes in particular the concepts of "good folder" and "designable structures", as explored by Sali *et al.* (1994), Shacknovich *et al.* (1998), and Li *et al.* (1996, 1997, 1998).

An interesting feature of hydrophobic–polar models is that they are so simple that they allow exhaustive scanning of all (or a large fraction of) structures and sequences for polymers that are not too large. Thus one can address the designability of a particular structure as follows (Li *et al.*, 1998, 2002).

> **Designability of structure** = number of sequences that have the structure as a unique ground state.

Presumably, highly designable structures have a large energy gap between the best sequence and the non-folding sequences; they are also more robust to mutations (Tiana *et al.*, 2001). Thus they should be seen more often in Nature. And, trivially, structures that are not designable at all should not occur. Thus designability is a criterion to limit the structure space, and also opens a way to generate clusters of sequences that fold into the same structure. By complete enumeration Li *et al.* (1996) found that the most designable structure of a 36 mer hydrophobic–polar sequence was the one shown in Fig. 5.12. In general it was found that the more designable structures are the ones with the highest symmetry.

We now return to the actual calculation of equilibrium properties for a given primary sequence. When dealing with long polymers, where exhaustive sampling is impossible, one instead samples a representative subset of possible configurations. To do this in a reasonable way one should not just sample random configurations

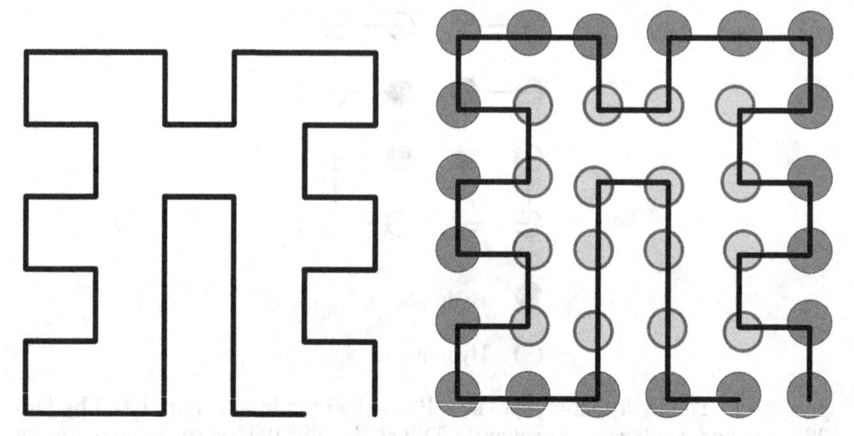

Figure 5.12. The most designable sequence of a 36 mer, when one allows only compact structures (from Li *et al.*, 1998). The right-hand panel shows a sequence that has this structure as the ground state: PPHHHHHHPPPPHHPPHHPPPPHHPPPPHHPPHH. (P represents a polar residue, and H represents a hydrophobic residue.)

from scratch. Doing that makes it nearly impossible to obtain compact polymer states. Instead one should use Metropolis sampling (Metropolis *et al.*, 1953). This is an algorithm to generate a representative sequence of structures, given interactions and temperature. At each iteration one performs the following steps.

- Given the current structure, with energy $E_{\text{old}}$, generate a neighbor structure and calculate its energy $E_{\text{new}}$.
- If the new structure has lower energy, $E_{\text{new}} < E_{\text{old}}$, it is accepted as the current structure.
- If the new structure has higher energy, $E_{\text{new}} > E_{\text{old}}$, the new structure is accepted with probability $\exp((E_{\text{old}} - E_{\text{new}})/k_B T)$.
- If the new structure was accepted, it replaces the old one. If the new structure is not accepted, one retains the old one.

The "art" in Metropolis sampling is to define a "move set", which is efficient and allows exploration of the full space of structures. The move set defines which neighboring structures a given structure has. A suitable set of moves for polymers on a cubic lattice is discussed by Socci & Onuchic (1994), and illustrated in Fig. 5.13. Simulating the system at any given temperature, one obtains a sample of sequences, where high-energy sequences are automatically suppressed together with their corresponding Boltzmann factor. Then one may calculate any thermodynamic quantity by "time" averaging. For example, for a set of structures $\{i\}$ the average energy and its second moment are

$$\langle E \rangle = \frac{\sum_i E_i}{\sum_i 1} \qquad \langle E^2 \rangle = \frac{\sum_i E_i^2}{\sum_i 1} \tag{5.18}$$

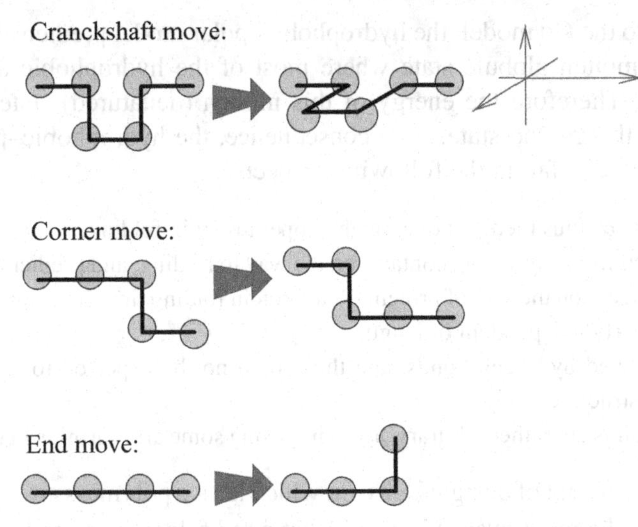


Figure 5.13. The three different local moves that allow a polymer on a cubic lattice to visit all configurations. In each case the move is accepted only when the new structure does not overlap with any part of the polymer. Notice also that in each case we have included some part of the chain that does not participate. These parts may be in different configurations than shown, as long as they do not interfere with the move.

In this way one can simulate the behavior as a function of temperature and, in particular, can obtain the heat capacity $C = (\langle E^2 \rangle - \langle E \rangle^2)/k_B T^2$ as a function of temperature, and thus examine the cooperativity of the melting transition through Eqs. (5.3) and (5.4).

Maybe the most striking lesson from the simple hydrophobic–polar model, as well as its related cousins based on more elaborate interaction alphabets, is that approaches that treat amino acids as independently interacting entities tend to give a folding that is far from being two state (see Kaya & Chan, 2000). Thus these models do not have sufficient cooperativity to reproduce the calorimetric findings of Privalov and collaborators.

Possible improvements to introduce cooperativity in such simple models involve higher-order interactions (Chan & Kaya, 2000). The conceptually simplest of these is the Go model (Takatomi *et al.*, 1975). The Go model assumes that only those amino acid pairs that interact in the ground state interact at all. Through this assumption, an unfolded protein has very little interaction energy, and the energy difference between the ground state and the unfolded states becomes sufficiently large to give two-state behavior. One may speculate whether this relative success of the Go model implies that interactions between real amino acids are very dependent on the local environment, including both the water and the positions of other amino acids.

In contrast to the Go model, the hydrophobic–polar model predicts that proteins unfold into a molten globule state where most of the hydrophobic amino acids remain buried. Therefore the energy of this molten (denatured) state is not that different from the ground state. As a consequence, the hydrophobic–polar model (and similar models) fail in the following respects.

- They give large $\alpha$, thus there is not enough cooperativity in folding.
- They give frustrations by wrong contacts on the way to folding, and predict folding times that depend hugely on the size of protein. Real protein folding times (for simple two-state folders) are nearly independent of length.
- They lack directed hydrogen bonds, and thus could not be expected to teach us much about protein structure.
- Lattice constraints are rather arbitrary, again imposing some arbitrariness in the structure.

Apart from the concept of designability for which lattice polymers are a particularly suitable model, the main overall lesson is that good folders need to have a ground state energy that is well separated from all the other non-native states. As we shall see in the next section, this result can be understood from analyzing the random energy model.

### *Questions*

(1) Apply the Metropolis algorithm to a two-state system with energies $E_1 = 0$ and $E_2 = 1$ and simulate the average energy and heat capacity as functions of temperature.

(2) How many different configurations are there for a homopolymer of length 4 on a two-dimensional square lattice? Consider all hydrophobic–polar sequences of length 4 (sequences of four amino acids where each element is characterized solely on the basis of being either hydrophobic or polar). How many designable sequences are there, what are they and what is their energy, when one assigns $-1$ unit of energy for each hydrophobic–polar contact and $-2$ for each hydrophobic–hydrophobic contact?

## ***Random energy model***

One main lesson from the lattice models has been that a good folding is associated with a big gap between the ground state energy, and the typical energies of partially folded states. This main result can be seen in a much simpler model, called the random energy model (REM). This is usually represented by an energy function (also called a Hamiltonian) that for each state takes a value (Derrida, 1980) of

$$H = \sum (\text{random numbers}) \tag{5.19}$$

with random numbers that all change value to new random non-correlated numbers for each change of state. This reflects that any change in structure of a compact polymer implies that many local interactions also change, and thus take new values (see Bryngelson & Wolynes, 1987). That is, when the energy is given by the sum of many local interactions that all change when one changes, the REM Hamiltonian becomes a reasonable first-order guess on an energy function.

The REM has a number of states with energy $E$ given by

$$\mathcal{N}(E) = \left(\frac{C-1}{e}\right)^N \exp\left(-\frac{E^2}{2\sigma^2}\right) \tag{5.20}$$

because the sum of random variables has a Gaussian distribution; the pre-factor takes into account the number of compact states of the polymer with coordination number $C$ ($C \sim 6$ for proteins). This energy spectrum (see Fig. 5.14) is valid as long as $\mathcal{N}(E) \gg 1$, and would then be identical for widely different proteins. It signals a Hamiltonian where any small adjustment in variables changes many energies in random directions. For an energy $E$ the corresponding temperature $T$ is determined from

$$\frac{1}{T} = \frac{\mathrm{d}S}{\mathrm{d}E} = -k_\mathrm{B}\,\frac{E}{\sigma^2} \tag{5.21}$$

i.e. the energy $E = -\sigma^2/k_\mathrm{B}T$ decreases with diminishing $T$ (we used the relation $S(E) = k_\mathrm{B}\ln(\mathcal{N}(E))$).

Motion in the associated potential is very difficult, characterized by multiple energy barriers, and slow transitions between states. In fact the typical rate for moving between states at temperature $T$ would be (see Kramers' escape in the Appendix)

$$\text{rate} \propto \mathrm{e}^{-E(T)/k_\mathrm{B}T} \tag{5.22}$$

where the energy is the barrier. If this energy is temperature independent, we would have an escape *rate* with temperature dependence $\ln(\text{rate}) \propto -1/T$. However, in the case of a glassy system described by the random energy model the typical barrier is given by the difference between the average energy $E$ of the system and the energies corresponding to most states of the system (in the middle of the Gaussian, i.e. at $E = 0$ in our parametrization). From Eq. (5.21) this defines a barrier $E(T) \propto 1/T$, and the rate for moves between typical energy states at a given temperature $T$ is therefore

$$\ln(\text{rate}) \propto -1/T^2 \tag{5.23}$$

which means that the rates change much more with $T$ than one would expect for the case where relevant barriers are independent of $T$. This behavior is characteristic of "glasses". Such behavior is indeed observed in the temperature dependence of the viscosity for certain liquids. In fact, we define the glass transition temperature

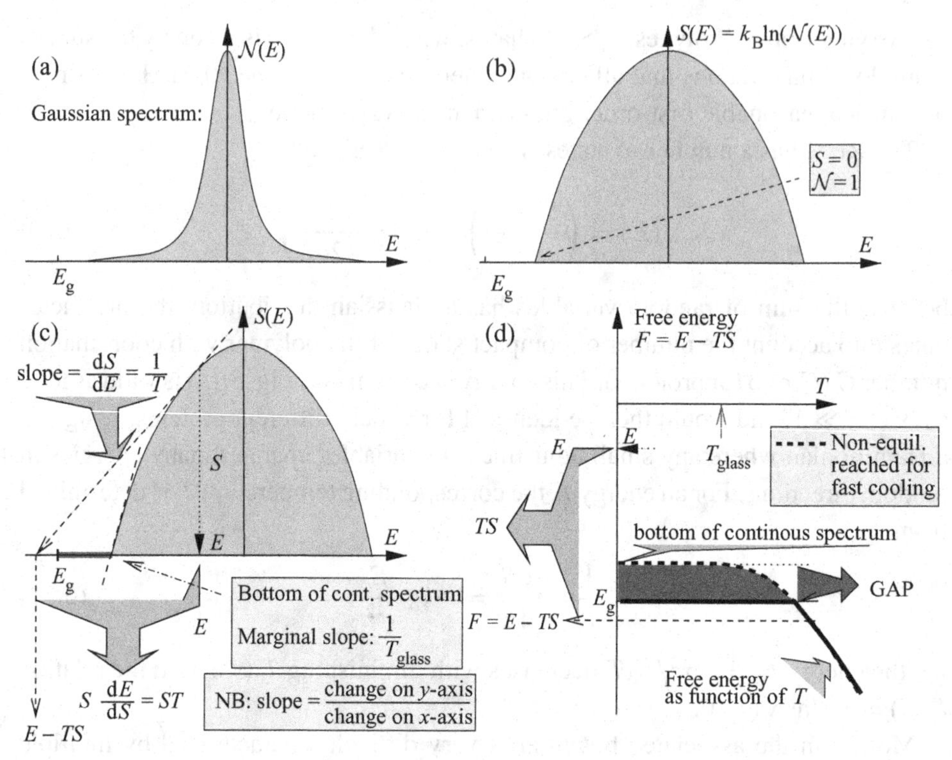


Figure 5.14. Energy spectrum (a) and solution of the random energy model (c, d). The energy spectrum from (a) gives the entropy as a function of the energy, shown in (b). Panel (c) shows how the derivative $\mathrm{d}S/\mathrm{d}E = 1/T$ at energy $E$ can be used to calculate the free energy $F = E - TS$. The parabolas in both (b) and (c) predict a continuous spectrum with a large number of states for all energies where $\mathcal{N}(E) > 1$. When, for decreasing $T$, the continuous spectrum runs out of states ($S(E) = 0$) the system freezes. However, as shown in (c), at some higher temperature (we could call it $T^*$) the $F$ given by the parabolic part of the spectrum may become larger than the ground state energy $E_g$. At this transition temperature $T^*$ the equilibrium system will select a single isolated state over all the states in the continuum spectrum. The resulting free energy as a function of temperature, $F = F(T) = E - TS$, is shown by the thick solid line in (d). If equilibrium is not reached, the system will instead follow the dashed thick line to the bottom of the continuous spectrum.

as the highest $T$ where the system can be at the bottom of the continuous spectrum (see Fig. 5.14d).

In practice one does not always reach the low energy states of the continuum, because a few states of even lower energy may take over. In fact, in terms of heteropolymer models, the individuality of each particular heteropolymer is associated with the behavior of exactly these low energy states. Thus the specificity is associated with the discrete states, and the possibility of folding to a unique ground state is associated with the behavior of the lowest energy state. To elaborate on this, see Fig. 5.14 where we compare the free energy $F_{\text{continuum}}$ of being somewhere in the continuum spectrum, with the ground state energy, $E_g$. We see that the more $E_g$ is separated from the continuum, the higher the temperature where it becomes thermodynamically favorable to be in the ground state.

In the canonical ensemble the free energy associated to the sub-part of states in the continuum is

$$F_{\text{continuum}} = E - TS = -\frac{1}{2}\frac{\sigma^2}{k_B T} - k_B T \cdot \ln\left(\left(\frac{C}{e}\right)^N\right) \tag{5.24}$$

When this continuum free energy is smaller than the free energy associated with being in the single discrete ground state, so that

$$F_{\text{ground state}} = E_g \tag{5.25}$$

the system stays in the continuum. Upon decreasing the temperature $T$ the continuum free energy increases and at one point becomes larger than the ground state energy. The shift from the continuum to the ground state defines a transition temperature for folding. When $T$ is lower than this the system prefers the ground state over the many states in the continuum part of the spectrum (see Fig. 5.14).

Now we can return to the difference between good and bad folders. The basic observation is that when the gap between the lowest and the next-lowest state is large, then one has a good folder. This is because a low $E_g$ allows a jump to this state at rather high temperatures, as indicated in Fig. 5.14. Therefore at these high temperatures it will be easier to cross the barrier between some high energy trap and the ground state. Furthermore, it requires that there is no other low-energy state, as the presence of such competing states may lead to a system that gets stuck in a relatively deep minimum.

Overall, the random energy model and its lattice counterparts focus on the energy part of the protein folding problem, emphasizing that the folding of random heteropolymers will be slowed down by the large friction associated with many large energy barriers. Models with a rough energy landscape predict a folding that is fastest at the highest temperature, whereas the experimental data show the maximal folding rate for physiological temperatures and a slowing down for larger $T$ (see the investigation of the protein CI2 by Oliveberg *et al.*, 1995). This, in part, may reflect an energy landscape that is temperature dependent. However, it is also indicative of a folding limited by entropy barriers. In the following sections we will explore models for this latter possibility.

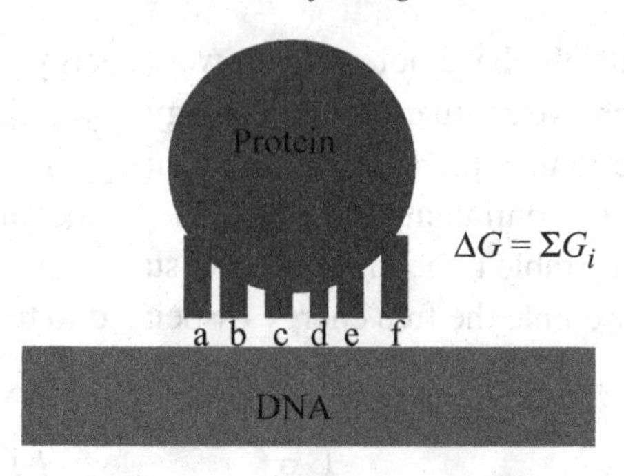


Figure 5.15. Protein–DNA binding where the interaction at each site is a sum of interactions between exposed amino acids and parts of the DNA. For non-specific sites this sum is effectively a sum of random numbers (Gerland *et al.*, 2002), making the random energy model a good basis for understanding the overall ensemble of binding to all possible non-specific parts of the DNA.

*Question*

(1) The random energy model (REM) can also be applied to analyze the unspecific binding of proteins to DNA, see Fig. 5.15 and Gerland *et al.* (2002). The basic observation of Stormo & Fields (1998) is that the binding energy of a DNA-binding protein at a position is specified by $j = 1, \ldots L$: $E_j = \sum_{i=j}^{j+l} \mathcal{E}(i)$, where $i$ is the base pair and $l$ is the length of the protein–DNA binding region ($\sim$15–20). Consider the energy spectrum for a protein binding to *E. coli* DNA with length $L \sim 5 \times 10^6$. Set $l = 15$ and assume that each $\mathcal{E}(i)$ is drawn from a distribution with average 0 and spread 1 kcal/mol. What is the average binding energy? What is the average free energy? Calculate the position of the lower end of a continuous spectrum.

## *The topomer search model*

Figure 5.16 show a number of two-state proteins, ordered by folding rate. One can notice a clear correlation between structure and folding, such that proteins with a large fraction of β-sheets are slower folders. This effect may be associated with the fact that formation of sheets requires contacts between amino acids that are far from each other along the peptide backbone. In fact, it may indicate that folding rate is limited by entropy, rather than by energy barriers. This can be quantified through Plaxco's observation (Plaxco *et al.*, 1998) of a correlation between folding time and the topology of the native state (see Fig. 5.17). Thus the folding rate $k_f$ of a single-domain protein is strongly correlated with the number of times, $Q$, that the peptide backbone crosses itself in the native state:

$$k_f \propto p^Q \tag{5.26}$$

where $p$ is some number smaller than 1, and $Q \sim$ 5–100 is the number of $C_\alpha$ in amino acids that are separated by more than 12 amino acids along the peptide

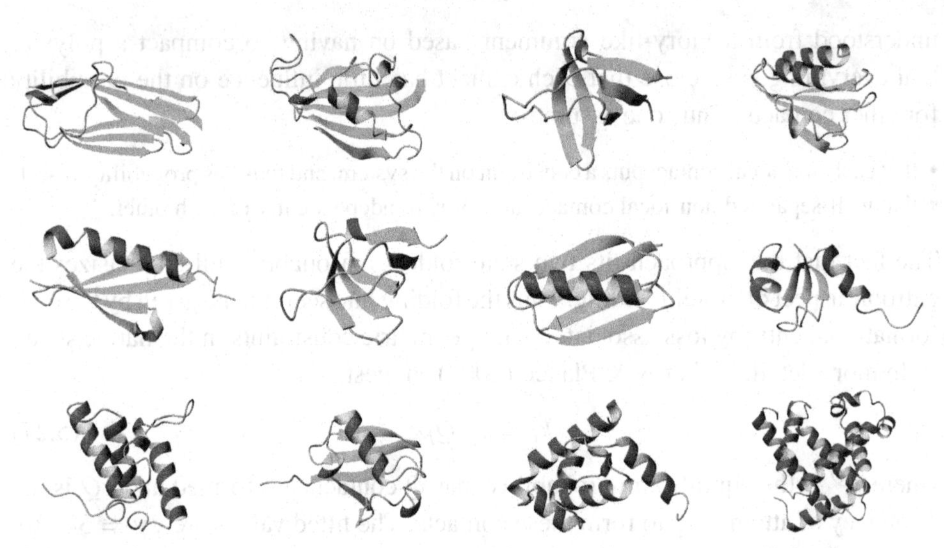

Figure 5.16. Twelve two-state proteins, ordered such that the slowest folders are in the upper left-hand corner. The folding rate is correlated with the fraction of sheet-to-helix content. The β-sheet is a non-local secondary structure, and the fraction of sheets is thus nearly equivalent to the number of non-local contacts. Figure prepared by J. Bock.

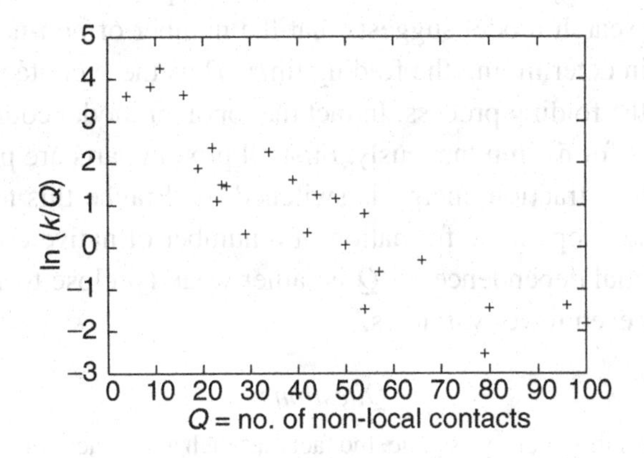


Figure 5.17. Correlation between folding rates and topology of the ground state of a number of single-domain proteins (data from K. W. Plaxco, private communication). The folding rate is observed to decrease exponentially with the number of native non-local contacts $Q$. The abscissa examines $\ln(k/Q)$, but a plot of $\ln(k)$ would exhibit very similar correlations.

backbone and that are closer than 6 Å from each other in the native state. Thus ln(folding rate) decreases linearly with $Q$. Figure 5.17 show $\ln(k/Q)$ versus $Q$ for a number of two-state folders.

The motivation for Eq. (5.26) is based on numerical simulations on how often a homopolymer crosses itself (Makarov & Plaxco, 2003). Theoretically it is understood from a Flory-like argument based on having so compact a polymer that everything is so close that each contact has little influence on the possibility for other contacts. Thus one assumes:

- that each non-local contact puts a constraint on the system, and thus has probability $p < 1$;
- that well-separated non-local contacts are formed independently of each other.

The fact that this approach fits two-state folders reasonably well emphasizes the entropic aspect of protein folding. Thus the folding rate seems to be given by the conformational entropy loss associated with geometric constraints in the native state.

In more detail, Makarov & Plaxco (2003) suggest

$$k_{\mathrm{f}} = \kappa \gamma Q p^{Q} \tag{5.27}$$

where $p^Q$ is the equilibrium probability that $Q$ contacts are formed and $\kappa Q$ is the frequency of attempting to form these contacts. The fitted values are $\kappa\gamma = 3800/\mathrm{s}$ and $p \approx 0.85$. The rate of collision between distant amino acid pairs is suggested to be given by $\kappa = 3D/d^2 \sim 10^8/\mathrm{s}$ with the diffusion constant $D = 40\ \mu\mathrm{m}^2/\mathrm{s}$. Here $d \sim 0.5$–$1$ nm is the distance at which an amino acid pair can form a contact; $\gamma$ is the entropy barrier to the first contact, and is estimated to be $\gamma \sim 4 \times 10^{-5}$. Thus $\gamma$ is the probability that the polymer is in a state that is compatible with a folding attempt.

The topomer search model suggests that the number of non-neighbor contacts $Q$ is important in determining the folding time. Thus these contacts should be the key contacts in the folding process. In fact the topomer model counts entropy as if all these contacts form simultaneously. First all protein parts are put into the right position, then the attraction energy is switched on. Maybe this teaches us about simultaneous and cooperative formation of a number of native contacts. The fact that the exponential dependence on $Q$ is rather weak ($p$ close to 1) indicates that $Q$ is a proxy for even fewer variables.

### *Questions*

(1) Argue for, and subsequently assume, the fact that each amino acid of a collapsed protein diffuses freely inside a volume $V = 4/3\pi d^3$ to estimate the rate of encounter between two random amino acids. Estimate prefactors in the topomer model.

(2) Experiments show that folding times change when some amino acids are changed in a protein. Can this be compatible with the topomer model?

## ***Pathway versus cooperativity of protein folding***

A folding protein consists of at least 50–100 amino acids that all need to be brought into the correct positions relative to one another. However, these amino acids are close together, and effectively the real independent number of degrees of freedom is probably smaller. Here we summarize some of the schematic models in terms of contact variable between the ruling degrees of freedom in a protein that folds. Thus, following Hansen *et al.* (1998) we introduce an event-based parametrization of protein folding in terms of a set of contact variables $\{\phi_i\}$ in the protein. Each can take one of two values:

$$\phi_i = 0 \text{ (unfolded)} \quad \text{and} \quad \phi_i = 1 \text{ (native contact)} \tag{5.28}$$

Typically there will be one correct native contact, and several unfolded ones for each contact variable $\phi_i$. If contact 2 can be made only when another contact 1 has already been formed, it means that the Hamiltonian (i.e. energy as function of the variable $\phi$) contains a term

$$H = \cdots - \phi_1\phi_2 - \cdots \tag{5.29}$$

and does not include any product terms containing $\phi_2$ but not $\phi_1$ (see Fig. 5.18). Therefore the Hamiltonian will be ordered according to what depends on what. A particularly simple example is the zipper model discussed in Chapter 3; see also Fig. 5.18. Using the logical binary variable by Hansen *et al.* (1998) the Hamiltonian for the zipper reads:

$$H = -\mathcal{E}_0(\phi_1 + \phi_1\phi_2 + \phi_1\phi_2\phi_3 + \cdots) \tag{5.30}$$

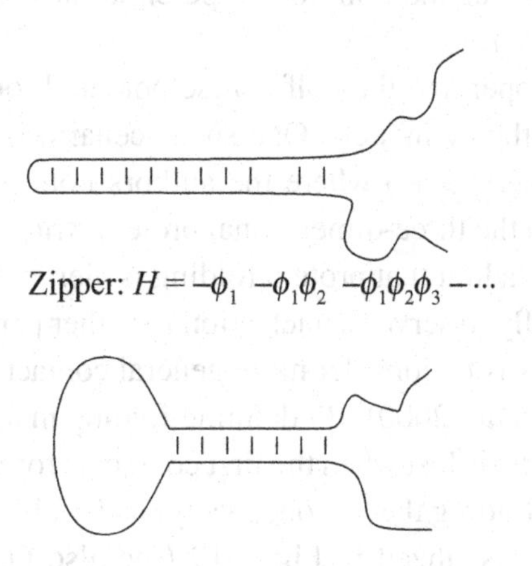

Zipper: $H = -\phi_1 - \phi_1\phi_2 - \phi_1\phi_2\phi_3 - \cdots$
Zipper with nucleation threshold:
$H = -\phi_1\phi_2\phi_3\phi_4\phi_5 - \phi_1\phi_2\phi_3\phi_4\phi_5\phi_6 - \phi_1\phi_2\phi_3\phi_4\phi_5\phi_6\phi_7$
Entropic barrier


Figure 5.18. The zipper approach to the folding of polymers, originally developed to describe DNA melting. The same model can be applied to protein folding, where the variables parametrize contacts along a folding pathway. Hydrophobic zippers for proteins have been discussed by Dill *et al.* (1993).

corresponding to the simplest hierarchical folding: a single pathway where any fold depends on a previous one being correctly in place. Another possible Hamiltonian is the "golf course" potential:

$$H = -N\mathcal{E}_0\,\phi_1\phi_2\phi_3\cdot\cdots\cdot\phi_N \tag{5.31}$$

where energy is first gained where exactly everything is in place at the same time. This is what is meant by "golf course": everything is equal throughout the configuration space, except for a very small hole.

The golf course potential is obviously wrong because each residue has maybe $C = 6$–$8$ different orientations. Thus there are $\sim 8^N$ states with energy 0 in the above potential, which makes the chance to find the correct state extremely small. Thus the folding time would be extremely long ($\sim 8^N$ time steps). This property of a featureless energy landscape is normally called Levinthal's paradox, and it emphasizes that protein folding should be guided by a potential (Levinthal, 1968). For example, if the variables are organized as a zipper, then folding consists of $N$ consecutive steps, where for each step the protein has to find only one in eight states. This will take only $8N$ time steps, thus removing Levinthal's paradox. Both of the above models can be solved analytically, and both predict a first-order transition at a melting temperature $T = \mathcal{E}_0/\ln(C+1)$. However, the zipper model has a van't Hoff coefficient $\alpha = 12$, whereas the golf course potential has $\alpha = 4$ when $N$ is large (see the Questions below).

Going beyond both the zipper and the golf course potential, one may consider more generalized folding pathway models. One such scenario is the hierarchical folding of Baldwin & Rose (1999a,b) where the authors notice that the folding polymer stays mostly local in the three-dimensional protein structure, and does not mix across it. This is taken as a hint that protein folding is hierarchical in the sense that some parts first fold locally to serve as nucleation for other parts. A formalistic way to play with such models is to consider more general contact models than the zipper model (Dommersnes *et al.*, 2000). By defining folding in terms of contacts, and assuming that most entropy is lost when the first contact is formed, whereas the main energy is gained while folding the last one, we typically obtain Hamiltonians with an energy spectrum as visualized in Fig. 5.19 (see also Questions below). In contrast to the random energy model, the thereby constructed "random entropy model" has a folding time that is limited solely by an entropic barrier.

## *Questions*

(1) What is the van't Hoff coefficient for the zipper model for $N \to \infty$?
(2) What is the van't Hoff coefficient for the golf course potential for $N \to \infty$?
(3) The native (folded) state has energy $E = -N$. How does energy where most contacts have been formed scale with the total number of contacts $N$ in the random

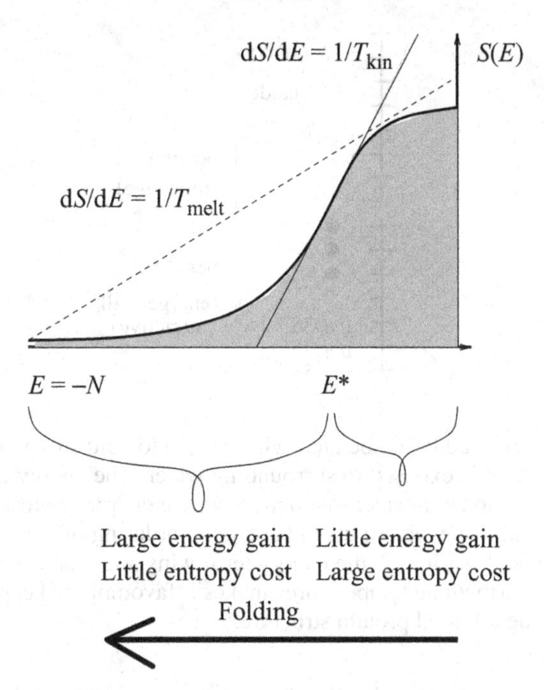


Figure 5.19. Entropy as a function of energy for the random entropy model (Dommersnes *et al.*, 2000), defined by a Hamiltonian given by a sum of $N$ attractive terms (all negative), each given by a product of $n = 1, 2, \ldots$ up to $n = N$ folding variables. As $S(E)$ bends upwards, the constructed Hamiltinian has a first-order phase transition, at a temperature $T_{\mathrm{melt}}$.

zipper model, defined by a random Hamiltonian of type: $H = -\phi_{\mathrm{r}(\{i\})} + \phi_{\mathrm{r}(\{i\})}\phi_{\mathrm{r}(\{i\}} + \phi_{\mathrm{r}(\{i\})}\phi_{\mathrm{r}(\{i\})}\phi_{\mathrm{r}(\{i\})} + \cdots$ with up to $N$ terms in the sum and where each $\mathrm{r}(\{i\})$ denotes a function that draws a random variable from $i = 1, \ldots N$? (An example of such a Hamiltonian would be the zipper model; another would be $H = -\phi_1 + \phi_2\phi_7 + \phi_8\phi_2\phi_4 + \cdots$)

### *Modelling cold unfolding – the effect of water*

In all the above models $\phi = 1$ signals a contact, and $\phi = 0$ a lack of contact. Lack of contact implies exposure to the surrounding solvent. To take the solvent into account, one may modify the Hamiltonian according to Hansen *et al.* (1998):

$$H = -\mathcal{E}_0\phi + (1 - \phi)w \tag{5.32}$$

where $\phi$ is the protein variable and $w$ the water variable. In principle $\phi$ is a combination of many folding variables, $\phi_i$, one for each degree of the peptide chain, with $\phi_i = 1$ representing the folded state and $\phi_i = 0$ representing the unfolded state of the variable. Similarly $w$ is a combination of water variables, each associated to one of the folding variables $\phi_i$. For the discussion of cold unfolding we do not

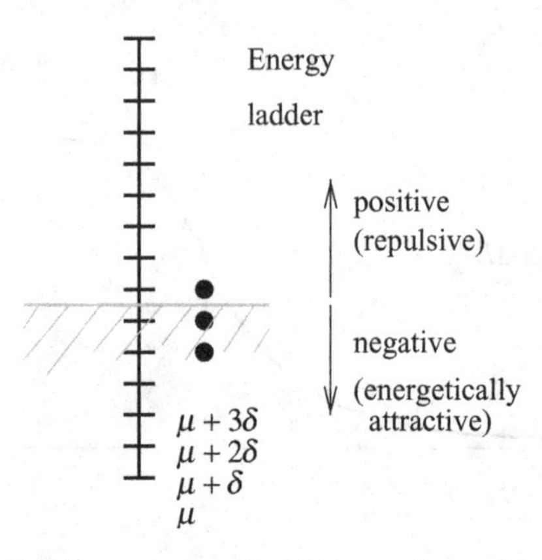


Figure 5.20. Energy ladder associated with an amino acid that is interior in the native state, but here is exposed to surrounding water. The energy ladder is associated with the amino acid–water interaction, where energetic attraction between the amino acid and some water puts the energetic ordering of water further away from the amino acid. At low $T$ the energy term wins and water tears the protein apart, whereas intermediate temperatures makes it favorable to keep hydrophobic amino acids inside a folded protein structure.

need to discuss the organization of the $\phi_i$, and thus we discuss $\phi$ and $w$ organized only as in Eq. (5.32). In this the water variable $w$ should produce the constant heat capacity, because the contribution of water to protein heat capacity seems fairly constant with temperature. That means $w$ has an equally spaced energy spectrum (we leave it to the reader to prove that), as illustrated in Fig. 5.20.

$$w \in \{\mu,\ \ \mu+\delta,\ \ \mu+2\delta,\ \ \mu+3\delta, \ldots \mu+(g-1)\delta\} \tag{5.33}$$

Here $\mu$ (<0) is the interaction energy between the exposed amino acid and water at low temperature, $\delta$ is the energy difference between subsequent steps on the ladder, and $g$ is the maximum number of states of the water that can be influenced by exposure to a $\phi$ variable.

Obviously when $\phi = 0$ the water is constrained to the lower parts of the ladder. For $\phi = 1$, however, all water levels become degenerate, and the water contributes with an entropy $S_w = k_B \ln(g)$. This represents a very simple model for the hydrophobic effect, which mainly consists of the entropy reduction of the water upon exposure to a hydrophobic group.

The partition function for the above variable is

$$Z = C \cdot e^{-\mu/k_BT} \cdot \sum_{j=0}^{g-1} e^{-j\delta/k_BT} + e^{\mathcal{E}_0/k_BT} \cdot g \tag{5.34}$$

where the first term counts the weight of the $\phi = 0$ situation. The $C$ coefficient is to remind the reader about the phase space from the polymer degree of freedom.

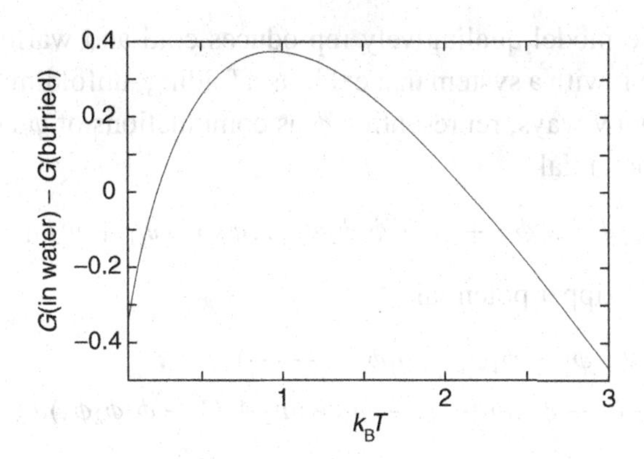


Figure 5.21. Free energy difference for a residue exposed or not exposed to the water, as modeled by an energy ladder for the residue–water interaction. The parameters chosen were $C = 2, g = 50$ and energy parameters $\delta = 0.1, \mu = -1.5$. All energies are in units of $\mathcal{E}_0$.

The second term in the equation corresponds to the native contact formed, $\phi = 1$. The first and last parts of this partition function, respectively, read

$$Z(\text{exposed}) = C \cdot \mathrm{e}^{-\mu/k_B T} \frac{1 - \mathrm{e}^{-g\delta/k_B T}}{1 - \mathrm{e}^{-\delta/k_B T}} \tag{5.35}$$

$$Z(\text{buried}) = e^{\mathcal{E}_0/k_B T} \cdot g \tag{5.36}$$

As a result, for $\delta \ll 1$ and $g\delta \gg 1$ one can write

$$\Delta G = -k_B T \Delta \ln \left( \frac{Z(\text{exposed})}{Z(\text{buried})} \right) = \mathcal{E}_0 + \mu - k_B T \ln \left( \frac{C}{g} \right) - k_B T \ln \left( \frac{k_B T}{\delta} \right) \tag{5.37}$$

which has the same non-monotonous form ($T - T \ln(T)$) as the one obtained from assuming[4] a constant $\Delta C_p$ in the Questions on p. 101. In Fig. 5.21 we show $\Delta G$, to be compared with Fig. 5.4 (left-hand panel). If $\mu < -\mathcal{E}_0$ then the unfolded state (first term, exposed residue) wins at low temperature. As $T$ increases, the water entropy term $g$ can dominate sufficiently over the few lower-energy terms in the water ladder, and this leads to folding (burial of residues in a hydrophobic core). At even higher $T$ more terms in the ladder contribute, and the effect of hydrophobicity decreases. In fact it is known that hydrophobic forces vanish at sufficiently high temperature (around 140 °C). However, before that limit is reached, the entropy of the polymer degrees ($\phi$) of freedom favors the unfolded state, and the protein melts ($\phi \to 0$).

[4] Where the sign of the first term $\propto T$ comes because $g > C$: the entropy of water is larger than the entropy of the polymer degree of freedom.

Thus the above model qualitatively reproduces cold and warm unfolding. We can now combine it with a system that exhibits a folding–unfolding transition. This can be done in many ways, representing $\phi$ as combinations of $\phi_i$, e.g. in the form of a golf course potential

$$H = -N\mathcal{E}_0\phi_1\phi_2\phi_3\ldots\phi_N + (1 - \phi_1\phi_2\phi_3\ldots\phi_N)\cdot(w_1 + w_2 + \cdots w_N) \quad (5.38)$$

or in the form of a zipper potential

$$\begin{aligned} H = {} & \mathcal{E}_0(-\phi_1 - \phi_1\phi_2 - \phi_1\phi_2\phi_3 - \cdots) \\ & + (1-\phi_1)w_1 + (1-\phi_1\phi_2)w_2 + (1-\phi_1\phi_2\phi_3)w_3 + \cdots \end{aligned} \quad (5.39)$$

or a combination. In any case we obtain a model with potentially two-phase transitions, a cold and a warm unfolding, and with a gap in heat capacity between the folded and unfolded states. The sharpness of the phase transition will depend sensitively on the way these variables interact with each other, and this, in fact, we know amazingly little about. Thus the protein models presented here are qualitative, maybe grasping a hint of the true interplay between local interactions and global cooperativity. However, they typically fail at quantitative tests, including protein cooperativity and protein folding times, as well as in predicting the real native structure of real sequences.

## *Questions*

(1) Solve the golf course model with water interactions: $\mathcal{E}_0 = 1$, $\mu = -2$ and $\delta = 0.1$. Find the minimal $g$ that gives two phase transitions.

(2) Solve the zipper model with water interactions, with $\mathcal{E}_0 = 1$, $\mu = -2$ and $\delta = 0.1$. Find the minimal $g$ that gives two phase transitions.

## References

Baldwin, R. L. & Rose, G. D. (1999a). Is protein folding hierarchic? I. Local structure and peptide folding. *Trends Biochem. Sci.* **24**, 26–33.

(1999b). Is protein folding hierarchic? II. Folding intermediates and transition states. *Trends Biochem. Sci.* **24**, 77–84.

Bryngelson, J. D. & Wolynes, P. G. (1987). Spin glasses and the statistical mechanics of protein folding. *Proc. Natl Acad. Sci. USA* **84**, 7524.

Carrion-Vazquez, M. *et al.* (2003). The mechanical stability of ubiquitin is linkage dependent. *Nature Struct. Biol.* **10**, 738.

Chan, H. S. & Kaya, H. S. (2000). Energetic components of cooperative protein folding. *Phys. Rev. Lett.* **85**, 4823.

Derrida, B. (1980). Random energy model: limit of a family of disordered models. *Phys. Rev. Lett.* **45**, 2.

Dill, K. A. (1985). Theory for the folding and stability of globular proteins. *Biochemistry* **24**, 1501.

Dill, K. A., Fiebig, K. M. & Chan, H. S. (1993). Cooperativity in protein folding kinetics. *Proc. Natl Acad. Sci. USA* **90**, 1942.
Dill, K. A., Bromberg, S., Yue, K., Fiebig, K. M., Yee, D. P., Thomas, P. D. & Chan, H. S. (1995). Principles of protein folding – a perspective from simple exact models. *Protein Sci.* **4**, 561.
Dommersnes, P. G., Hansen, A., Jensen, M. H. & Sneppen, K. (2000). Parametrization of multiple pathways in proteins: fast folding versus tight transitions (cond-mat/0006304).
Gerland, U., Moroz, J. D. & Hwa, T. (2002). Physical constraints and functional characteristics of transcription factor–DNA interactions. *Proc. Natl Acad. Sci. USA* **99**, 12 015.
Guerois, R. & Serrano, L. (2000). The SH3-fold family: experimental evidence and prediction of variations in the folding pathways. *J. Mol. Biol.* **304**, 967.
Hansen, A., Jensen, M. H., Sneppen, K. & Zocchi, G. (1998). Statistical mechanics of the folding transitions in globular proteins. *Eur. Phys. J.* B**6**, 157–161.
Jackson, S. E. & Fersht, A. R. (1991). Folding of chymotrypsin inhibitor 2.1. Evidence for a two-state transition. *Biochemistry* **30**, 10 428–10 435.
Kaya, H. & Chan, H. S. (2000). Polymer principles of protein calorimetric two-state cooperativity. *Proteins* **40**, 637.
Levinthal, C. (1968). Are there pathways for protein folding? *J. Chim. Phys.* **65**, 44.
Li, H., Helling, R., Tang, C. & Wingreen, N. (1996). Emergence of preferred structures in a simple model of protein folding. *Science* **273**, 666.
Li, H., Tang, C. & Wingreen, N. (1997). Nature of driving force for protein folding – a result from analyzing statistical potential. *Phys. Rev. Lett.* **79**, 765.
(1998). Are protein folds atypical? *Proc. Natl Acad. Sci. USA* **95**, 4987.
(2002). Designability of protein structures: a lattice-model study using the Miyazawa-Jernigan Matrix. *Proteins: Struct. Funct. Genet.* **49**, 403–412.
Makarov, D. E. & Plaxco, K. W. (2003). The topomer search model: a simple, quantitative theory of two state protein folding kinetics. *Protein Science* **12**, 17–26.
Makarov, D. E., Wang, Z., Thompson, J. & Hansma, H. G. (2002). On the interpretation of force extension curves of single protein molecules. *J. Chem. Phys.* **116**, 7760.
Musacchio, A., Noble, M. E. M., Pauptit, R., Wierenga, R. & Saraste, M. (1992). Crystal structure of a Src-homology 3 (SH3) domain. *Nature* **359**, 851.
Metropolis, N., Rosenbluth, A., Rosenbluth, M., Teller, A. & Teller, E. (1953). Equation of state calculations by fast computing machines. *J. Chem. Phys.* **21**, 1087.
Oliveberg, M., Tan, Y. & Fersht, A. R. (1995). Negative activation enthalpies in the kinetics of protein folding. *Proc. Natl Acad. Sci. USA* **92**, 8926.
Plaxco, K. W., Simons, K. T. & Baker, D. (1998). Contact order, transition state placement, and the refolding rates of single domain proteins. *J. Mol. Biol.* **277**, 985.
Privalov, P. L. (1979). Stability of proteins. Small globular proteins. *Adv. Protein Chem.* **33**, 167.
(1990). Cold denaturation of proteins. *CRC Crit. Rev. Biochem. Mol. Biol.* **25**, 281.
Privalov, P. L. & Potekhin, S. A. (1986). Scanning microcalorimetry in studying temperature-induced changes in proteins. *Methods Enzymol.* **131**, 4.
Sali, A., Shacknovich, E. I. & Karplus, M. (1994). How does a protein fold? *Nature* **369**, 248.
Schlierf, M. *et al.* (2004). The unfolding kinetics of ubiquitin captured with single-molecule force-clamp techniques. *Proc. Natl Acad. Sci. USA* **101**, 7299.

Shacknovich, E. I. *et al.* (1998). Protein design: a perspective from simple tractable models. *Fold Design* **3**, R45.
Socci, N. D. & Onuchic, J. N. (1994). Folding kinetics of protein-like heteropolymers. *J. Chem. Phys.* **101**, 1519.
Stormo, G. D. & Fields, D. S. (1998). Specificity, free energy and information content in protein–DNA interactions. *Trends Biochem. Sci.* **23**, 178.
Takatomi, H., Ueda, Y. & Go, N. (1975). Studies on protein folding, unfolding and fluctuations by computer simulations. 1. The effect of specific amno acid sequence represented by specific inter-unit interactions. *Int. J. Pept. Protein. Res.* **7**, 445.
Tiana, G., Broglia, R. A. & Shakhnovich, E. I. (2000). Hiking in the energy landscape in sequence space: a bumpy road to good folders. Proteins: *Struct. Funct. Genet.* **39**, 244.

## Further reading

Anfinsen, C. B., Haber, E., Sela, M. & White, F. H. (1961). The kinetics of formation of native ribonuclease during oxidation of the reduced polypeptide chain. *Proc. Natl Acad. Sci. USA* **47**, 1309.
Baldwin, R. L. (1986). Temperature dependence of the hydrophobic interaction in protein folding. *Proc. Natl Acad. Sci. USA* **83**, 8069–8072.
(1994). Matching speed and stability. *Nature* **369**, 183–184.
Bryngelson, J., Onuchic, J. N., Socci, N. & Wolynes, P. G. (1995). Funnels, pathways and the energy landscape in protein folding. *Proteins: Struct. Funct. Genet.* **21**, 167.
Chothia, C. (1992). One thousand families for the molecular biologist. *Nature* **357**, 543–544.
Dill, K. A. (1990). The meaning of hydrophobicity. *Science* **250**, 297.
Dill, K. A. & Shortle, D. (1991). Denatured states of proteins. *Ann. Rev. Biochem.* **60**, 795–825.
Dill, K. A. & Sun Chan (1997). From Levinthal to pathways to funnels. *Nature Struct. Biol.* **4**, 10.
Dobson, C. M. (1999). Protein misfolding, evolution and disease. *Trends Biochem. Sci.* **24**, 329–332.
Dobson, C. M. & Karplus, M. (1999). The fundamentals of protein folding: bringing together theory and experiment. *Curr. Opin. Struct. Biol.* **9**, 92.
Fan, K., Wang, J. & Wang, W. (2001). Modeling two-state cooperativity in protein folding. *Phys. Rev.* E **64**, 041907.
Fersht, A. (1991). *Structure and Mechanism in Protein Science: A Guide to Enzyme Catalysis and Protein Folding*. New York: W. H. Freeman.
(1997). Nucleation mechanisms in protein folding. *Curr. Opin. Struct. Biol.* **7**, 3–9.
Fersht, A. R., Jackson, S. E. & Serrano, L. (1993). Protein stability: experimental data from protein engineering. *Phil. Trans. R. Soc. Lond.* A**345**, 141–151.
Fisher, D. (1994). List of representative protein structures. http://www.lmmb.ncifcrf.gov/nicka/lrs.html
Flory, P. J. (1949). The configuration of real polymer chains. *J. Chem. Phys.* **17**, 303.
Frauenfelder, H. & Wolynes, P. G. (1994). Biomolecules: where the physics of complexity and simplicity meet. *Physics Today*, pp. 58–64.
Frauenfelder, H., Sligar, S. G. & Wolynes, P. G. (1991). The energy landscapes and motions of proteins. *Science* **254**, 1598.
Fulton, K. F., Main, E. R., Daggett, V. & Jackson, S. E. (1999). Mapping the interactions present in the transition state for unfolding/folding of FKBP12. *J. Mol. Biol.* **291**, 445.

Gillespie, B. & Plaxco, K. W. (2000). Non-glassy kinetics in the folding of a simple, single domain protein. *Proc. Natl Acad. Sci. USA* **97**, 12 014.
Go, N. (1983). Theoretical studies of protein folding. *Ann. Rev. Biophys. Bioeng.* **12**, 183–210.
Hartl, F. U. (1996). Molecular chaperones in cellular protein folding. *Nature* **381**, 571–580.
Heggie, M. I., Latham, C. D., Maynard, S. C. P. & Jones, R. (1996). Cooperative polarisation in ice Ih and the unusual strength of the hydrogen bond. *Chem. Phys. Lett.* **249**, 485.
Helling, R., Li, H., Miller, J. *et al.* (2001). The designability of protein structures. *J. Mol. Graph. Model.* **19**, 157.
Irback, A. & Sandelin, E. (2000). On hydrophobicity correlations in protein chains. *Biophys. J.* **79**, 2252.
Jackson, S. E. (1998). How do small, monomeric proteins fold? *Folding Des.* **3**, R81.
Kaya, H. & Chan, H. S. (2003). Simple two-state protein folding kinetics requires near-Levinthal thermodynamic cooperativity. *Proteins* **52**, 510.
Keutsch, F. N. & Saykally, R. J. (2001). Water clusters: untangling the mysteries of the liquid, one molecule at a time. *Proc. Natl Acad. Sci. USA* **98**, 10 533.
Klimov, D. K. & Tirumalai, D. (1996). Factors governing the foldability of proteins. *Proteins* **26**, 411.
Luck, W. A. P. (1998). The importance of cooperativity for the properties of liquid water. *J. Mol. Struct.* **448**, 131.
Main, E. R. G., Fultond, F. & Jackson, S. E. (1998). Stability of FKBP12: the context-dependent nature of destabilizing mutations. *Biochemistry* **37**, 6145.
Makarov, D. E., Keller, C. A., Plaxco, K. W. & Metiu, H. (2002). How the folding rate constant of simple, single-domain proteins depends on the number of native contacts. *Proc. Natl Acad. Sci. USA* **99**, 3535.
Myers, J. K., Pace, C. N. & Scholtz, J. M. (1995). Denaturant m values and heat capacity changes: relation to changes in accessible surface areas of protein unfolding. *Protein Sci.* **4**, 2138.
Mayor, U., Johnson, C. M., Daggett, V. & Fersht, A. R. (2000). Protein folding and unfolding in microseconds to nanoseconds by experiment and simulation. *Proc. Natl Acad. Sci. USA* **97**, 13 518.
Ptitsyn, O. B. & Rashin, A. A. (1975). A model of myoglobin self-organization. *Biophys. Chem.* **3**, 120.
Privalov, P. L. (1989). Thermodynamic problems of protein structure. *Ann. Rev. Biophys. Biophys. Chem.* **18**, 47.
(1996). Intermediate states in protein folding. *J. Mol. Biol.* **258**, 707–725.
Privalov, P. L. & Gill, S. J. (1988). Stability of protein structure and hydrophobic interaction. *Adv. Protein Chem.* **39**, 191.
Shimizu, S. & Chan, H. S. (2000). Temperature dependence of hydrophobic interactions: a mean force perspective, effects of water density, and non-additivity of thermodynamic signatures. *J. Chem. Phys.* **113**, 4683.
Silverstein, K. A. T., Haymet, A. D. J. & Dill, K. A. (2000). The strength of hydrogen bonds in liquid water and around nonpolar solutes. *J. Am. Chem. Soc.* **122**, 8037.
Smith, T. F. (1995). Models of protein folding. *Science* **507**, 959.
Tiana, G. & Broglia, R. A. (2000). Statistical analysis of native contact formation in the folding of designed model proteins. (cond-mat/0010394).
Tiana, G., Broglia, R. A. & Provasi, D. (2001). Designability of lattice model heteropolymers. (cond-mat/0105340).

Thompson, J. B., Hansma, H. G., Hansma, P. K. & Plaxco, P. W. (2002). The backbone conformational entropy of protein folding: experimental measures from atomic force microscopy. *J. Mol. Biol.* **322**, 645–652.
Yee, D. P., Chan, H. S., Havel, T. F. & Dill, K. A. (1994). Does compactness induce secondary structure in proteins? *J. Mol. Biol.* **241**, 557.
Wickner, S., Maurizi, M. R. & Gottesman, S. (1999). Posttranslational quality control: folding, refolding, and degrading proteins. *Science* **286**, 1888.
Wetlaufer, D. B. (1973). Nucleation, rapid folding, and globular intrachain regions in proteins. *Proc. Natl Acad. Sci. USA* **70**, 697 701.
Wolynes, P. G. (1997). As simple as can be. *Nature Struct. Biol.* **4**, 871.
Wolynes, P. G., Onuchic, J. N. & Thirumalai, D. (1995). Navigating the folding routes. *Science* **267**, 1619.

# 6

# Proteins in action: molecular motors

Kim Sneppen

## Molecular motors: energy from ATP

Proteins can be grouped into a few broad categories with respect to their function. Some are regulatory, some are enzymes, some are structural, and some proteins do mechanical work. It is this latter group that we now discuss. Molecular motors include kinesin, myosin, dynein, the motors connected to DNA replication, to gene transcription and to translation. The motors are mostly driven by ATP hydrolysis: ATP→ ADP + P, a process with $\Delta G \approx 13$ kcal/mol for typical conditions in the cell. Exactly how the free energy difference from ATP hydrolysis is converted into directed motion and mechanical work is a most interesting question, which is not resolved. In many cases the conformational changes of the protein are known in considerable detail from structural studies. The sequence of events associating conformational changes and substrate binding and release is also known. Nonetheless, the actual physical mechanism by which the motor works is not obvious. Thermal noise and diffusion certainly play a role, making this "soft" machine qualitatively different from a macroscopic motor. In the next section we elaborate on these ideas through some models.

The most studied motors include myosin and kinesin, which move along the polymers that define the cytoskeleton. Kinesin walks on microtubules (Fig. 6.1), whereas myosin walks on polymerized actin. Microtubules and actin fibers are long (μm) polymers where the monomer units are proteins. Microtubules are very stiff; actin fibers more flexible. Kinesin motors work independently of each other, and are associated with the transport of material (vesicles) inside the eukaryotic cell. A myosin protein works in coordination with other myosin proteins, and is associated with muscle movements. Myosin and kinesin are structurally related; they both have a head, which directs the action, and they both go through an ATP burning cycle, which involves binding to the substrate (actin or microtubule), a forced power stroke, and a detachment phase for each cycle of the motion. Figure 6.1 illustrates

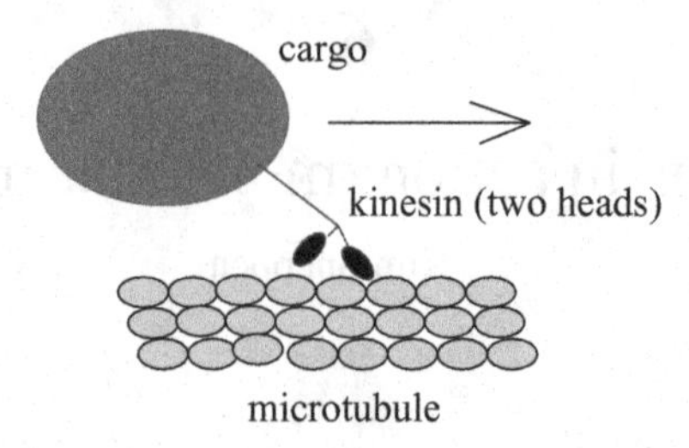


Figure 6.1. Kinesin walking on a microtubule. Kinesin forms dimers, consisting of two globular heads, a stretched stalk about 80 nm long, and a tail. The head and the tail domains contain the microtubule- and the cargo-binding sites, respectively. Microtubules are 25 nm thick and about 5–20 μm long hollow cylindrical fibers that are formed by tubulin dimers. Microtubules are polar; there are kinesins moving from + to −, and there are kinesins designed to move the opposite way. Each tubulin dimer is 8 nm long, and the kinesin moves in steps of 8 nm along the surface of microtubules.

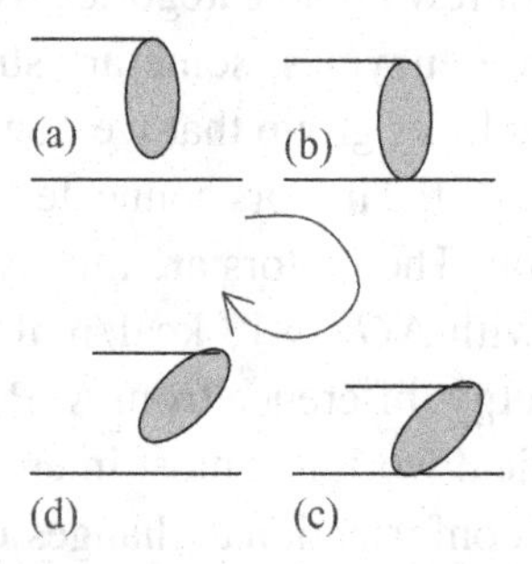


Figure 6.2. Schematic view of some states during one cycle of a motor protein that walks on a substrate. The protein may, for example, be kinesin, in which case the substrate is a microtubule. Or it may be myosin walking on actin. The stroke state (b) → (c) is associated with release of the products P and possibly also ADP from the motor. State (c) is called the rigor state, because this is the state a motor without ATP ends in. The release of the head from substrate (c) → (d) is associated to binding of ATP. For myosin, the energetics of all involved transitions are tabulated by Smith & Geeves (1995a,b). ATP bind weakly to attached rotated myosin (c) but strongly to de-attached myosin (d,a) with $\Delta G = -16$ kcal/mol, and myosin with ATP/ADP bind to actin with $\Delta G = -6$ kcal/mol (b). ATP → ADP + P releases ~ 9 kcal/mol inside the myosin head (a) or (b), and when P and ADP is released from the attached and rotated myosin there is a further gain of about 7–8 kcal/mol, coupled to the power stroke (b) → (c) (see also Rayment, 1996).

kinesin motion along a microtubule. Figure 6.2 gives a cartoon visualization of the steps involved in the movements of the individual heads.

Contemporary experiments on kinesin show that the motion takes place in discrete 8 nm steps (corresponding to the spacing of tubulin dimers along the microtubule). There is one ATP hydrolysis per step, and the efficiency is large (maybe >50%) at low load. The stall force (maximum force the motor can move against) is about 5–6 pN (Meyhofer & Howard, 1995). The maximum speed of a kinesin motor is about 1–2 μm/s, which means that one step of 8 nm in total takes about 0.005 s. The speed depends on ATP concentration when this is low. In vitro, maximum speed is reached above ~1 mM concentration of ATP.

On a more detailed level we now examine the energy source, namely the ATP→ ADP + P reaction. In the cell ATP is found at a concentration of 1 mM. Thus each potential reaction point in the cell will be bombarded with $\sim 10^5$ ATP molecules per second (see the Questions on p. 131). Thus the limit imposed by the ATP capturing rate is very high when compared with the overall time it takes kinesin to move one step. Thus it can be ignored: the motor typically works under conditions with ample energy supply. Energetically, the reaction

$$\mathrm{ATP} \rightleftharpoons \mathrm{ADP} + \mathrm{P} \tag{6.1}$$

is characterized by the dissociation constant

$$K = \left( \frac{[\mathrm{ADP}][\mathrm{P}]}{[\mathrm{ATP}]} \right)_{\mathrm{eq}} = [\mathrm{M}] \cdot \mathrm{e}^{\Delta G_0 / k_B T} \tag{6.2}$$

where the concentrations are at equilibrium and $\Delta G_0 = 7.3$ kcal/mol is the *standard free energy change of the reaction*. This means that $\Delta G_0$ is defined as the free energy of ATP relative to ADP + P when all reactants are present at 1 M concentration ([M]). Having all reactants at 1 M is not equilibrium; in fact if [ADP] = [P] = 1 M, the equilibrium [ATP] concentration

$$[\mathrm{ATP}]_{\mathrm{eq}} = \mathrm{e}^{-\Delta G_0 / k_B T} \mathrm{M}^{-1} [\mathrm{ADP}][\mathrm{P}] = \mathrm{e}^{-7.3/0.617} \mathrm{M} \sim 10^{-5} \mathrm{M} \tag{6.3}$$

is very low. This reflects the fact that [ATP] is in an unlikely high-energy state. This high-energy state of ATP is exactly what makes it a good way to store energy in the living cell.

The equilibrium concentration of [ATP] is the one where it is not possible to extract energy from the hydrolysis. In the cell, concentrations are not at equilibrium, and it is therefore possible to extract work from the reaction in Eq. (6.1). The in vivo free energy that ATP has stored relative to ADP+P in the living cell, $\Delta G = G(\mathrm{ATP}) - G(\mathrm{ADP} + \mathrm{P})$, is (see derivation in Question 6 on p. 131):

$$\Delta G_{\text{in vivo}} = k_B T \ln \left( \frac{([\mathrm{ADP}][\mathrm{P}]/[\mathrm{ATP}])_{\mathrm{eq}}}{([\mathrm{ADP}][\mathrm{P}]/[\mathrm{ATP}])_{\text{in vivo}}} \right) \tag{6.4}$$

$$= \Delta G_0 - k_B T \ \ln ([\mathrm{ADP}][\mathrm{P}]/[\mathrm{ATP}])_{\text{in vivo}} \tag{6.5}$$

$$= 7.3 \text{ kcal/mol} - k_B T \ln(0.000\,02) \approx 13 \text{ kcal/mol} \tag{6.6}$$

where, in the last equality, we insert the actual in vivo concentrations: [ATP] ~1 mM, [ADP] ~20 μM, and [P] ~1 mM. Equation (6.5) simply states that the free energy release due to hydrolysis is equal to what it would be at 1 M concentrations of all reactants, $\Delta G_0$, plus a contribution due to the entropy gain of diluting/concentrating the reactants to the actual in vivo concentrations. Thus to a large extent it is the high concentration of ATP relative to ADP and P concentrations that brings us up to the hydrolysis energy of 13 kcal/mol in the living cell.

For the motor myosin (or kinesin) the ATP hydrolysis takes place in a small pocket deeply buried inside the protein. This pocket has a size of about 1 nm. Confining the reactants to this 1 $nm^3$ cavity in fact corresponds to concentrations of order 1 M. Thus the last of the two terms in Eq. (6.6) means that about half the in vivo free energy of hydrolysis is associated to the entropy gained by moving the ADP and P far away from this reaction volume after the hydrolysis. Figure 6.2 illustrates the cycle of steps that the motor goes through during capture, hydrolyses and release of fuel.

## *Questions*

(1) Consider ATP $\rightarrow$ ADP + P with 7.3 kcal/mol energy released in the cavity of 1 $nm^3$ inside the myosin head. If we ignore thermal conductivity, and other ways to store energy chemically, what is the maximum temperature increase one may have in this small volume?

(2) A kinesin molecule takes a step of 8 nm and can sustain a load of 6 pN. What work does can it do per ATP cycle? What is its efficiency in percent?

(3) Convert 1 $k_B T$ to pN · nm units.

(4) A myosin motor has a power stroke of 3–5 pN. How many myosin molecules are needed to carry a man of 80 kg? The speed that myosin can induce is of order 0.5–4 μm/s, dependent on the myosin. How does one obtain macrosocopic speeds of, say, 1 m/s? How many myosins should be used to move a human body of 80 kg 1 m/s vertically? How much do these weigh, when each myosin molecule is 500 000 Da? Muscles contain about 7–10% myosin, so what is the lowest muscle mass that can sustain us?

(5) Find, by dimensional analysis, a formula for the rate at which diffusing molecules visit a given region of space, depending on the concentration $c$, diffusion constant $D$, and size $s$ of the target region. Show that a small molecule like ATP, at mM concentrations, visits a reaction center ~$10^5$ times per second.

(6) We here consider an alternative derivation of Eq. (6.5), using the formalism derived in the λ-phage chapter. Argue that statistical weight of an ATP in the in vivo cell is $Z_{ATP} = [ADP][P]\, e^{-\Delta G_0/k_B T}$, and that the statistical weight of the de-hydrolyzed state ADP+P is $Z_{ADP+P} = [ATP]$. Show that the maximum work that one can obtain from the hydrolysis, $W_{max} = k_B T \ln(Z_{ADP+P}/Z_{ATP})$, is given by Eq. (6.5).

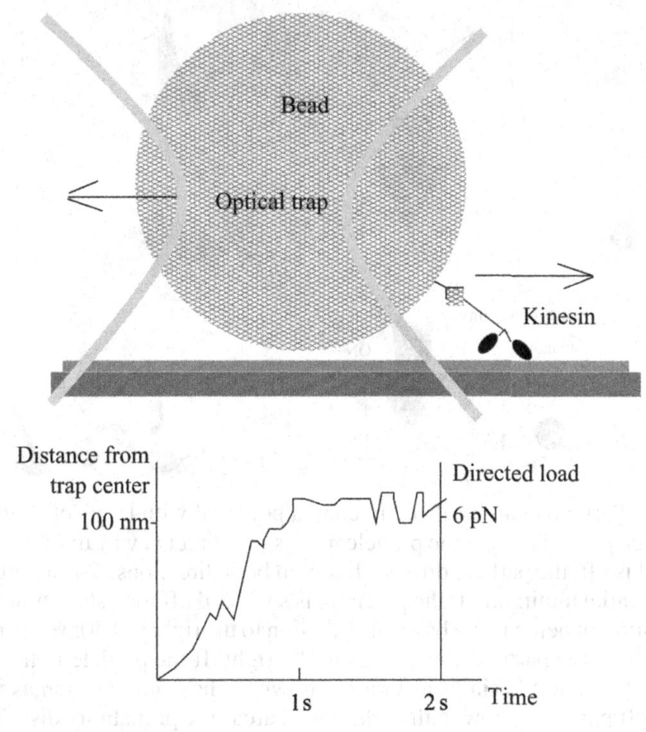


Figure 6.3. Optical tweezer measurement of the kinetic properties of kinesin. The kinesin tail is attached to the bead with some antigen. The bead is large, about 1 μm; thus the determination of displacement steps of 8 nm is not trivial! The lower part of the figure illustrates that kinesin stops when the restoring force from the optical trap on the bead in about 5–6 nm.

## Molecular motors: ratchet mechanism

In this brief discussion we do not venture into the large literature describing the many beautiful experiments on molecular motors. During the past decade, many of these experiments have been based on single molecule techniques. In Fig. 6.3 we illustrate one particular experiment, the setup of Block and coworkers, which uses optical trapping interferometry to determine the force needed to stop the movement of kinesin. The reader can find abundant material on these types of experiment in the reference list (see Block, 1996, 1998; Block *et al.*, 1990). Here we limit our discussion to simple models that introduce some physical concepts that are probably relevant for understanding the conversion from chemical to mechanical energy.

One attractive idea for modeling comes from considering the motion of Brownian ratchets (Julicher *et al.*, 1997). In the model we follow the coordinate of a

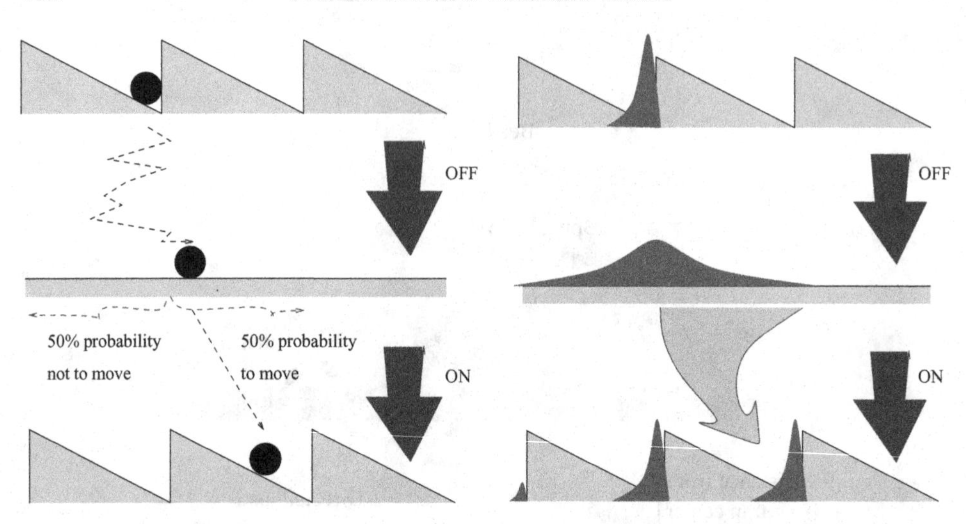


Figure 6.4. Particle in an idealized ratchet: a potential with large left–right asymmetry. When potential is on, the particle moves in a directed way until it is trapped. If potential is off, the particle diffuses freely in both directions. Consider the particle in a potential minimum. If the potential is switched off for a short time interval, the probability of being in the basin of attraction to the right is 0.50. When potential is switched on, the particle then moves to the right. If the particle in the off state has moved to the left basin, it just moves down to the same position as it started in. In the left panel we show with dark shaded areas the probability distribution of the particle at various steps of the process.

particle, which may be seen as an idealized state coordinate for the motor protein. Directed motion of the particle is achieved by changing the potential experienced by the particle. In a simple ratchet, the change is between a state where diffusion is non-directed, to a state where the particle's motion is mostly directed. Switching the potential requires energy from the outside, which in the motor protein could be provided by ATP hydrolysis. In Fig. 6.4 we show an idealized ratchet, and illustrate how directed motion can occur, owing to the interplay between random motion in the "off" state, and directed motion in the "on" state of the potential.

We now want to illustrate the ratchet idea in terms of the steps associated to the sequence of events that characterize one cycle in the motion of a motor protein, see Fig. 6.2. In Fig. 6.5 this is done through the two binary variables $(\phi_1, \phi_2)$, from Hansen *et al.* (2000). Here $\phi_1$ describes whether the motor protein is bound to the substrate ($\phi_1 = 1$) or not bound ($\phi_1 = 0$), and $\phi_2$ describes whether the head of the motor protein is relaxed and randomly diffusing (then $\phi_2 = 0$), or whether it is stretched ($\phi_2 = 1$). Thus each of the $\phi$ variables is equal to 1 when it is bound or forced, and 0 when it is relaxed.

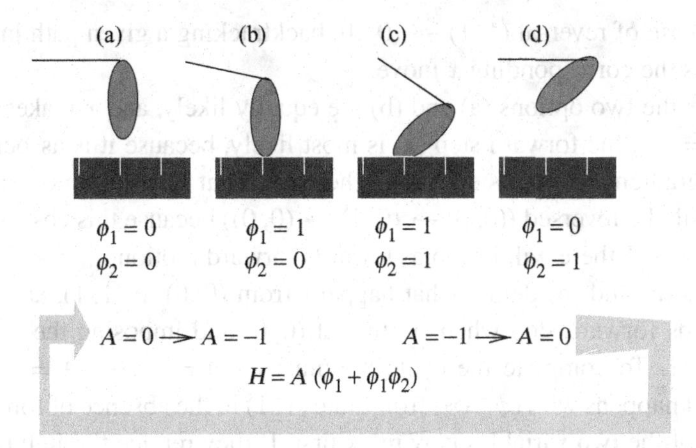


Figure 6.5. Schematic view of some states during one cycle of the discrete ratchet model discussed in the text. Below the four states that are labeled as in Fig. 6.2, we show the parametrization of the states in terms of two binary variables. By making movement of $\phi_2$ dependent on the state of $\phi_1$ as shown by the energy function $H$, one can drive the cycle of states into a directed motion by externally changing the potential through the parameter $A$. As with the continuous ratchet, this model is also at maximum 50% effective, because state (1,1) can go to state (1,0) or (0,1), respectively, with equal probability when $A = 0$. In the first case, $(1, 1) \rightarrow (1, 0)$, we are back in state (b), and have not performed any motion. Here, (c) is the rigor state and the $A = -1 \rightarrow A = 0$ transition corresponds to the ATP binding transition.

The two $\phi$ variables are controlled through one external variable $A$, which may take the value $A = 0$ or $A = -1$, and is associated to the hydrolysis state of the motor. The control is enforced through the total energy function:

$$H = A(\phi_1 + \phi_1\phi_2)\mathcal{E}_0 \qquad (6.7)$$

As there are more non-specific states than specific (bound) ones, the degeneracy of the $\phi = 0$ states is larger than that of the $\phi = 1$ states. One may set degeneracy of the $\phi_1 = 0$ and of the $\phi_2 = 0$ states to be $g = 10$, whereas we assume only one specific state for $\phi_1 = 1$ and $\phi_2 = 1$. Then, for big $g$ and $A = 0$, the state $(0, 0)$ is by far the most likely state, and the $(1, 1)$ state will decay to the $(0, 0)$ situation. For $A = -1$, on the other hand, then for small enough temperature the $(1, 1)$ state is favoured.

For each value of $A$ and starting conditions for $(\phi_1, \phi_2)$ one may consider the trajectory for $(\phi_1, \phi_2)$. Starting at $(0, 0)$ we have the options:

(a) forward motion $x \rightarrow x + 1$, associated with the sequence of events where binding is followed by stroke: $(0, 0) \rightarrow (1, 0) \rightarrow (1, 1)$;
(b) backward motion $x \rightarrow x - 1$, associated with the sequence where stroke happens before binding: $(0, 0) \rightarrow (0, 1) \rightarrow (1, 1)$.

Further, in case of reversal $(1, 1) \to (0, 0)$, backtracking a given path implies that one reverses the corresponding $x$ move.

If $A = 0$, the two options (a) and (b) are equally likely, and $x$ makes a random step. If $A = -1$, the forward step (a) is most likely, because it is associated with an energy gradient on both steps along the path. That is, a first move along step (b) may easily be reversed $((0, 0) \to (0, 1) \to (0, 0))$ because this costs no energy. Thus for $A = -1$ there will be a bias towards forward motion.

The steps (a) and (b) define what happens from $(0, 0)$ to $(1, 1)$, and we saw a trend towards forward step when starting at $(0, 0)$ and imposing the switch $A = 0 \to A = -1$. To complete the cycle we must let $A = -1 \to A = 0$. Thus we retrace what happens with $(\phi_1, \phi_2)$ from state $(1, 1)$ in the absence of forces. In that case either of the two variables may relax first. If they retrace through the reverse of path (a), then $x \to x - 1$, and there is no net motion. However, if they retrace as the reverse of path (b), then the system in fact progresses even further. As a result the $A = -1 \to A = 0$ change induces a move from $(1, 1)$ to $(0, 0)$ that makes no average motion.

To study the dynamics in a discrete simulation, we shift between $A = 0$ and $A = -1$ at some frequency defined by a certain preset number of updates in each $A$ situation. For each $A$ the model is simulated in a Metropolis-like algorithm, where at each timestep (= update) we select randomly either $\phi_1$ or $\phi_2$ and try to change it. The factor $g > 1$ in degeneracy of the $\phi = 0$ states means that for zero energy difference ($A = 0$, or considering $\phi_2$ for $\phi_1 = 0$) a transition $0 \to 0$, or $1 \to 0$ is $g$ times more likely than a transition $0 \to 1$ or $1 \to 1$. When imposing forces by letting $A = -1$, then any move where the energy is increased by one unit is accepted with a probability proportional to $e^{-\mathcal{E}_0/k_B T}$. Thereby, with $A = -1$, the reversal from midway in path (a) is suppressed, whereas reversal midway in path (b) is not. The smaller the value of $T$, the more suppression of wrong moves, and the more efficient the ratchet. Figure 6.6 shows the simulated movement of $x$ for two different temperatures.

In reality there are several states of the ATP–motor association, for ATP binding, hydrolysis, and the subsequent release of the products (see Smith & Geeves, 1995a, b). Thus this type of simplified scenario where the "off state" is without any directed motion should be understood only as a zero-order version of a possible true ratchet mechanism.

## *Questions*

(1) Consider the thermal ratchet in Fig. 6.4. Let the diffusion constant of the particle be $D = 5\mu m^2/s$ and the distance between subsequent wells be $d = 8$ nm. What is the time it takes the particle to diffuse the distance $d$ if the potential is off?

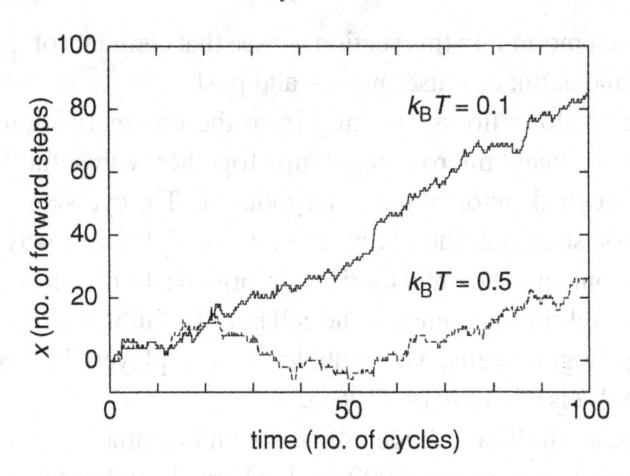


Figure 6.6. Movement in the discrete ratchet of Fig. 6.5, $H = A(\phi_1 + \phi_1\phi_2)\mathcal{E}_0$ for $A$ cycling between 0 and $-1$ and a degeneracy factor $g = 10$ for the detached state ($\phi = 0$). When the temperature increases, the state $\phi_1, \phi_2 = 1, 1$ becomes thermodynamically suppressed even when $A = -1$, and the "motor" performs a random walk (any trajectory between the (0, 0) and the (1, 1) states becomes equally likely). Note that $k_BT$ is measured in units of $\mathcal{E}_0$.

(2) Consider the ratchet problem above and implement it on a one-dimensional discrete lattice with spacings of 1 nm. Use computer simulations to estimate the average speed as a function of the switching rate of the potential. What is the timescale of the simulations, in units of real time (use the diffusion constant from Question (1) and consider the system at room temperature where $k_BT = 4\ \text{pN} \cdot \text{nm}$)? Discuss energy consumption under the assumption that each switch of potential costs one ATP unit of energy.

(3) Consider the discrete ratchet described by the Hamiltonian $H = A(\phi_1 + \phi_1\phi_2)$ at various temperatures. For $A = -1$ and degeneracy of 0 states, $g = 10$, what temperature is $(\phi_1, \phi_2) = (1, 1)$ the lowest free energy state? Simulate the ratchet, by switching appropriately from $A = 0$ to $A = -1$ at $T = 0.2$ and $T = 1.0$ (in same units as $A$).

(4) Consider the setup from Question (4), but with the addition of an external drag: $H = A(\phi_1 + \phi_1\phi_2) + F\phi_2$, with $F > 0$ being a drag force. Assume that each $\phi$ variable has one state where $\phi = 1$ and 10 states where $\phi = 0$ (thus the probability that a state moves from 0 to 0 is a priori 10 times the probability that a state moves from 0 to 1, as in Fig. 6.6). Simulate the behavior at finite $F$ (at say $T = 0.1$), and determine the stall force.

## The cytoskeleton: motion by polymerization

The motors in the cell often work on substrates. Because work is directed motion the motor needs to have at least one end fixed to something in order to move something else. For RNA polymerase, the motion is along the DNA. For kinesin the movement is along a microtubule. However, motors are not the only molecules capable of directed motion. In this section we see that elements of the cytoskeleton, such as tubulin and actin, can also move – and push.

Microtubules are long fibers resulting from the polymerization of the protein tubulin; they can be many micrometers long. Together with actin they are a major component of the cytoskeleton of the eukaryotic cell. The cytoskeleton provides the cell with stiffness, structure, the ability to move, and directed ways to reorganize the cellular environment. The most striking reorganization is during cell division, where the duplicated chromosomes of the cell must be faithfully separated into each of the emerging daughter cells. Microtubules are key players in this separation, as first observed by Weisenberg *et al.* (1968).

A microtubule is a hollow cylinder about 25 nm in diameter. It is very stiff, with persistence length $l_p \sim 5$ mm $= 5000$ μm! Along the microtubule axis, tubulin heterodimers are joined end-to-end to form protofilaments. Each heterodimer consists of an $\alpha$-sub-unit and a $\beta$-sub-unit. A staggered assembly of 13 protofilaments yields a helical arrangement of tubulin heterodimers in a cylindrical wall. Microtubules are not only structural elements, they are dynamical in the sense that they can both grow and shrink (see Fig. 6.7). This dynamical behavior requires energy, which is supplied in the form of GTP.

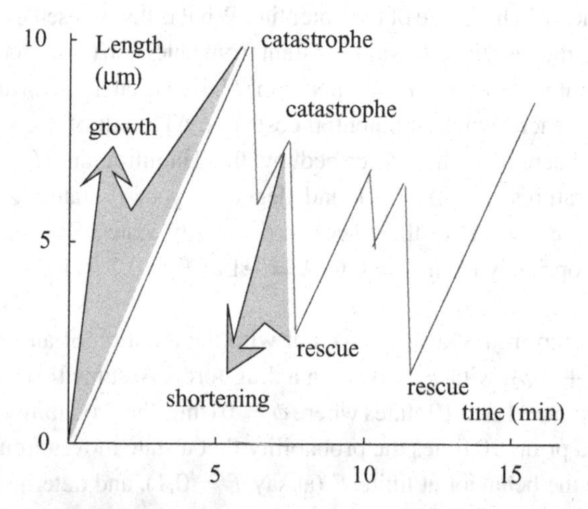


Figure 6.7. Typical in vitro trajectory of a growing and shrinking microtubule (D. K. Fygenson, private communication). Notice the alternating phases of growth and collapse, allowing the possibility of retracing from growth attempts that do not result in attachment of the microtubule end. This trial and error strategy may be behind a number of reorganization phenomena of living cells, including the formation of neuron contacts. In the figure only the "plus" end is monitored; the "minus" end is fixed. The speeds of growth and collapse depend on the conditions: if glycerol is added, the collapse speed reduces whereas the growth speed remains unchanged.

α-Tubulin has a bound GTP molecule, that does not hydrolyze. β-Tubulin may have bound GTP or GDP, and changes properties upon hydrolysis. Heterodimers with GTP-charged β-tubulin can polymerize into microtubules. At high enough concentrations they may even spontaneously polymerize into microtubules, as seen in in vitro experiments (Fygenson *et al.*, 1995). However, the most interesting aspect of microtubules is their ability to alternate between a slowly growing state, and a rapidly disassembling state (Mitcheson & Kirshner, 1984a, b). Both phases can be directly monitored both in vitro and also inside a living (Xenopus) cell, where one finds that the growth rate is of order 8 μm/min and the shortening rate is of order 20 μm/min (Shelden & Wadsworth, 1993; Fygenson *et al.*, 1994; Vasquez *et al.*, 1997). The ability of microtubules to grow and shrink allows reorganization of cellular environments, features that can be implemented in vitro (Dogterom & Yurke, 1998). Moreover, microtubules can exert forces as they grow.

Actin filaments are another major component of the cytoskeleton. An actin polymer consist of two actin strands that form a helix. The diameter of actin, the protein, is 4 nm; the diameter of polymerized actin is 7 nm. The actin polymer has a persistence length $l_p \sim 10$ μm. The actin monomer has a binding site for ATP at its center, and actin with ATP polymerizes easily. Hydrolysis of ATP to ADP destabilizes the polymer, which results in a new type of polymer dynamics that we will discuss later. Apart from its association with the myosin motor, actin polymerization in itself is thought to produce forces associated with cell motility (see review by Theriot (2000)). Cells that crawl across solid substrates use two types of force: protrusion force to extend the leading edge of the cell margin forward and traction forces to translocate the cell body forward (Elson *et al.*, 1999; Mitchison & Craner, 1996). Actin polymerization is also utilized by parasites; in particular, bacteria like *Listeria monocytogenes* and *Shigella flexneri* that propel themselves inside a eukaryotic cell by polymerizing the actin on one side, and depolymerizing it on the other side (Theriot, 2000; Dramsi & Cossart, 1998).

The mechanism for how actin polymerization may perform work was originally examined by Hill & Kirschner (1982). Here we present a simplified version (see Figs. 6.8 and 6.12). Assume that actin polymerizes with a rate $k(\text{on}) = k_{\text{on}}C$, where $C$ is the free concentration of actin monomers, and depolymerizes with rate $k(\text{off}) = k_{\text{off}} e^{\Delta G/k_B T}$, where $\Delta G$ is the free energy associated to the binding of one actin monomer to the end of the filament (large negative $\Delta G$ means strong binding). Thus at the concentration $C_{\text{crit}}$, where $k_{\text{on}} C_{\text{crit}} = k(\text{off})$, the actin filament will not grow on average. Above this concentration actin tends to polymerize. However, when the growth is limited by something else, say a bead that is pushed towards it by an external force $F$ (see Fig. 6.8), the growth rate decreases. In fact, most of the time the distance between the end of the growing filament and the bead is

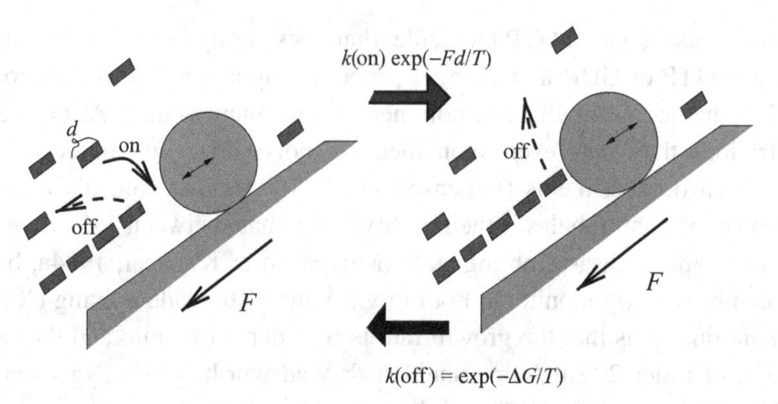


Figure 6.8. A polymerizing actin filament that pushes a bead against a force $F$. The bead is exposed to random Brownian movement, confined towards the growing polymer. When the bead is sufficiently separated from the polymer, a new unit may be added to the polymer; it grows, and effectively pushes the bead one unit to the right. The tilted plane illustrates the potential, and not the effect of gravity (gravity is insignificant on the molecular scale).

less than the minimal distance $d$ that allows a new actin monomer to bind. The fraction of time where one may add a monomer is the fraction of time where this distance is $>d$. This is given by the Boltzmann weight associated to the energy $F \cdot d$:

$$\frac{t(F)}{t(F=0)} = \mathrm{e}^{-Fd/k_{\mathrm{B}}T} \tag{6.8}$$

where $t(F=0)$ is the time where the exposure of the tip is as if there were no external limits to the growth. Thus the "on-rate" at force $F$ is

$$k(\mathrm{on}, F) = k(\mathrm{on}, 0)\mathrm{e}^{-Fd/k_{\mathrm{B}}T} \tag{6.9}$$

whereas for simplicity we assume that the off rate is independent of $F$. The average growth rate is then

$$r = k(\mathrm{on}, F) - k(\mathrm{off}) = k(\mathrm{on}, 0)\mathrm{e}^{-Fd/k_{\mathrm{B}}T} - k(\mathrm{off}) \tag{6.10}$$

$$= k(\mathrm{on}, 0)\mathrm{e}^{-Fd/k_{\mathrm{B}}T} - k(\mathrm{off}) = Ck_{\mathrm{on}}\mathrm{e}^{-Fd/k_{\mathrm{B}}T} - k_{\mathrm{on}}C_{\mathrm{crit}} \tag{6.11}$$

which is $>0$ when the concentration $C > C_{\mathrm{crit}}\mathrm{e}^{Fd/k_{\mathrm{B}}T}$. Thus the maximum force that a polymerizing filament can generate at concentration $C$ is

$$F_{\max} = \frac{k_{\mathrm{B}}T}{d}\ln\left(\frac{k_{\mathrm{on}}C}{k(\mathrm{off})}\right) = \frac{k_{\mathrm{B}}T}{d}\ln\left(\frac{C}{C_{\mathrm{crit}}}\right) \tag{6.12}$$

For a single actin filament in a solution with active actin at concentration $C = 50\ \mu\mathrm{M}$ the stall force is $F_{\max} \sim 9$ pN (Peskin *et al.*, 1993).

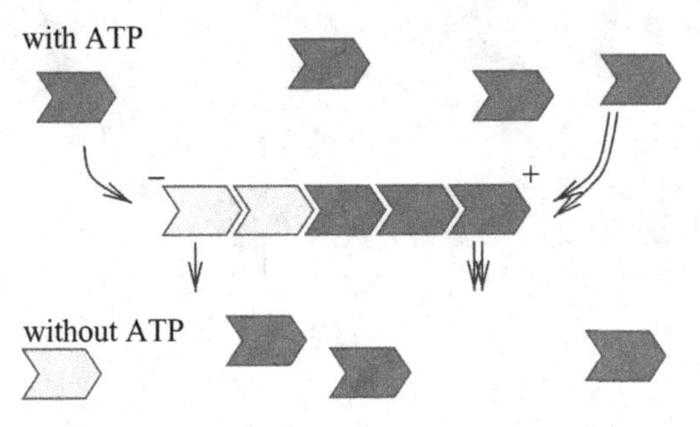


Figure 6.9. Polymerization of a dipolar filament, with different rates at the different ends. If hydrolysis does not take place, the ratio in the plus end $k(\text{in}, +)/k(\text{off}, +)$ must be equal to the ration of rates in the minus end $k(\text{in}, -)/k(\text{off}, -)$. However, because hydrolysis takes place in the filament, the polymer tends to grow at the plus end, and shrink at the minus end.

Force can be generated by any polymerization process. To be useful, however, it must be possible to repeat the process, such that movements can be generated beyond the distances set by the number of molecules in the cell. This recycling is associated with ATP hydrolysis taking place within the monomers that build the cytoskeleton. Through ATP hydrolysis the monomers can be in at least two states, one that is "charged" by, say, ATP (for actin), and one that is not. Directed movements of the filament are made possible by the following three properties (Wegner, 1976; Theriot, 2000).

(a) The polymer is not front–back symmetric, and the proteins polymerize with distinctly different rates at the plus end and at the minus end of the growing filament. The plus end grows faster.
(b) Monomer binding at the end of the filament depends on whether the monomer is ATP charged or not. If ATP is hydrolyzed in the actin monomer, the monomer leaves the filament when it reaches its end.
(c) Charging of monomers with ATP takes place in the bulk, not when the monomers are bound to the filament (where instead ATP is hydrolyzed faster). Thus monomers in the filament have on average less ATP bound than the free monomers, and because of (a) they have least ATP in the negative end of the filament.

In Fig. 6.9 we illustrate the mechanism by which the plus end grows, while the polymer simultaneously shrinks at the minus end. Figure 6.10 shows that, as a consequence, the polymer effectively moves across a surface (see also Fig. 6.11). This phenomenon can be monitored in vitro by fluorescent-labeled actin. The end of the polymer may perform work by pushing something or, alternatively, if the polymer is confined between two barriers the net difference in growth will make monomers move from the confining barrier in the growing end to the barrier in the shrinking end.

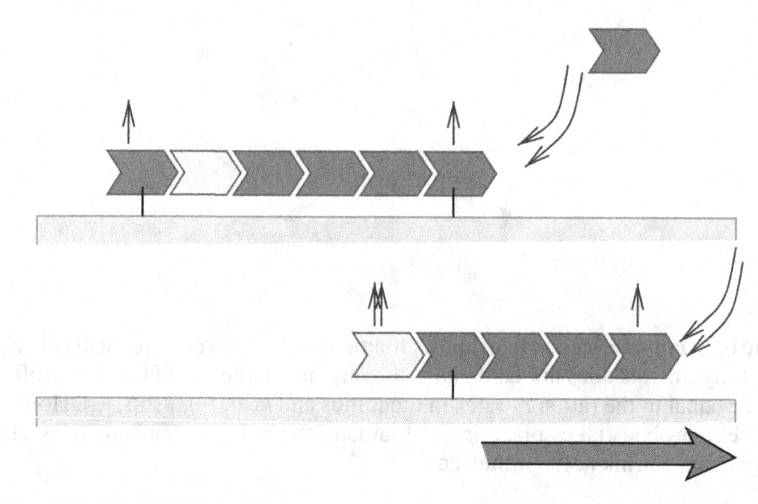

Figure 6.10. Treadmilling, a phenomenon originally suggested by Wegner (1976). Treadmilling has been seen in vitro for actin and demonstrated in vivo for tubulin (Rodinov & Borisy, 1997).

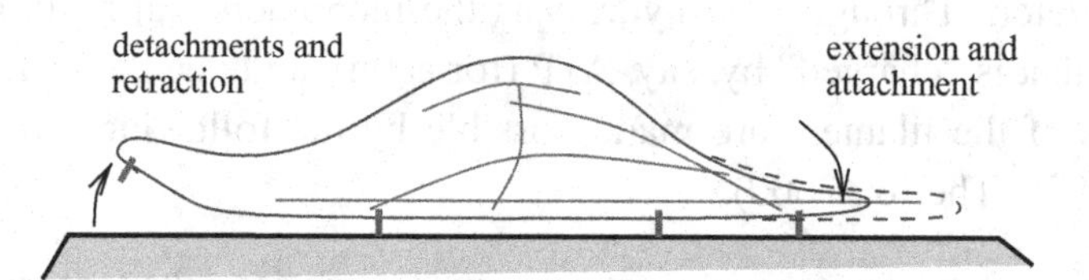


Figure 6.11. A single cell moving along a surface. A crucial element is the ability to control polymerization of actin in the front, and depolymerization in the back.

The polymerization of a single filament is an idealization; in real life the polymers consist of several filaments. Thus an actin polymer consists of two actin filaments, whereas a microtubule polymer consists of 13 filaments forming a rod. In addition different polymers may help each other's growth by subsequently securing the space for the neighbor polymers to grow (Mogilner & Oster, 1999).

The use of polymerization for directed cell movements not only involves the polymer building blocks and ATP/GTP, but also utilizes proteins that facilitate polymerization or depolymerization, and also the initial nucleation of the polymers. Thus actin depolymerization is increased by a factor 30 when the ADF protein binds to it. Other proteins, like the Arp2/3 complex, serve as a nucleation center for growing a second actin polymer on an already existing polymer, and thus allow the formation of a cross-linked cytoskeleton, which gives the cell internal structure, and upon which molecular motors may work.

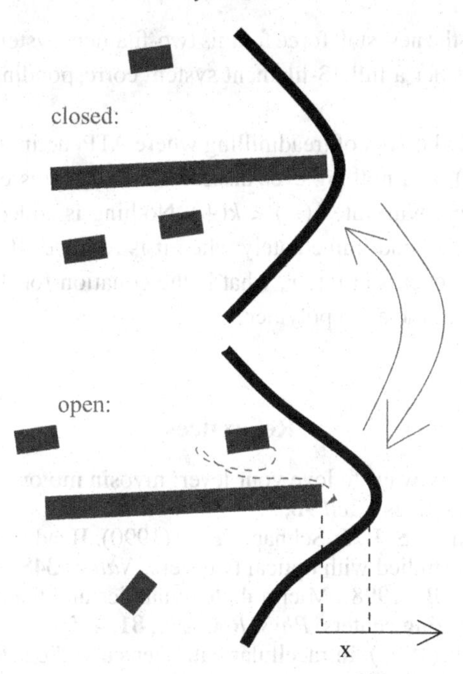


Figure 6.12. Figure for Questions (1)–(4), where polymerization of a protein filament pushes against a membrane, which may be the cell wall. Because of Brownian motion the membrane position $x$ fluctuates in front of the growing filament. When the gap $x$ is large enough, there is space for a new subunit. If this is attached, the whole scenario will be moved one step to the right. This resembles the ratchet mechanism from Fig. 6.4.

## Questions

(1) Consider the growth of a single microtubule filament against a membrane (see Fig. 6.12). At any point the membrane prefers to stay close to the tubule (at position $x = 0$). The membrane has thermal fluctuations. If the restoring force is $F = -k < 0$, what is the distribution of spacings between the membrane and the end of the microtubule? If $k = 4$ pN, what is the probability that the spacing exceeds 8 nm?

(2) Consider the setup from Fig. 6.12 and Question (1) above. Assume that the density of tubulin dimers is 1 mM. A tubulin dimer can bind to the end of the growing microtubule if this is longer than $d = 8$ nm from the membrane. Assume that the binding affinity of tubulin to tubule is infinite, and estimate the rate at which this microtubule will grow against the membranes. What happens to the growth rate if restoring force $F = -k$ is doubled from 4 pN to 8 pN?

(3) Describe qualitatively what happens when the binding affinity of tubulin in a microtubule is finite, and what happens if $k(\text{off}) > k(\text{on})$.

(4) Consider Question (2) above with a finite $k_{\text{on}}$ that is of a size that sets the stall force for the single filament to 4 pN. Consider a polymer consisting of two separate filaments. The building blocks still have length $d = 8$ nm, but are displaced by 4 nm relative to

each other. What is the new stall force for this two-filament system? What would the stall force be if we consider a full 13-filament system corresponding to a real microtubule rod?

(5) Consider a simplified model of treadmilling where ATP–actin is added only at the plus end with rate $k(+)$, and nothing ever dissociates at the plus end. Further ATP–actin leaves the minus end with rate $k(-) < k(+)$. Nothing is added at the minus end, but ADP–actin leaves this end immediately when it is reached. If conversion from ATP–actin to ADP–actin occurs at rate $r$, what is the equation for the growth and possible steady-state length of the actin polymer?

## References


Block, S. M. (1996). Fifty ways to love your lever: myosin motors. *Cell* **87**, 151.
(1998). Kinesin: what gives? *Cell* **93**, 5.

Block, S. M., Goldstein, L. S. B. & Schnapp, B. J. (1990). Bead movement by single kinesin molecules studied with optical tweezers. *Nature* **348**, 348.

Dogterom, M. & Yurke, B. (1998). Microtubule dynamics and the positioning of microtubule organizing centers. *Phys. Rev. Lett.* **81**, 485.

Dramsi, S. & Cossart, P. (1998). Intracellular pathogens and the actin cytoskeleton. *Ann. Rev. Cell. Dev. Biol.* **14**, 137.

Elson, E. L., Felder, S. F., Jay, P. Y., Kolodney, M. S. & Pastemak, C. (1999). Forces in cell locomotion. *Biochem. Soc. Symp.* **65**, 299.

Fygenson, D. K., Braun, E. & Libchaber, A. (1994). Phase diagram of microtubules. *Phys. Rev. E* **50**, 1579.

Fygenson, D., Flyvbjerg, H., Sneppen, K., Leibler, S. & Libchaber, A. (1995). Spontaneous nucleation of microtubules. *Phys. Rev.* E **51**, 5058.

Hansen, A., Jensen, M. H., Sneppen, K. & Zocchi, G. (2000). Modelling molecular motors as unfolding–folding cycles. *Europhys. Lett.* **50**, 120.

Hill, T. L. & Kirschner, M. W. (1982). Bioenergetics and kinetics of microtubule and actin filament assembly–disassembly. *Int. Rev. Cytol.* **78**, 1.

Julicher, F., Ajdari, A. & Prost, J. (1997). Modeling molecular motors. *Rev. Mod. Phys.* **69**, 1269.

Meyhofer, E. & Howard, J. (1995). The force generated by a single kinesin molecule against an elastic load. *Proc. Natl Acad. Sci. USA* **92**, 574.

Mitchison, T. J. & Cramer, L.P. (1996). Actin based cell motility and cell locomotion. *Cell* **84**, 359.

Mitchison, T. & Kirschner, M. (1984a). Microtubule assembly nucleated by isolated centrosomes. *Nature* **312**, 232.
(1984b). Dynamic instability of microtubule growth. *Nature* **312**, 237.

Mogilner, A. & Oster, G. (1999). The polymerization ratchet model explains the force–velocity relation for growing microtubules. *Eur. Biophys. J.* **28**, 235.

Peskin, C. S., Odell, G. M. & Oster, G. F. (1993). Cellular motions and the thermal fluctuations: the Brownian ratchet. *Biophys. J.* **65**, 316.

Rayment, I. (1996). The structural basis of the myosin ATPase activity. *J. Biol. Chem.* **271**, 15 850.

Rodinov, V. I. & Borisy, G. G. (1997). Microtubule treadmilling in vivo. *Science* **275**, 215.

Shelden, E. & Wadsworth, P. (1993). Observation and quantification of individual microtubule behavior in vivo: microtubule dynamics are cell-type specific. *J. Cell Biol.* **120**, 935.
Smith, D. A. & Geeves, M. A. (1995a). Strain-dependent cross-bridge cycle for muscle. *Biophys. J.* **69**, 524.
(1995b). Strain-dependent cross-bridge cycle for muscle. II. Steady-state behavior. *Biophys. J.* **69**, 538.
Szilard, L. (1929). Uber die entropieverminderung in einem thermodynamischen system bei eingriffen intelligenter wesen. *Z. Phys.* **53**, 840.
Theriot, J. A. (2000). The polymerization motor. *Traffic* **1**, 19.
Vasquez, R. J., Howell, B., Yvon, A. M., Wadsworth, P. & Cassimeris, L. (1997). Nanomolar concentrations of nocodazole alter microtubule dynamic instability in vivo and in vitro. *Mol. Biol. Cell.* **8**, 973.
Wegner, A. (1976). Head to tail polymerization of actin. *J. Mol. Biol.* **108**, 139.
Weisenberg, R. C., Borisy, G. G. & Taylor, E. W. (1968). The colchicine-binding protein of mammalian brain and its relation to microtubules. *Biochemistry* **7**, 4466.

## Further reading

Astumian, R. D. (1997). Thermodynamics and kinetics of a Brownian motor. *Science* **276**, 917.
Brockelmann, U., Essavaz-Roulet, B. & Heslot, F. (1997). Molecular stick-slip motion revealed by opening DNA with piconewton forces. *Phys. Rev. Lett.* **79**, 4489.
Cameron, L. A., Footer, M. J., van Oudenaarden, A. & Theriot, J. A. (1999). Motility of ActA protein-coated microspheres driven by actin polymerization. *Proc. Natl Acad. Sci. USA* **96**, 4908.
Coy, D. L., Wagenbach, M. & Howard, J. (1999). Kinesin takes one 8-nm step for each ATP that it hydrolyzes. *J. Biol. Chem.* **274**, 3667.
Davenport, R. J., Wuite, G. J. L., Landick, R. & Bustamante, C. (2000). Single-molecule study of transcriptional pausing and arrest by *E-coli* RNA polymerase. *Science* **287**, 2497.
Duke, T. & Leibler, S. (1996). Motor protein mechanics: a stochastic model with minimal mechanochemical coupling. *Biophys. J.* **71**, 1235.
Elston, T. & Oster, G. (1997). Protein turbines I: the bacterial flagellar motor. *Biophys. J.* **73**, 703.
Elston, T., Wang, H. & Oster, G. (1997). Energy transduction in ATP synthase. *Nature* **391**, 510.
Endow, S. A. & Waligora, K. W. (1998). Determinants of kinesin motor polarity. *Science* **281**, 1200.
Frey, E. (2001). Physics in cell biology: actin as a model system for polymer physics. *Adv. Solid State Phys.* **41**, 345.
(2002). Physics in cell biology: on the physics of biopolymers and molecular motors. *Chem. Phys. Chem.* **3**, 270.
Fygenson, D. K., Elbaum, M., Shraiman, B. & Libchaber, A. (1997a). Microtubules and vesicles under controlled tension. *Phys. Rev. E* **55**, 850.
Fygenson, D. K., Marko, J. F. & Libchaber, A. (1997b). Mechanics of microtubule-based membrane extension. *Phys. Rev. Lett.* **79**, 4497–4450.
Fygenson, D. K., Needleman, D. J. & Sneppen, K. (2002). Structural distribution of Amino Acid Variability in tubulin. *Biophys. J.* **82**, 2015.

Gelles, J. & Landick, R. (1998). RNAP as a molecular motor. *Cell* **93**, 13.
Hancock, W. O. & Howard, J. (1999). Kinesin's processivity results from mechanical and chemical coordination between the ATP hydrolysis cycles of the two motor domains. *Proc. Natl Acad. Sci. USA* **96**, 13 147.
Harmer, G. P. & Abbott, D. (1999). Losing strategies can win by Parrondo's paradox. *Nature* **402**, 864.
Higgs, H. N. & Pollard, T. D. (1999). Regulation of actin polymerization by Arp2/3 complex and WASp/Scar proteins. *J. Biol. Chem.* **274**, 32 531.
(2001). Regulation of actin filament network formation through Arp2/3 complex: activation by a diverse array of proteins. *Ann. Rev. Biochem.* **70**, 649.
Holy, T. E., Dogterom, M., Yurke, B. & Leibler, S. (1997). Assembly and positioning of microtubule asters in microfabricated chambers. *Proc. Natl Acad. Sci. USA* **94**, 6228.
Houdusse, A., Szent-Gyrgyi, A. G. & Cohen, C. (2000). Three conformational states of scallop myosin S1. *Proc. Natl Acad. Sci. USA* **97**, 11 238.
Howard, J. (1998). How molecular motors work in muscle. *Nature* **391**, 239.
Hunt, A. J., Gittes, F. & Howard, J. (1994). The force exerted by a single kinesin molecule against a viscous load. *Biophys. J.* **67**, 766.
Huxley, A. F. & Simmons, R. M. (1971). Proposed mechanism of force generation in striated muscle. *Nature* **233**, 533–538.
Ishijima, A., Doi, T., Sakurada, K. & Yanagida, T. (1991). Sub-piconewton force fluctuations of actomyosin in vitro. *Nature* **352**, 301.
Ishijima, A., Kojima, H., Funatsu, T. *et al.* (1998). Simultaneous observation of individual ATPase and mechanical events by a single myosin molecule during interaction with actin. *Cell* **92**, 161–171.
Kitamura, K., Tokunaga, M., Iwane, A. H. & Yanagida, T. (1999). A single myosin head moves along an actin filament with regular steps of 5.3 nanometres. *Nature* **397**, 129.
Kojima, H., Muto, E., Higuchi, H. & Yanagida, T. (1997). Mechanics of single kinesin molecules measured by optical trapping nanometry. *Biophys. J.* **73**, 2012.
Kron, S. J. & Spudich, J. A. (1986). Fluorescent actin filaments move on myosin fixed to a glass surface. *Proc. Natl Acad. Sci. USA* **83**, 6272.
Leibler, S. & Huse, D. (1993). Porters versus rowers: a unified stochastic model of motor proteins. *J. Cell Biol.* **121**, 1357.
Magnasco, M. O. (1993). Forced thermal ratchets. *Phys. Rev. Lett.* **71**, 1477.
Molloy, J. E., Burns, J. E., Kendrick-Jones, J., Tregear, R. T. & White, D. C. S. (1995). Movement and force produced by a single myosin head. *Nature* **378**, 209.
Moritz, M. & Agard, D. A. (2001). g-Tubulin complexes and microtubule nucleation. *Curr. Opin. Struct. Biol.* **11**, 174.
Nogales, E. (2000). Structural insights into microtubule function. *Ann. Rev. Biochem.* **69**, 277–302.
Noji, H., Yasuda, R., Yoshida, M. & Kinosita, K. Jr (1997). Direct observation of the rotation of F-1-ATPase. *Nature* **386**, 299–302.
Parrondo, J. M. R. (1998). Reversible ratchets as Brownian particles in an adiabatically changing periodic potential. *Phys. Rev. E* **57**, 7297.
Prost, J., Chauwin, J. F., Peliti, L. & Ajdari, A. (1994). Asymmetric pumping of particles. *Phys. Rev. Lett.* **72**, 2652.
Rieder, C. L., Faruki, S. & Khodjakov, A. (2001). The centrosome in vertebrates: more than a microtubule-organizing center. *Trends Cell Biol.* **11**, 413.

Rodionov, V. I., Nadezhdina, E. & Borisy, G. G. (1999). Centrosomal control of microtubule. *Proc. Natl Acad. Sci. USA* **96**, 115.

Sellers, J. R. & Goodson, H. V. (1995). Motor proteins 2: myosin. *Protein Profile* **2**, 1323.

Shingyoji, C., Higuchi, H., Yoshimura, M., Katayama, E. & Yanagida, T. (1998). Dynein arms are oscillating force generators. *Nature* **393**, 711.

Simon, A. & Libchaber, A. (1992). Escape and synchronization of a Brownian particle. *Phys. Rev. Lett.* **68**, 3375.

Song, Y. H., Marx, A., Muller, J. *et al.* (2001). Structure of a fast kinesin: implications for ATPase mechanism and interactions with microtubules. *Embo J.* **20**, 6213.

Tawada, K. & Sekimoto, K. (1991). Protein friction exerted by motor enzymes through a weak-binding interaction. *J. Theor. Biol.* **150**, 193.

Vale, R. D., Funatsu, T., Pierce, D. W., Romberg, L., Harada, Y. & Yanagida, T. (1996). Direct observation of single kinesin molecules moving along microtubules. *Nature* **380**, 451.

Vilfan, A., Frey, E. & Schwabl, F. (1999). Force-velocity relations of a two-state crossbridge model for molecular motors. *Europhys. Lett.* **45**, 283.

Walczak, C. E. (2000). Microtubule dynamics and tubulin interacting proteins. *Curr. Opin. Cell Biol.* **12**, 52.

Walczak, C. E., Mitchison, T. J. & Desai, A. (1996). XKCM1: a Xenopus kinesin-related protein that regulates microtubule dynamics during mitotic spindle assembly. *Cell* **84**, 37.

Wang, H.-Y., Elston, T., Mogliner, A. & Oster, G. (1998). Force generation in RNAP. *Biophys. J.* **74**, 1186.

Wang, M. D., Schnitzer, M. J., Yin, H., Landick, R., Gelles, J. & Block, S. M. (1998). Force and velocity measured for single molecules of RNA polymerase. *Science* **282**, 902.

Yasuda, R., Noji, H., Kinosita, K. & Yoshida, M. (1998). F-1-ATPase is a highly efficient molecular motor that rotates with discrete 120 degrees steps. *Cell* **93**, 1117.

Yin, H., Wang, M. D., Svoboda, K., Landick, R., Block, S. M. & Gelles, J. (1995). Transcription against an applied force. *Science* **270**, 1653.

# 7

# Physics of genetic regulation: the λ-phage in *E. coli*

Kim Sneppen

Biological molecular systems work inside living cells. As a cell prototype we consider the prokaryote *Escherichia coli*. This may be viewed as a small bag of DNA, RNA and proteins, surrounded by a membrane. The bag has a volume of about 1 $\mu m^3$. This volume varies as the cell grows and divides, and also varies in response to external conditions such as osmotic pressure. The interior of the cell is a very crowded environment, with about 30% to 40% of its weight in proteins and other macromolecules, and only about 60% as water. Further, the water contains a number of salts, in particular $K^+$, $Cl^-$ and $Mg^{2+}$, each of which influences the stability of different molecular complexes.

The dry weight of *E. coli* consists of 3% DNA, 15% RNA and 80% proteins. The genome of *E. coli* is a single DNA molecule with $4.6 \times 10^6$ base pairs (total length of about 1.5 mm). It codes for 4226 different proteins and a number of RNA molecules. However, the information content of the genome is larger than that corresponding to the structure of the coded macromolecules. Important information is hidden and resides in the regulation mechanisms that appear when these proteins interact with the DNA and with each other. Some proteins, called transcription factors, regulate the production of other proteins by turning on or off their genes. Figure 7.1 shows two ways by which a transcription factor can regulate the transcription of a gene. The figure also shows a specific example of a regulatory protein bound to the DNA: the CAP protein.

In this chapter we describe in detail one regulation system, the λ-phage switch. We also use this example to illustrate a number of physical concepts related to genetic regulation. These include the relation of chemical equilibrium to statistical mechanics, cooperativity in binding, and timescales and rates associated with diffusion and chemical binding events in the living cell. We will thereby show how one builds quantitative models of gene regulation. We will also discuss the concept of robustness in relation to genetic switches, both in view of how it has been observed in mutants of the λ-phage, and in the variations of genetic switch mechanisms that

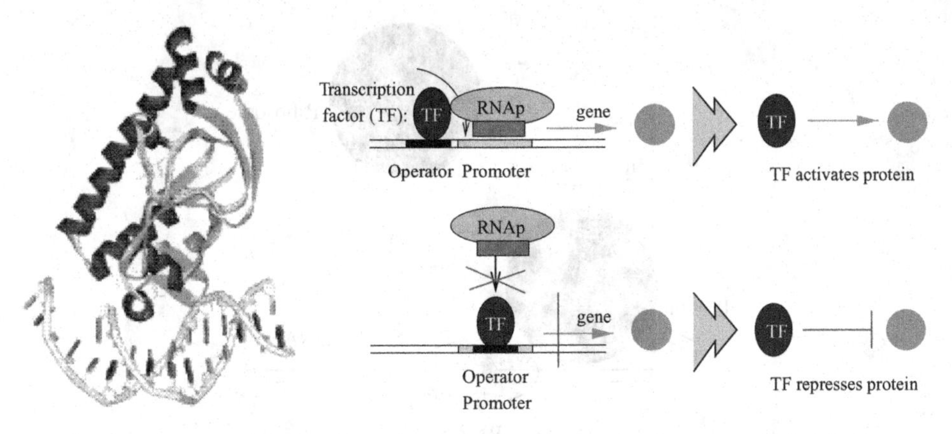


Figure 7.1. In the left-hand panel we show an example of a transcription factor, the protein CAP, that binds to a piece of DNA. In the right-hand panel we illustrate positive and negative regulation, respectively, by a transcription factor (TF). The TF is a protein that binds to a region on the DNA called an operator (the dark region on the DNA). Positive regulation is through the binding between the TF and the RNAp (RNA polymerase); this increases the chance that RNAp binds to the promoter, which is shown as the medium dark region on the DNA strand. Negative regulation occurs when the operator is placed such that the bound TF prevents the RNAp from binding to the promoter. On the far right we show how one typically draws the elementary regulations in a regulatory network.

one finds in other vira that infect bacteria (bacteriophages). But before discussing these concepts, we will introduce the general properties of the transcription and translation processes.

## Transcription and translation in numbers

In this section we will describe the basic production apparatus of the cell; the mechanisms of transcription and translation. These processes are the action of huge molecular motors, which work their way along a one-dimensional string, using this as a template, for generating another one-dimensional string. This, of course, demands energy. Energy is, however, abundant in the cell in form of huge amounts of ATP and GTP. Also, production demands the presence of some building blocks: the bases to make RNA units, and the 20 amino acids from which one can build the protein polymer string. We will now describe the processes involved in transcription and translation in more detail, and put some numbers on this overall production machinery.

### *Transcription*

When DNA is transcribed, a messenger RNA (mRNA) is produced (see Fig. 7.2). First RNA polymerase (RNAp) binds to the DNA at certain binding sites, called

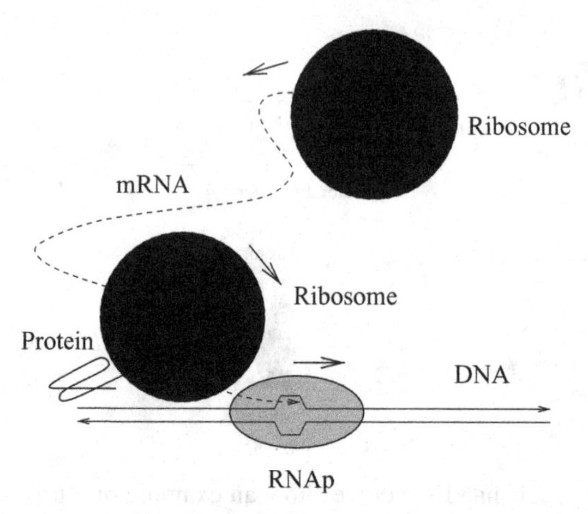


Figure 7.2. RNA polymerase (RNAp) moves along the DNA while transcribing it into a mRNA string: that is, it opens the DNA double strand, and makes a complementary single-stranded RNA polymer. This mRNA is subsequently translated into a protein by a ribosome, here shown in black. The same mRNA may easily be translated into several copies of the protein, as shown here where two ribosomes act simultaneously on the mRNA. Notice that translation may be initiated before transcription is fully finished. However, one should then be aware that the RNAp rotates around the DNA, implying that the translating ribosome has to spiral around the DNA. It is thus fortunate that translation speed is only a fraction of transcription speed.

promoters. Then RNAp may transcribe the DNA in a direction determined by the orientation of the promoter, see Fig. 7.1. Each gene has one, or more, corresponding promoter, and some promoters control more than one gene.

The polymerase machine RNAp is a protein complex that consists of at least four protein sub-units and has a total mass of about 400 kDa. There are about 2000–3000 RNAp molecules in an *E. coli* cell (Pedersen *et al.*, 1978). This is about one per promoter. At any time nearly all RNAp enzymes are engaged in transcribing genes. At any given time only a fraction of the genes are being transcribed. In fact, in a rich medium about half of all RNAp activity is associated with transcription of genes associated with production of ribosomes. As a result, the free RNAp concentration is expected to be only about 30–50 nM. Thus there are only about 30 RNAp complexes available at any time in the 1 $\mu m^3$ *E. coli* cell volume. We remind the reader that 1 nM corresponds to 0.6 molecules per cubic micrometer.

Whether an RNAp can bind to a promoter may depend on the presence or absence of other proteins, called transcription factors, at nearby locations (operator sites) on the DNA. This represents one major control mechanism of transcription. In general such control mechanisms are called genetic regulation. Figure 7.1 illustrates how one protein may increase or decrease, respectively, the transcription rate of a given gene.

If RNAp can bind, then the transcription initiation frequency of RNAp is a result of three consecutive steps (Hawley & McClure, 1982).

(1) An RNAp binding rate: this is facilitated by RNAp binding to nearby DNA and the subsequent sliding to the promoter site. The recognition part of the promoter stretches from −55 bp to −6 bp, where position +1 is the starting site for the RNA production, and + numbers refer to subsequent nucleotides for the transcription. Especially important for binding are the short sequences around the −10 bp (TATAAT) and the −35 bp (TTGACA) positions. These regions are separated by a spacing region 17 ± 1 bp long.
(2) A DNA opening rate (isomerization step): this is the unwinding of about 12–18 bp of double-stranded DNA to form an open bubble that exposes the base pairs for transcription initiation. The bubble is around the region −9 bp to +3 bp. The bound RNAP occupies the region from −55 bp to +20 bp.
(3) The time it takes RNAp to initiate successful transcription downstream along the DNA on the 5′ to 3′ DNA strand: notice that the RNAP must rotate along the helix.

Total initiation rates of mRNA synthesis (the three steps above) vary from ∼1/s (for ribosomal genes) to 1/(18 s) for promoter PR in the λ-phage, and 1/(400 s) for the unstimulated promoter PRM in the λ-phage (Ian Dodd, personal communication). Thus the initiation frequency is important in determining the concentrations of the proteins in the cell. The RNAp is quite a large molecular machine. When it binds to the promoter it is believed to occupy about 75 bp of DNA. After transcription initiation, the RNAp transcribes the DNA with a speed that varies between ∼ 30 bp/s and 90 bp/s, where the ribosomal genes are the ones that are transcribed fastest (Liang *et al.*, 1999; Vogel *et al.*, 1992; Vogel & Jensen, 1994). The transcription is stopped at terminator signals, which in *E. coli* is a DNA sequence that at least codes for an mRNA sequence that forms a short hairpin.

## Translation

After an mRNA is generated, it can subsequently be translated by a ribosome into a peptide heteropolymer, which folds to form a protein (see Fig. 7.2). The ribosome is a large molecular complex that consists of 50 different protein sub-units plus some fairly large RNA sub-units. In bacteria it has a mass of $2.5 \times 10^6$ Da, of which one-third is proteins. There are about 10–15 000 ribosomes in an *E. coli* cell, of which about 750 are free at any time (Vind *et al.*, 1993).

As already stated, production of ribosomes occupies up to 50% of all RNAp activity in the *E. coli* cell. In a rich medium the production of ribosomal RNA (rRNA) represents about half of all RNA production in the cell. And at any time rRNA occupies 80% to 90% of the total RNA content in the cell. Ribosomes are huge, and are produced in huge quantities. Ribosome production is controlled, and balanced. In particular, if ribosomal proteins are not occupied in ribosomes by binding to ribosomal RNA, they down-regulate production of more ribosomal proteins. Ribosomal RNA (rRNA) on the other hand can be up-regulated by both ribosomal proteins and other factors (Voulgaris *et al.* (1999a,b), and references therein).

It is inside the ribosome that the information world of DNA/RNA is merged with the machine world of proteins. Ribosomes thus translate the nucleotide sequence to amino acids, by using the genetic code. Each amino acid is attached to a specific tRNA molecule that contains the triplet codon corresponding to that amino acid. In the ribosome this triplet is matched to the mRNA triplets, and its attached amino acid is linked covalently to the previous amino acid in the growing peptide chain. Thereby a sequence of codons is translated to a sequence of amino acids.

Once initiated, the mRNA-to-protein translation in a ribosome proceeds at a rate between 6 codons/s and 22 codons/s; see Pedersen *et al.* (1978). Each mRNA is translated between 1 and 40 times before it is degraded. This number depends primarily on the start of the mRNA sequence, in particular whether this is a good ribosome binding site. The best binding site is the Shine–Delgarno sequence (AG-GAGGU), located about 10 base pairs upstream from the translation start signal (Ringquist *et al.*, 1992). For optimal sequences one expects the previously mentioned ~40 fold translation for each mRNA in an *E. coli*; mRNA degradation factors also play a role in controlling this copy number (Rauht & Klug, 1999). This degradation is presumably also under control, in particular through the RNase enzymes, which actively degrade the mRNA. Typical half times of mRNA in *E. coli* are found to be 3–8 min (Bernstein *et al.*, 2002).

## *Questions*

(1) One *E. coli* cell has a volume of 1 μm$^3$, and consists of about 30% protein by weight. An amino acid weighs on average 100 Da. Assuming that amino acids have the same density as water, how many amino acids are there in an *E. coli* cell?

(2) If an average protein consists of 300 amino acids, how many proteins are there in a cell?

(3) One bacterial generation takes 30 min, and there are about 10 000 ribosomes in a cell. What is the average rate of translation in the cell (in units of codons per second per ribosome)?

(4) There are about 2000–3000 RNAp in an *E. coli*. If one assumes that transcription and translation occur at the same rates, how many proteins are produced per mRNA on average?

(5) Consider the transcription initiation kinetics:
RNAp + P ↔ RNAp − P(closed) → RNAp − P(open), where P is the promoter, and transition from the closed to open complex is irreversible. Prove that, under certain conditions, the rate for open complex formation can be written as $k_{\text{obs}} = k_{\text{f}}[\text{RNAp}]/$

($K_B$ + [RNAp]). Identify the expressions in terms of constants for the transcription initiation process, and give the conditions for the validity of this equation.

(6) If $K_B = 2 \times 10^7$ M$^{-1}$, $k_f = 0.05$/s and [RNAp] = 30 nM, calculate the mean time between transcription initiations.

(7) For the $P_R$ promoter in the λ-phage, we can approximately take $K_B = 1 \times 10^8$ M$^{-1}$ and $k_f = 0.01$/s. How many mRNA transcripts can be produced in one cell generation (30 min)?

(8) Consider the setup in Question (5) but with the addition that the RNAp binding site could be blocked with one repressor at some concentration [Rep] and a binding equilibrium constant $K_{rep}$. Derive the equation for the transcription initiation rate.

(9) How long does it takes to clear a promoter site before a new RNAp can bind? What is the maximal activity of a promoter?

(10) Consider the following production/feedback system for balancing ribosomal rRNA (r) and ribosomal protein (p) production: $dr/dt = \eta_p(t) + \Theta(p - r) - r/\tau$ and $dp/dt = c + \eta_p(t) - \Theta(p - r) - p/\tau$, where the $\eta$s are random noise terms with mean 0, and the function $\Theta$ reflects a feedback from an overabundance of total $p$ relative to the total rRNA ($\Theta(x) = 1$ for $x > 0$ and $= 0$ elsewhere; in the smoothed version one may set $\Theta(x) = x^2/(x^2 + 1)$). Note that $c$ is a constant that secures the presence of $p$. Simulate the equations and compare the behavior with that of similar systems without feedback. Investigate the response to a sudden doubling of $c$. In terms of real timescales, an effective lifetime $\tau$ for ribosomes given by dilution due to the doubling rate of the bacteria is ~30 min. Notice that we have not defined the noise term properly, and the result may depend on the size of the noise and the timestep in integration.

## The genetic switch of the λ-phage

With about $5 \times 10^{30}$ prokaryotes on Earth, these relatively simple single-celled organisms are a dominating life form. However, not even for bacteria is the world a safe place! The bacteria are exposed to their own parasites, the bacteriophages. The number of phages exceed the number of bacteria by several orders of magnitude, and their diversity and ability to manipulate genetic information probably makes these simple information carriers an important force in shaping the global ecosystem. Molecular biologists have selected a few of these phages for detailed study, and this has been a major inspiration for developing our understanding of how genetic regulation works. In particular, the bacteriophage λ has been studied in great detail. This is not because the λ is particularly abundant in nature, but rather because, by accident, it was the first bacteriophage that was found to have a genetic switch.

The λ-phage displays one of the simplest examples of computation in living systems: in response to a sensory input it decides between one of two states (Johnson *et al.*, 1981; Ackers *et al.*, 1982). Thus, with the same genome, it can be in two different states, a phenomena called epigenetics. Epigenetics is defined as "a heritable

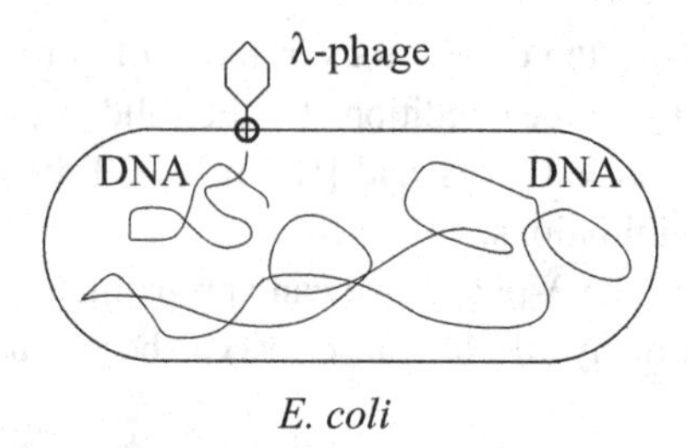


Figure 7.3. *E. coli* cell with a λ-phage injecting its DNA through the membrane (i.e. through the maltose receptor of the *E. coli* cell). The *E. coli* cell has dimensions of about 1 μm, the λ-phage head is 55 nm in diameter, whereas the full length of λ-head plus tail is 0.2 μm.

change in phenotype in the absence of any change in the nucleotide sequence of the genome" (Bestor *et al.*, 1994). Epigenetics is inherent to cell differentiation, and thus essential for multicellular life. Epigenetics comes about because of a positive feedback that forces the system to differentiate into well-separated states (Thomas, 1998). In an eukaryote, such states would correspond to cell types. In bacteria, it allows proliferation in widely different environments (Casaderus & D'ari, 2002). In the λ-phage, epigenetics allows the phage to adopt two widely different survival strategies.

The λ-phage is a bacterial virus (=phage) that uses a specific receptor protein on the bacterial surface to enter the cell. If there are no maltose receptors on the surface, the *E. coli* is immune to the λ-phage. With the receptor present, the phage can bind and inject its DNA into the cell, as sketched in Fig. 7.3.

After infection of an *E. coli* cell the λ-phage enters either into an explosive *lytic* state, where finally the bacteria bursts and many copies of λ are released, or the λ integrates its DNA into the host cell DNA (see Fig. 7.4). The latter case leads to the *lysogenic* state, in which the phage DNA can be passively replicated for many generations of the *E. coli*. A phage that has this ability to enter into lysogeny is called *temperate*.

The initial decision, whether to integrate or to lyse, is taken through the interplay between two proteins both of which have their production initiated from promoter PR (see Fig. 7.5). The protein Cro, which degrades slowly (~30 min; see Pakula *et al.*, 1986), and the protein CII whose degradation is faster (~5 min; see Hoyt *et al.*, 1982). CII degradation depends on a number of cell-dependent factors; in particular its degradation is slower when the cell is starving, because the protease that degrades CII is repressed at starvation. Also CII is stabilized at lower temperature (Obuchowski *et al.*, 1997). Cro favors lysis whereas CII favors lysogeny.

As a result, the lysogeny state is primarily selected when the cell has a slow metabolism (starvation increases lysogeny frequency by a factor of 100; Kourilsky,

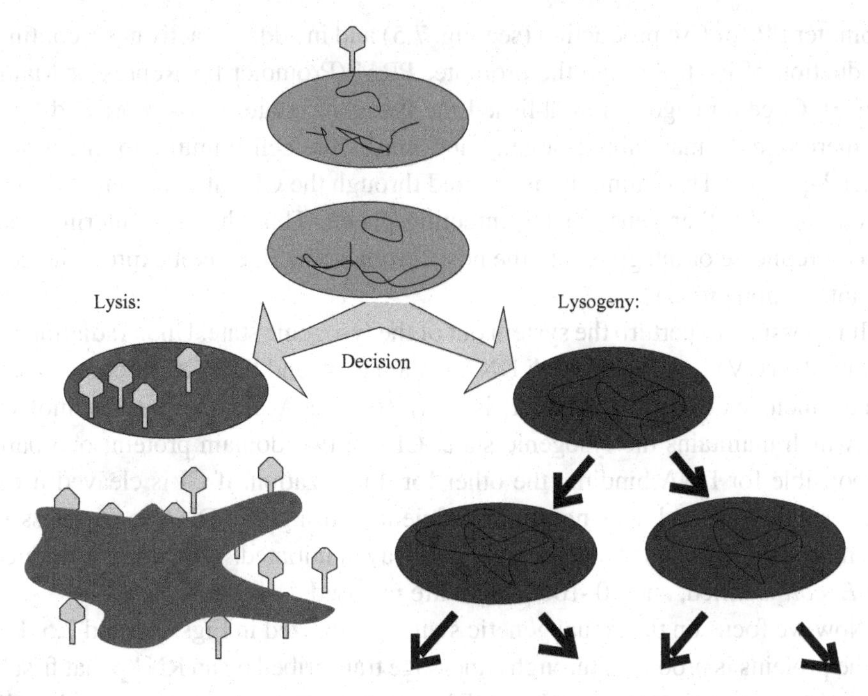


Figure 7.4. *E. coli* cell after infection by a λ-phage enters either the lysis pathway (left), or the lysogenic pathway (right). Lysis leads to death plus 50–100 released phages, whereas lysogeny lets the phage integrate into the *E. coli* chromosome, and be passively replicated for many (maybe millions) of generations.

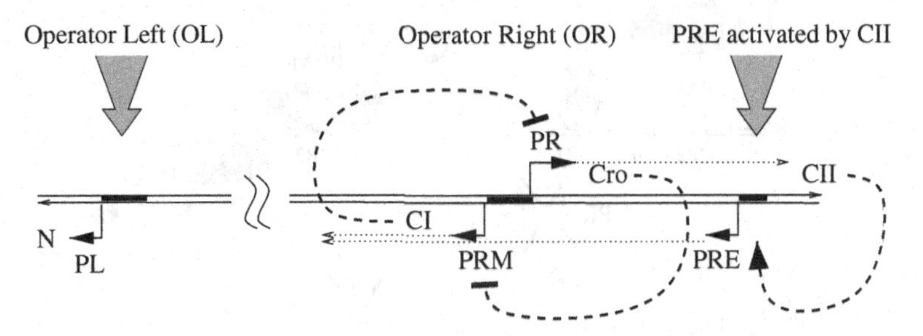


Figure 7.5. The location of the λ-switch and its immediate neighborhood in the λ-genome. Right of OR, after the Cro gene, there is a promoter site PRE for left-directed transcription of anti-Cro message and CI. The promoter PRE is activated by CII, which thereby initiates CI production. This is important in the early phase of infection, and PRE refers to the promoter for repressor establishment. Left of OR, after the CI gene, one finds the operator OL that controls the gene for protein N, which also has a regulatory role in the PR–PRE competition during infection.

1973; Friedman, 1992) or when there are many λ-phages trying to infect the cell simultaneously (Kourilsky, 1973). CII activates the strong promoter PRE and initiates production of the protein CI that stabilizes the lysogenic state (PRE is short for Promoter for Repressor Establishment). CI stabilizes lysogeny by repressing the promoter PR for Cro production (see Fig. 7.5) and in addition activates a continous production of itself through the promoter PRM (Promoter for Repressor Maintenance). Once a phage has established the lysogenic state, PR is repressed by CI. CI thereby both maintains lysogeny and makes the cell immune to infection by other λ-phages. This immunity is secured through the CI that will repress Cro and essentially all other genes of the infecting phage. Thus the new entering phage cannot replicate or integrate into the host chromosome (it cannot express the genes for integration either).

It is possible to perturb the system out of the lysogenic state. Upon radiation with ultraviolet (UV) light, the *E. coli* DNA is damaged, and the SOS rescuing system, which includes the protein RecA, is activated. RecA cleaves the key molecule CI, which maintains the lysogenic state. CI is a two-domain protein; one part is responsible for DNA binding, the other for dimerization. If CI is cleaved it does not form dimers, and does not bind sufficiently strongly to DNA to suppress PR. Thereby Cro is produced and the lytic pathway is initiated. When lysis is induced, the *E. coli* is killed, and 50–100 phages are released.

Now we focus on the actual genetic switch visualized in Figs. 7.6 and 7.5. Each of the proteins is produced through a message transcribed by an RNAp that first has to bind to a corresponding promoter. The main promoter for the *cI* gene is called PRM, and the promoter for the *cro* gene is called PR. They are both controlled by

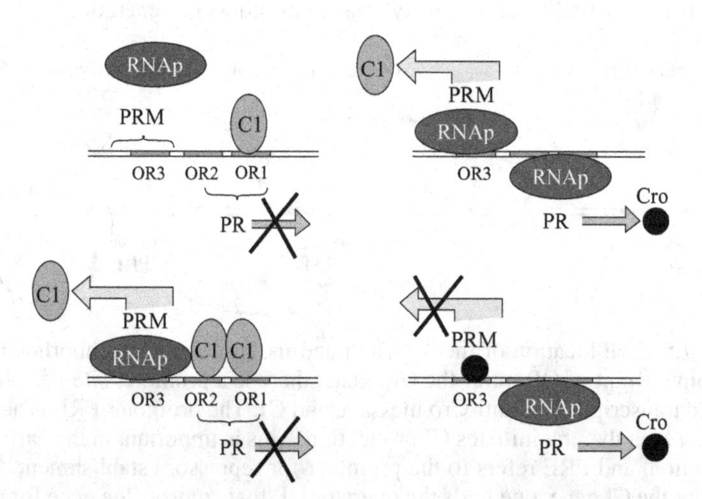


Figure 7.6. Basic mechanisms of the λ-phage switch. Arrows indicate transcription directions. Transcription of the CI gene is thus in the opposite direction (and on opposite DNA strand) to transcription of the Cro gene. The geometry of operator sites relative to the two promoters makes CI repress Cro, and Cro repress CI. Thus a mixture of CI and Cro is unstable and there is a positive feedback driving the system away from the mixed CI–Cro state. This results in either a CI-dominated state (lysogeny), or a Cro-dominated state (lysis).

3′ end 5′ end

TATCACCGCAAGGGATAAATATCTAACACCGTGCGTGTTGACTATTTTACCTCTGGCGGTGATA

ATAGTGGCGTTCCCTATTTATAGATTGTGGCACGCACAACTGATAAAATGGAGACCGCCACTAT

5′ end 3′ end

OR3 OR2 OR1

Figure 7.7. Detailed view of switch where individual base pairs are shown. Each operator covers 17 base pairs, indicated by bold letters. Further, each operator is fairly palindromic. For example on OR3 the upper 3′ end starts at TATC, which is also found on the lower 3′ end of OR3 (on the opposite strand). This means that the protein binding is mirror symmetric, which is natural as both Cro and CI bind as dimers.

the operator called OR ("operator right") that is located in the middle of the $\sim$40 k base λ-DNA. When CI binds to the right part of this operator, OR1 or OR2, it represses the *cro* gene and thereby represses lysis. The cell is then in the lysogenic state, and CI will be constantly expressed from the promoter PRM. CI is often just referred to as the repressor.

CI and Cro are encoded in opposite directions along the DNA. The promoters for the two proteins are controlled by the operator sites that form OR, as seen in Fig. 7.6. OR consists of three operator sites, OR1, OR2 and OR3 as illustrated in Fig. 7.6. Each of these operator sites is a binding site for both Cro and CI, but with different affinities. Qualitatively CI binds first to OR1 and OR2, and then to OR3. Cro binds first to OR3 and then to OR1 and OR2. As a result either Cro, or CI, but not both, can be simultaneously produced from promoters around OR. If CI is produced, the cell is in a lysogenic state. If Cro is produced the cell is on the way to lysis.

In Fig. 7.7 we show the DNA sequence for the operators OR. In the real system the DNA is of course not linear. It is a double helix with a period of 10.4, whereas the operators consists of 17 base pairs, separated by 6–7 base pairs. Thus the center-to-center distances are about 23 base pairs, and therefore the binding of two CI dimers to two consecutive operators will place the two CI dimers on nearly the same side of the DNA, thus facilitating interactions between the CI dimers. Notice also that the operators' sequences are similar; in fact CI and Cro bind to each operator with energy $\Delta G$ of about $-10$ kcal/mol. The order of binding is set by the rather small differences between the operators, with differences in free energies of $\Delta\Delta G = \Delta(\Delta G) \sim$ 1–3 kcal/mol. This is not much in energy difference, but is up to a factor $\exp(\Delta G/T) \sim 100$ in the chemical binding constant. In the following sections we will explore the biochemistry and physics involved in this control system. First, however, we discuss how to make simple biological measurements.

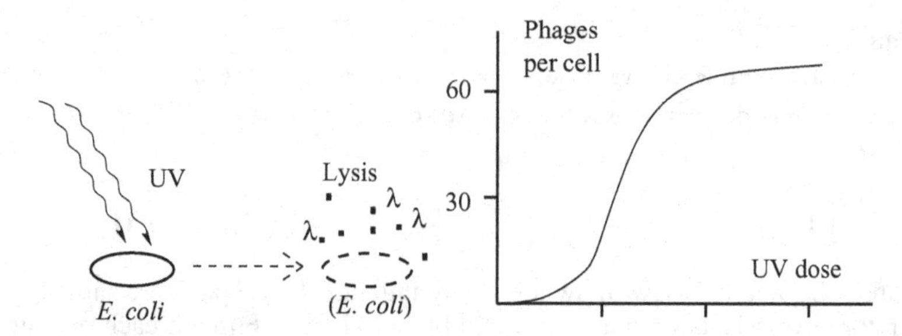


Figure 7.8. Left: when exposing *E. coli* to UV light, some are killed and each of these releases 50–100 λ-phages. The ability to lyse is an inherited trait. Right: response of the λ-phage to UV radiation. We observe a threshold-like behavior, where phages lyse when UV is at a level that kills the bacteria. Thus the *y*-axis is proportional to probability to lyse.

### *Questions*

(1) Why is it a good strategy for the λ to favor lysogeny when a bacterium is infected simultaneously by many λs?

(2) A λ-lysogen is infected with another λ-phage. Why does the new phage not induce lysis?

## Using bacteria to count phages

One nice feature of molecular biology is that it facilitates its own exploration. Often one can use the control of some parts to explore other parts. One example is the counting of phage release, which is an important quantitative tool of most phage research. In Fig. 7.8 we show the result of an experiment where λ-infected *E. coli* is exposed to different levels of UV radiation. Fig. 7.9 illustrates how one counts the number of released phages.

Let us say that we want to measure the stability of the lysogenic state. Thus we want to measure how many phages are released per bacterium in a given situation. The overall experiment is sketched in Fig. 7.9. The experiment proceed as follows. First we need a strain of *E. coli* that is infected with the λ but does not have the λ-receptor. Such strains have been prepared. Bacteria from this strain are grown into a culture overnight. In the morning we will have a large sample of bacteria, plus some phages that were released from bacteria that underwent lysis. We want to count both the number of bacteria, and the number of released phages. Notice that because none of the bacteria has the maltose receptor, released phages cannot enter another bacterium. Therefore only first-generation lysis events are counted.

To count bacteria is easy. One dilutes the solution by a large factor and distributes it homogeneously on a culture dish (plate it), allows it to grow overnight and subsequently counts the number of colonies on the dish. Each bacterium gives rise

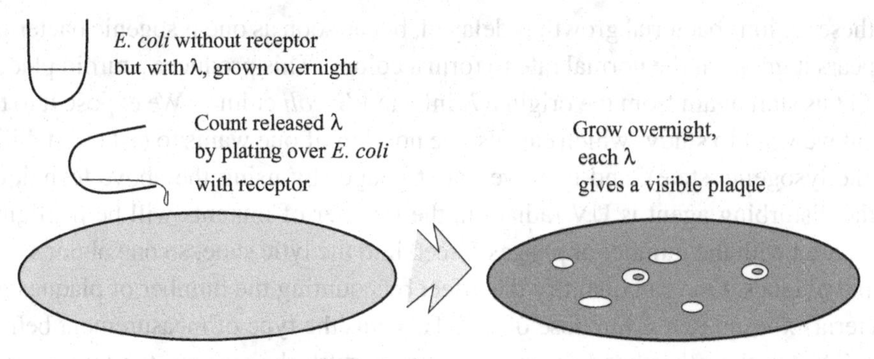


Figure 7.9. Measurement of phage release. This figure shows spontaneous phage release (similar experiments can done for the UV-induced release shown in Fig. 7.8). The primary experiment is done in a test tube; the counting of what occurred is subsequently done by plating the contents of the test tube on a growing strain of bacteria which is ready for λ-infection. Each λ will infect and kill its host, leading to a plaque that can be seen with the naked eye after a night's incubation. The resulting number of plaques is then counted manually.

to one colony. If the solution of bacteria was sufficiently diluted the colonies are non-overlapping and thus countable by naked eye. One can then easily recalculate the original number of bacteria per mL.

To count phages we remove the *E. coli* from the solution, and we are left with a solution containing the released phages. An appropriate dilution of this solution is poured over a detector strain of *E. coli* that has receptors for the λ. This plate is grown overnight. Each phage will then initiate a local plaque. The reason is as follows. The phage will infect a bacterium and in most cases (~99%) it enters lysis. Then new phages are produced and they infect and kill the nearby bacteria. This is seen as a small circular area where bacterial growth is reduced: a plaque. By counting the number of plaques one obtains the original number of phages. The dynamic range of this type of measurement is very large: if there are many phages one dilutes the fraction poured over the plate accordingly. One can measure phage release in a range from 100 phages per *E. coli* to one phage per $10^7$ *E. coli*.

Actually there is more information that one can extract from this experiment. It turns out that one observes two types of plaque: clear and turbid. In both types the bacterial growth is reduced compared with the growth in the surrounding phage-free area, but for the clear plaques it is reduced to zero. The clear plaques correspond to phages that cannot establish lysogeny, and thus kill all the bacteria they infect. These phages are virulant mutants. The turbid plaques originate from phages that are able to establish lysogeny (in a small percentage of their hosts). The bacteria where the phage has established lysogeny are the only ones to survive; all others are killed. This is because lysogenic bacteria cannot be re-infected by another λ. Thus in these regions bacterial growth is delayed, but as soon as one lysogenic bacterium appears it grows at the normal rate to form a colony. This produces a turbid plaque.

Let us start again from the original λ-infected *E. coli* culture. We expose it to the agent we want to study (which can also be nothing, if one wants to test the stability of the lysogenic state), and then we count phages by using the above technique. If the disturbing agent is UV radiation, the number of mutants will be negligible compared with the number of phages forced into the lytic state, so one obtains only turbid plaques. One can quantify the effect by counting the number of plaques per bacteria exposed to a given dose of UV. This was the type of measurement behind Fig. 7.8. On the other hand, if one does not perturb the system, the transition to lysis is mostly caused by mutations. Thus there are more clear plaques than turbid ones, and we can quantify the mutation rate by counting the clear plaques (about 1 in $10^6$ phages mutate to become virulent).

*Questions*

(1) Simulate a plaque on a growing bacterial culture plated on a two-dimensional disk. Key parameters are the diffusion rate of λ-particles, the capture cross section for a phage to enter the bacteria, and the number of phages released per lysis. Assume for simplicity that all bacteria lyse.

(2) When a λ decides to lyse the bacteria it doubles its genome through a so-called rolling mode, where about one copy is generated each 30 s. Assume that an escaped λ instantly infects a new *E. coli* and initiates the lysis there. Discuss the optimal lysis strategy in terms of how many copies the λ should generate from each lysis event. (In principle there is material for generating about 2000 phage particles in one *E. coli* cell.)

## Basic chemistry – and cooperativity

We now want to turn to a more quantitative description of the possible processes involved in genetic regulation. An elementary process here is the binding–unbinding of proteins to each other and to the operator DNA that controls transcription initiation. Thus, for readers who are not familiar with chemistry, we now go through some basic chemistry and cooperativity. In this section we follow the standard approach quite closely, with on- and off-rates, and consequences of cooperativity. In the next section we will venture into a more detailed discussion of how to treat binding to a combination of operators, and through this we will demonstrate how statistical mechanics provides us with a generic framework.

The basis of chemistry can be understood from considering a titration experiment. In this we examine the bound fraction of a complex as a function of the total concentration of one of its constituents. Thus, consider a titration experiment where we investigate the reaction between repressor CI and an operator site O on a small piece of DNA:

$$CI + O \rightleftharpoons CIO \tag{7.1}$$

This reaction is in equilibrium when the rates of the reactions

$$CI + O \rightarrow CIO \quad \text{with} \quad rate_{\rightarrow} = k_{\rightarrow}[CI][O] \tag{7.2}$$

$$CI + O \leftarrow CIO \quad \text{with} \quad rate_{\leftarrow} = k_{\leftarrow}[CIO] \tag{7.3}$$

balance ($rate_{\rightarrow} = rate_{\leftarrow}$):

$$K = \frac{k_{\leftarrow}}{k_{\rightarrow}} = \frac{[CI][O]}{[CIO]} \tag{7.4}$$

$K$ is called the dissociation constant. Further we emphasize that all concentrations refer to free concentrations in the solution, and thus not total concentration. For example, the total concentration of operator sites is $[O_{total}] = [O] + [CIO]$, and fraction of occupied operator is therefore

$$\frac{[CIO]}{[O_{total}]} = \frac{[CI]}{K + [CI]} \tag{7.5}$$

Thus $K$ is the CI concentration at which [O] is half occupied, that is [O] = [CIO]. In molecular biology $K$ is typically found to be $\sim 10^{-8 \pm 2}$ M. In Fig. 7.10 we illustrate such an experiment. Notice that the free concentrations are less than the total concentrations, as $[CI_{total}] = [CI] + [CIO]$, and that the experiment is most

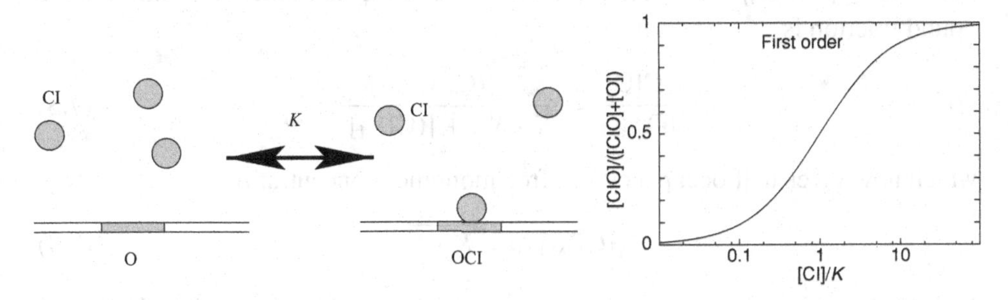


Figure 7.10. Left: elementary chemistry illustrating the first-order reaction between a repressor CI and one isolated operator site. Right: occupancy as a function of repressor concentration. This is measurable in a titration experiment, where one measures variation in the fraction of bound operator as a function of total repressor concentration (for the first-order reaction shown here: $[CIO]/([O_{total}]) = [CIO]/([CIO] + [O])$). The measurement can be made by separating bound and free DNA through the difference in displacement upon electrophoresis in a gel filter. More efficient, however, is "foot-printing", where parts of the DNA that are bound to the protein are protected against a DNAase enzyme that degrades all DNA that is not bound. The fraction of surviving DNA fractions is then measured by gel electrophoresis.

easily analyzed when operator concentration is kept much lower than all other concentrations ([O] $\ll$ [CI]), such that $[\mathrm{CI}_{\mathrm{total}}] \approx [\mathrm{CI}_{\mathrm{free}}]$. This situation is also typical inside the *E. coli* cell, where there is often only one operator of any particular type.

### *Cooperative chemistry*

A genetic switch should switch at a reasonable concentration, and be efficient in discriminating between its options: it should switch between its two states with only a small change in concentration of the controlling protein. In molecular biology this is obtained by an ordered sequence of bindings, the simplest being a dimerization step.

In the λ-switch, the DNA binding protein CI in fact binds significantly only to the DNA in the form of a dimer. CI is produced as a monomer, and binding to DNA goes through a two step process:

$$(\mathrm{CI})_\mathrm{M} + (\mathrm{CI})_\mathrm{M} \rightleftharpoons \mathrm{CI} \quad \text{with} \quad K_\mathrm{D} = \frac{[(\mathrm{CI})_\mathrm{M}]^2}{[\mathrm{CI}]} \tag{7.6}$$

with a dimerization constant $K_\mathrm{D}$, followed by the association

$$\mathrm{CI} + \mathrm{O} \rightleftharpoons \mathrm{CIO} \text{ with a bound fraction } \frac{[\mathrm{CIO}]}{[\mathrm{O}_{\mathrm{total}}]} = \frac{[\mathrm{CI}]}{K + [\mathrm{CI}]} \tag{7.7}$$

where $K = [\mathrm{CI}][\mathrm{O}]/[\mathrm{CIO}]$. Expressed in terms of the monomer concentration the bound fraction is

$$\frac{[\mathrm{CIO}]}{[\mathrm{O}_{\mathrm{total}}]} = \frac{[(\mathrm{CI})_\mathrm{M}]^2}{K \cdot K_\mathrm{D} + [(\mathrm{CI})_\mathrm{M}]^2} \tag{7.8}$$

which now gives half occupancy at a free monomer concentration

$$[(\mathrm{CI})_\mathrm{M}] \approx \sqrt{K \cdot K_\mathrm{D}} \tag{7.9}$$

Before plotting the behavior versus concentration one should be aware that in the above expression the $(\mathrm{CI})_\mathrm{M}$ represents the free concentration of CI monomers. If we want to express everything as a function of total concentration we have to solve Eq. (7.6):

$$\frac{2[(\mathrm{CI})_\mathrm{M}]^2}{[\mathrm{CI}_{\mathrm{total}}] - [(\mathrm{CI})_\mathrm{M}]} = K_\mathrm{D} \tag{7.10}$$

where the total concentration of CI molecules is given by the contribution from the free dimers [CI] plus the free monomers $[(\mathrm{CI})_\mathrm{M}]$, $[\mathrm{CI}_{\mathrm{total}}] = 2[\mathrm{CI}] + [(\mathrm{CI})_\mathrm{M}]$, when

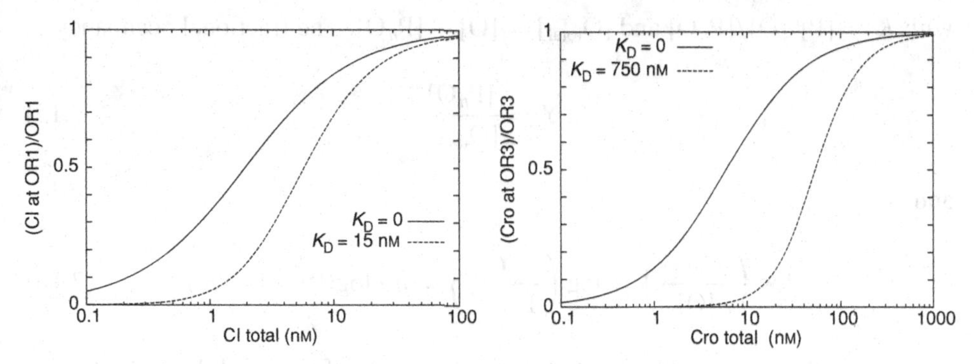


Figure 7.11. Left: occupancy of OR1 as a function of total CI concentration. The two curves illustrate the effect of the finite dimerization constant of CI. Dimerization makes the transition between on and off occur in a narrower CI interval. Right: a similar plot for Cro-OR3 association, demonstrating that cooperativity is important on both sides of the switch.

we ignore the few CI dimers bound to the operator. This gives:

$$(\mathrm{CI})_{\mathrm{M}} = -\frac{K_{\mathrm{D}}}{4} + \frac{K_{\mathrm{D}}}{4}\sqrt{1 + \frac{8}{K_{\mathrm{D}}}[\mathrm{CI}_{\mathrm{total}}]} \sim [\mathrm{CI}_{\mathrm{total}}] \tag{7.11}$$

where the last approximation requires small total concentrations $[\mathrm{CI}_{\mathrm{total}}] \ll K_{\mathrm{D}}$, where most of CI is monomers, and dimerization thus represents a barrier to the operator bound CI state. Depending on the value of total CI concentration the fractional occupancy of an operator is accordingly:

$$\frac{[\mathrm{CIO}]}{[\mathrm{O}_{\mathrm{total}}]} \approx \frac{[\mathrm{CI}_{\mathrm{T}}]^2}{K\,K_{\mathrm{D}} + [\mathrm{CI}_{\mathrm{total}}]^2} \quad \text{for} \quad [\mathrm{CI}_{\mathrm{total}}] \ll K_{\mathrm{D}} \tag{7.12}$$

$$\frac{[\mathrm{CIO}]}{[\mathrm{O}_{\mathrm{total}}]} \approx \frac{[\mathrm{CI}_{\mathrm{total}}]/2}{K + [\mathrm{CI}_{\mathrm{total}}]/2} \quad \text{for} \quad [\mathrm{CI}_{\mathrm{total}}] \gg K_{\mathrm{D}} \tag{7.13}$$

We leave Eq. (7.13) to be proven by the reader. In general we obtain cooperative effects when concentrations are smaller that $K_{\mathrm{D}}$, whereas high concentration implies that all CI effectively act as if they are always dimers. The maximum cooperativity is thus obtained when $K_{\mathrm{D}}$ is larger than $K$, which indeed is the case for the control of the OR complex (see Fig. 7.11).

In general one may consider an $n$th-order process of $n$ identical proteins P binding simultaneously to a molecule O:

$$n\mathrm{P} + \mathrm{O} \leftrightarrow \mathrm{P}_n\mathrm{O} \tag{7.14}$$

with $K = [P]^n[O]/[P_nO]$ and $[O_{total}] = [O] + [P_nO]$. The fractional occupancy

$$Y = \frac{[P_nO]}{[O_{total}]} \tag{7.15}$$

and

$$\log\left(\frac{[P_nO]}{[O]}\right) = \log\left(\frac{Y}{1-Y}\right) = n \cdot \log(P) - \log(K) \tag{7.16}$$

allow us to determine $n$ provided we can measure the functional dependence as a function of free P concentration [P]. The Hill plot of $\log(P_nO/[O])$ versus $\log(P)$ determines a slope $h$ that is called the Hill coefficient. A Hill coefficient $h > 1$ signals a cooperative process. The larger the Hill coefficient, the sharper (more cooperative) is the transition from off to on.

Let us now return to the CI binding to OR in the λ-phage switch. In the right-hand part of Fig. 7.12, we show a Hill plot for the process where the CI first dimerize, and then bind to a single operator. The two plotted curves show the behavior for moderate and relatively strong dimerization, compared with the binding to the operator.

In biology, very large Hill coefficients indeed exist: for example, the chemotaxis motor in *E. coli* responds to changes in the phosporylated molecules CheYp with a Hill coefficient $h \sim 11$.

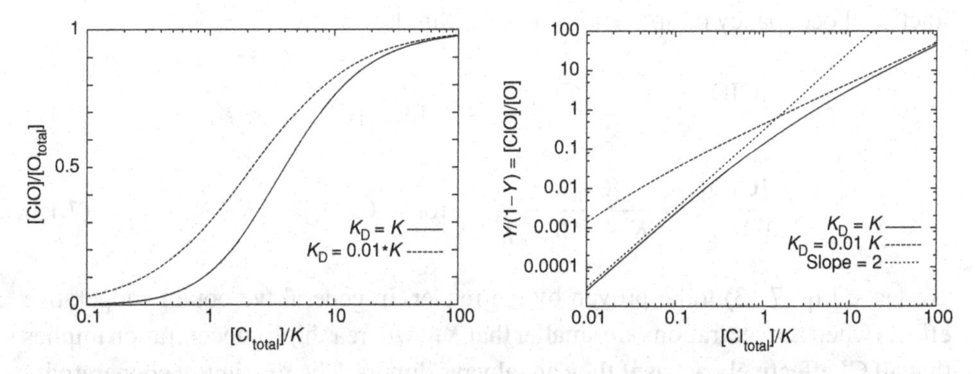


Figure 7.12. Fractional occupancy $Y = [CIO]/[O_t]$ of a single operator, where the CI molecules first dimerize, and then bind to the operator. When the dimerization constant $K_D$ is much smaller than $K$, the reaction is effectively first order. On the other hand, when $K_D > K$, the occupancy curve switches in a narrower concentration interval. On right panel we show the usual Hill plot of $Y/(1-Y) =$ [CIO]/[O], here again versus the total concentration of CI. This plot is on a log–log scale, thereby determining a slope called the Hill coefficient. In principle this Hill coefficient is usually defined as the slope at half saturation.

### *Questions*

(1) If a living bacterium had diameter 1 m, what would one expect for the typical binding affinities between proteins and an operator?

(2) Consider a titration experiment where one varies CI concentration (with fixed operator concentration) in order to probe the first-order chemical reaction $\mathrm{CI} + \mathrm{O} \leftrightarrow \mathrm{CIO}$. Compare the fraction of bound [CIO] to [O] as a function of free [CI] and $[\mathrm{CI_{total}}]$, respectively, for a case where $[\mathrm{O_{total}}] = 10K$. Repeat the comparison when $[\mathrm{O_{total}}] = K/10$.

## Chemistry as statistical mechanics

One molecule in a bacterial volume $V$ corresponds to a concentration

$$\rho = 1/V \approx 10^{15} l^{-1} = 10^{15}(1\mathrm{M}/6 \times 10^{23}) = 1.5 \times 10^{-9}\mathrm{M} \sim 1\ \mathrm{nM} \qquad (7.17)$$

where M is molar (moles per liter). We now repeat some elementary chemistry inside such a volume, in order to emphasize the relation between simple counting and chemical equilibrium. In keeping with the character of the book we choose to go through the basic derivations from a statistical-mechanical point of view, because in this approach the role of entropy is more transparent. Also this approach is easily generalized to more complex situations, as we will see in the subsequent applications.

Consider, for example, CI binding to DNA. The probability that a molecule is bound depends on the (Gibbs) free energy difference between the bound and unbound states. Call this difference $\Delta G' = G(\text{bound}) - G(\text{unbound})$. When $\Delta G'$ is negative the "on" state has lower free energy than the "off" state. Remember that the statistical weight (=unnormalized probability) of a state is given by the number of ways the state can be realized, multiplied by exp(−energy of state/$T$); see Appendix. Now we count by considering only two states for the $N$ molecules: an "off state" where all $N$ molecules are free, and an "on state" with $N - 1$ free molecules and 1 CI bound to the DNA. The statistical weight of the case with one molecule bound is then

$$Z(\text{on}) = \frac{1}{(N-1)!}\left(\int_V \int \frac{\mathrm{d}^3 r \mathrm{d}^3 p}{h^3} \mathrm{e}^{-p^2/(2mk_{\mathrm{B}}T)}\right)^{N-1} \mathrm{e}^{-\Delta G'/k_{\mathrm{B}}T} \qquad (7.18)$$

whereas the statistical weight of having no molecules bound to the site is

$$Z(\text{off}) = \frac{1}{N!}\left(\int_V \int \frac{\mathrm{d}^3 r \mathrm{d}^3 p}{h^3} \mathrm{e}^{-p^2/(2mk_{\mathrm{B}}T)}\right)^{N} \qquad (7.19)$$

Here the multiple integrals count all possible positions $r$ and moment $p$ of one molecule in the cell of volume $V$. Division by Planck's constant $h$ takes into account the discreteness of phase space imposed by quantum mechanics. Division by $(N - 1)!$ and $N!$, respectively, counts all permutations of molecules and is included because these molecules are indistinguishable. This is necessary in order to avoid double counting of any of the identical free molecules in the cell volume $V$. In principle there are other parts of the system, for example the water that surrounds the molecules. However, one typically assumes that everything else besides the $N$ molecules is independent of the state of the system (whether on or off), and therefore gives the same contribution to all weights. Thus one simply forgets it in the counting, but may include it later on by renormalizing the free energies.

Integrating (and using $\int_{-\infty}^{\infty} dy \exp(-y^2) = \sqrt{\pi}$), one obtains the result that the statistical weights of the state where one of the particles is bound (and $N-1$ are free) is:

$$Z(\text{on}) = \frac{\left(V[2mk_B T\pi/h^2]^{3/2}\right)^{N-1}}{(N-1)!} \exp\left(-\frac{\Delta G'}{k_B T}\right) \propto c^{N-1} \rho^{-(N-1)} \exp\left(-\frac{\Delta G'}{k_B T}\right) \tag{7.20}$$

where $\rho = N/V$ and $c = (2mk_B T\pi/h^2)^{3/2}$. The statistical weight that all $N$ particles are free is

$$Z(\text{off}) = \frac{\left(V[2mk_B T\pi/h^2]^{3/2}\right)^{N}}{N!} \propto c^N \rho^{-N} \tag{7.21}$$

In the last step we used Sterling's formula $x! \approx (x/e)^x$ and the approximation $(N-1)/N \approx 1$. This assumes that the density $\rho$ of the free particles does not change significantly when one molecule binds. The system defined by the $N$ particles in $V$ with either one bound (on) or none bound (off) has now a total statistical weight (partition function) of

$$Z = Z(\text{on}) + Z(\text{off}) \tag{7.22}$$

and the probability of the on state is

$$P_{\text{on}} = \frac{Z(\text{on})}{Z} \quad \text{and the ratio} \quad \frac{P_{\text{on}}}{P_{\text{off}}} = \frac{Z(\text{on})}{Z(\text{off})} = \frac{\rho}{c} e^{-\Delta G'/k_B T} \tag{7.23}$$

The counting determines the statistical weights associated with, for example, the chemical reaction of the DNA binding protein CI to its specific operator O:

$$\text{CI} + \text{O} \leftrightarrow \text{CIO} \tag{7.24}$$

Usually this is characterized by the dissociation constant

$$K = \frac{[\text{CI}][\text{O}]}{[\text{CIO}]} \quad \text{where} \quad [\text{O}] + [\text{CIO}] = [\text{O}_t] \tag{7.25}$$

where [O] is the free operator concentration, [CIO] is the bound operator concentration and $[O_t] = 1/V$ is the total operator concentration associated with one specific operator located in the bacterial volume.

Since the given single operator O can only be on or off, the probability of being on is $P_{\text{on}} = [\text{CIO}] \cdot \text{V}$, and off is $P_{\text{off}} = [\text{O}] \cdot \text{V}$, where $P_{\text{on}} + P_{\text{off}} = [O_t] \cdot V = 1$. From chemistry

$$\frac{P_{\text{on}}}{P_{\text{off}}} = \frac{[\text{CIO}]}{[\text{O}]} = \frac{[\text{CI}]}{K} \tag{7.26}$$

which is consistent with Eq. (7.23) when we set $[\text{CI}] = \rho$ and identify

$$K = c \cdot \mathrm{e}^{\Delta G'/k_B T} = [1\ \text{M}] \cdot \mathrm{e}^{\Delta G/k_B T} \tag{7.27}$$

where the last equality sign adopts the usual convention for $\Delta G$ by measuring the pre-factor in moles per liter (M). Notice that we occasionally use the association constant $K_A = 1/K$. Looking back to Eq. (7.20) the meaning of $\Delta G$ is that it reflects the correct statistical weight for the "on" state if $N = N_A$ (equals Avogadro's number $= 6 \times 10^{23}$) and $V = 1$ l in Eq. (7.20). In the following we will not write the [1 M] in our equations, thus always implicitly measuring all concentrations in moles per liter.

As a standard, in the following we will measure all statistical weights normalized in terms of $Z_{\text{off}}$. Thus Eq. (7.20) is

$$Z(\text{on}) = [\text{CI}]\mathrm{e}^{-\Delta G/k_B T} \quad \text{and} \quad Z(\text{off}) = 1 \tag{7.28}$$

At this point it is worthwhile to dwell a little on the meaning of the free energy difference: $\Delta G$ counts the free energy difference between the bound state and the free state when the concentration of CI in the free state is 1 M. If the concentration is smaller than 1 M, as indeed it is in the cell, then the entropy gain by going from on to off is larger, because the volume per molecule in the free state is larger. The free energy difference between the bound and the free state of one molecule at density [CI] can be deduced from Eq. (7.28) if we express [CI] in molar terms

$$\Delta G^* = k_B T\,(\ln(Z_{\text{off}}) - \ln(Z_{\text{on}})) = \Delta G - k_B T \ln([\text{CI}]) \tag{7.29}$$

corresponding to the fact that lower density favors the free state (remember that [CI] means [CI]/1 M).

Finally we want to conclude the association between statistical mechanics and chemical equilibrium by addressing the operational range of possible $K$ (and $\Delta G$ values). From Eq. (7.25) we obtain a bound fraction

$$\frac{[\text{CIO}]}{[\text{O}_t]} = \frac{[\text{CI}]}{[\text{CI}] + K} = \frac{Z_{\text{on}}}{Z_{\text{off}} + Z_{\text{on}}} = \frac{Z_{\text{on}}}{1 + Z_{\text{on}}} \tag{7.30}$$

where $Z_{\text{on}} = [\text{CI}]/K$ and $Z_{\text{off}} = 1$ (see Eq. (7.28)) and occupancy therefore switches from 0 to 1 when $[\text{CI}] \sim K = \mathrm{e}^{\Delta G/k_B T}$. Typically the system should be able to switch from on to off, or back, by changing the number of molecules from around 1/cell to around 100/cell. Because one molecule in $V$ corresponds to a concentration of 1 nM we expect functional $K$s of about $10^{-8\pm1}$ M. As $K = [1\ \text{M}] \cdot e^{\Delta G/k_B T}$ then the binding free energies $\Delta G \approx -14\,k_B T \rightarrow -18\,k_B T$. In the biological literature, free energies are often given in units of kcal/mol. Because $1\ k_B T = 0.617$ kcal/mol at room temperature, one expects $\Delta G \sim -10$ kcal/mol, as indeed is the case for typical binding between regulatory proteins and DNA, including CI and OR.

For the λ-phage, the system uses only a dimer to bind to the DNA. In order to increase further the sharpness of the switch several CI dimers bind cooperatively (i.e. strengthening each other) to the closely located operator sites on the DNA. We now analyze this sequence of reactions employing the statistical-mechanical analysis of Darling *et al.* (2000) (see also Fig. 7.13).

At OR in λ there are three operator sites, OR1, OR2 and OR3, that are adjacent to each other. Thus there are $2^3 = 8$ possible states of CI occupancy of the operators, and we have to know the $\Delta G$ for all possible occupation of these states (plus some states, which here we ignore, associated with Cro and RNAP bindings). Each of these states $s = (s_1, s_2, s_3)$ where $s_\sigma = 0$ if operator with number $\sigma$ is empty, and $s_\sigma = 1$ if OR$\sigma$ is occupied by CI dimer. Thus for a state characterized by $n_M$ free monomers, $n_D$ free dimers and an operator state $s$ where $i = i(s)$ dimers are bound

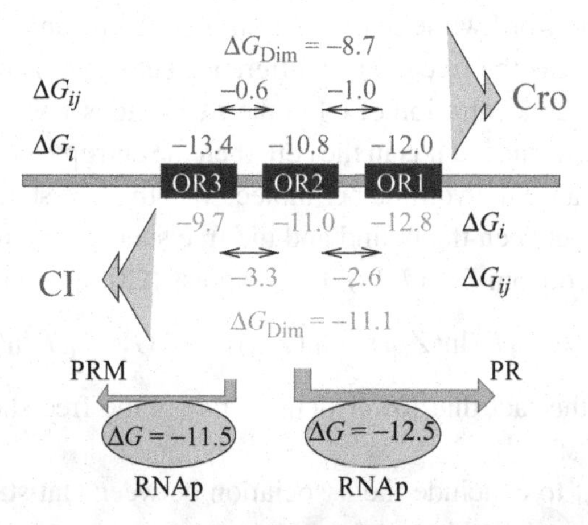


Figure 7.13. Illustration of data on $\Delta G$ (in kcal/mol) involved in OR regulation from Darling *et al.* (2000). The top part shows Cro bindings, with $G_i$ being individual bindings, and $\Delta G_{ij}$ being additional cooperative bindings. The lower part shows corresponding numbers for CI and RNAp respectively. Total $\Delta G$ for a state is obtained by summing individual contributions. Notice that the cooperative bindings are pairwise exclusive, such that only one of them counts if there is CI or Cro on all three operator sites. Also notice that when RNAp binds to PR there is no additional space to bind a CI or Cro on either OR1 or OR2. Free RNAp concentration in *E. coli* is about 30 nM. All energies are relative to the non-occupied reference state.

to operators, the exact statistical weight is

$$Z(s, n_{\mathrm{M}}) = \frac{V^{n_{\mathrm{D}}}}{n_{\mathrm{D}}!} \cdot \frac{V^{n_{\mathrm{M}}}}{n_{\mathrm{M}}!} \cdot \frac{1}{K_{\mathrm{D}}^{n_{\mathrm{D}}+i(s)}} \cdot \mathrm{e}^{-\Delta G(s)/k_{\mathrm{B}}T} \tag{7.31}$$

where $n_{\mathrm{D}} = (N - n_{\mathrm{M}} - 2i)/2$ because the number of free dimers in the cell is fixed by the conservation requirement: $N = n_{\mathrm{M}} + 2n_{\mathrm{D}} + 2i$ with $N$ the total numbers of CI proteins in the cell. Notice that each of the $n_{\mathrm{D}}$ dimers contributes with its dimerization binding free energy through $K_{\mathrm{D}} = \mathrm{e}^{-\Delta G_{\mathrm{D}}/k_{\mathrm{B}}T}$, whereas only the dimers bound to the operators contribute to $\Delta G(s)$. For CI the dimerization constant is $K_{\mathrm{D}} = 5 \times 10^{-7}\,\mathrm{M}$. $V$ is counted in $\mathrm{M}^{-1}$. In Eq. (7.31) $V^{n_{\mathrm{D}}}/n_{\mathrm{D}}!$ counts the number of states of the free dimers and $V^{n_{\mathrm{M}}}/n_{\mathrm{M}}!$ counts the number of states of the free monomers. Again the total partition function can be written as a sum over all states $(s, n_{\mathrm{M}})$ that the $N$ molecules can be in: $Z = \sum Z(s, n_{\mathrm{M}})$.

The probability of a state $s$ is then given by the appropriately normalized ratio

$$P(s) = \frac{\sum_{n_{\mathrm{M}}} Z(s, n_{\mathrm{M}})}{\sum_{s, n_{\mathrm{M}}} Z(s, n_{\mathrm{M}})} \tag{7.32}$$

For large $N$, say $N > 30$, the above calculation can be simplified by considering that the fraction of molecules bound to the operators is only a small perturbation on the monomer–dimer equilibrium. Then we can calculate [CI], from Eq. (7.6), and express approximately the statistical weights for the $s$ state in terms of dimer density [CI]:

$$Z(s) = [\mathrm{CI}]^{i(s)} \mathrm{e}^{-\Delta G(s)/k_{\mathrm{B}}T} \tag{7.33}$$

which is normalized such that the statistical weight of the state where nothing is bound is $Z(s = 0) = 1$. The concentration-dependent factor reflects the entropy loss in going from a freely moving dimer to a bound dimer. We have seen it already in Eqs. (7.18)–(7.28): in the case of $i(s)$ repressors bound, the exponent in Eq. (7.18) would be $N - i(s)$ instead of $N - 1$; correspondingly the concentration-dependent factor in Eq. (7.20) would be $\rho^{-(N-i(s))}$, and in Eq. (7.28) it would be $[\mathrm{CI}]^{i(s)}$ instead of [CI].

If we include both CI and possible RNAp states then Eq. (7.33) generalizes to (Shea and Ackers, 1985):

$$Z(s) = [\mathrm{RNAp}]^{j(s)} [\mathrm{CI}]^{i(s)} \mathrm{e}^{-\Delta G(s)/k_{\mathrm{B}}T} \tag{7.34}$$

where $j(s)$ is the number of bound RNAp molecules in state $s$. Similarly one can generalize to include the possibility of Cro binding to the operators (Darling *et al.*, 2000). This formalism is summarized in Fig. 7.14. The chemical binding free energies for CI and Cro to OR can be found in Fig. 7.13.

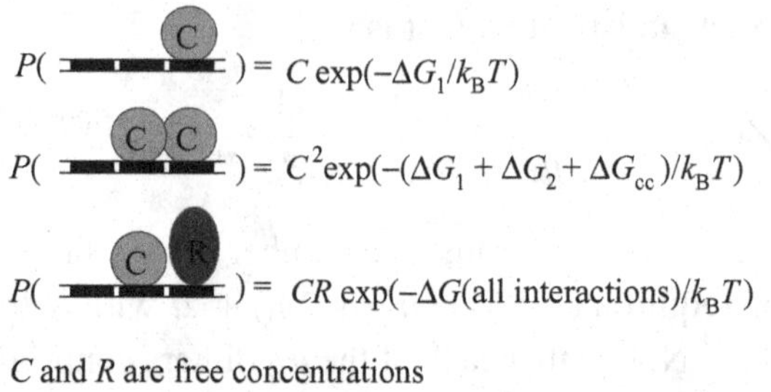

$P(\text{C on operator}) = C \exp(-\Delta G_1/k_B T)$

$P(\text{C C on operators}) = C^2 \exp(-(\Delta G_1 + \Delta G_2 + \Delta G_{cc})/k_B T)$

$P(\text{C and R on operators}) = CR \exp(-\Delta G(\text{all interactions})/k_B T)$

$C$ and $R$ are free concentrations

Figure 7.14. Statistical mechanics of genetic regulation. Any state of occupancy of operators can be assigned a statistical weight, or a probability, that is almost exactly proportional to free concentrations of the involved molecules in the cell multiplied by some binding energies $\Delta G$ (see also text).

Now all equilibrium properties to be calculated. In particular we can calculate the probability for any subset of the states to be bound as a function of the total amount of CI in the cell.

As an example let us calculate promoter activity of the two promoters for CI and Cro in a cell where there is no Cro. First let us consider the *cro* promoter PR. This promoter can only be accessed by RNAp when both OR2 and OR1 are free. Additionally in order for the promoter to be active, RNAp must first be bound. The probability for the corresponding state where RNAp is bound to PR is

$$P(0, \text{RNAp}) + P(1, \text{RNAp}) = \frac{Z(0, \text{RNAp}) + Z(1, \text{RNAp})}{\sum_s Z(s)} \tag{7.35}$$

where 0 means absence of CI on OR3 whereas 1 means CI bound to OR3. To calculate the $Z$s we make use of Eq. (7.34); the corresponding values of $\Delta G$ are given in Fig. 7.13. For example

$$Z(1, \text{RNAp}) = [\text{RNAp}][\text{CI}] \exp\left(-\frac{\Delta G(1, \text{RNAp})}{k_B T}\right) \tag{7.36}$$

$$= [\text{RNAp}][\text{CI}] \exp\left(-\frac{22.2 \text{ kcal/mol}}{0.62 \text{ kcal/mol}}\right) \tag{7.37}$$

where we inserted the appropriate binding energy from Fig. 7.13 and the room temperature $k_B T = 0.617$ kcal/mol. The right-hand panel in Fig. 7.15 shows the Cro promoter activity as function of number of CI calculated in this way. As CI bound to either OR1 or OR2 prevents RNAp from binding to PR, the probability of initiating PR decreases strongly with CI concentration.

Similarly one can calculate the activity of the CI promoter PRM, which is shown in the left-hand panel of Fig. 7.15. The CI promoter PRM is weak, but is strengthened by a factor of about 5–10 when OR2 is occupied by a CI dimer and OR3 is free. Thus PRM increases by such a factor when CI concentration becomes sufficient

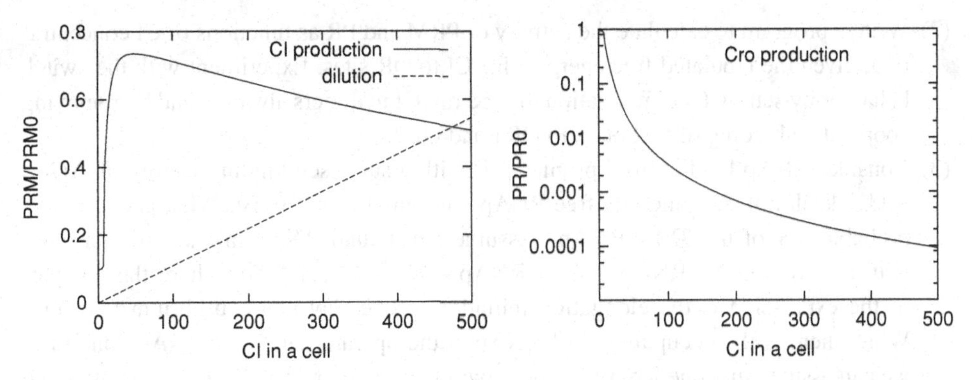


Figure 7.15. PRM and PR activity as a function of number of CI molecules in an *E. coli* cell that is assumed to have zero Cro molecules. PRM activity is calculated to be proportional to RNAp polymerase bound to OR3 ($= \sum_{i=0}^{1} \sum_{j=0}^{1} P(\text{RNAp}, i, j)$), whereas PR activity is proportional to one RNAp molecule occupying PR that covers both OR1 and OR2 ($= \sum_{i=0}^{1} P(i, \text{RNAp})$). RNAp binding to PRM is set to −11.5 kcal/mol, and to PR −10.5 kcal/mol, and CI is here assumed to bind only to operator DNA.

to make CI on OR2 likely. When OR3 is occupied by either CI or Cro, PRM is blocked. Further, binding of CI dimers to both OR1 and OR2 involves an additional binding between the CI dimers, as seen from the additional binding of the $s = (011)$ state compared with the sum of the (010) and the (001) states; see Fig. 7.13. All this is included in the PRM activity as a function of state $s$, and the probability of state $s$ as a function of CI. The PRM activity curve has in fact been measured, and Fig. 7.15 differs from the experiments in several ways, owing to effects that will be discussed in the next sections.

PR and PRM activities are not only functions of CI, but also of Cro in the cell. When Cro dimers are present in the cell, they bind first to OR3 (with $\Delta G(2, 0, 0) = -13.4$ kcal/mol), thereby blocking transcription of $cI$. In this way a lytic state can in principle, stabilize itself. It would, however, not lead to a very large amount of Cro, since Cro binds to OR1 and OR2, and thereby partially represses itself. The CI-dominated state is the lysogenic state with a total of about 250 CI molecules in an *E. coli* cell (Reichardt & Kaiser, 1971), whereas an artificially confined lytic state (anti-immune state) has about 300 Cro molecules in a cell (Reinitz & Vaisnys, 1990, from Pakula *et al.*, 1986).

## *Questions*

(1) If CI is cleaved by RecA, its dimerization is prevented. What concentration of CI monomers is needed for maintaining a similar probability of having OR1 occupied as 100 nM of CI dimers does? Assume that monomer–OR3 binding is half of the dimer CI–OR1 binding (= −12.8 kcal/mol).

(2) Write a program to calculate the activity of PRM and PR as functions of CI concentration, given the tabulated free energies for CI to OR sites. Experiment with the switch behavior by setting CI dimerization to $-\infty$ (all CI in dimers always), and by removing cooperative binding of CI bound to OR1 and OR2.

(3) Consider RNAp binding to a promoter $\mathcal{P}$ with a supposed binding energy of $\Delta G = -11.5$ kcal/mol. Assume that free RNAp concentration is 30 nM. What is occupancy probability $\theta$ of the $\mathcal{P}$ by RNAp? Assume that bound RNAp initiates transcription with rate $k_f = 0.1$/s: $\text{RNAp} + \mathcal{P} \leftrightarrow \text{RNAp} - \mathcal{P} \rightarrow \mathcal{P} \ldots \text{RNAp}$ (where the last part of the expression is the elongation initiation, a one-way non-equilibrium reaction). What, then, is the occupancy $\theta$ of RNAp at the operator site? For RNAp "on" rates we can assume that one RNAp in 1 $\mu\text{m}^3$ will have an on rate $k_{\text{on}} = 0.1$/s. (Hint: argue with $\mathrm{d}\theta/\mathrm{d}t = k(\text{on}) \cdot [\text{RNAp} - \text{free}] \cdot ([\mathcal{P}] - \theta) - k(\text{off}) \cdot \theta - k \cdot \theta = 0$, and use that $\mathrm{e}^{\Delta G/k_B T} = k(\text{off})/k(\text{on})$).

## Non-specific binding to DNA

Gene control involves transcription factors, i.e. proteins that bind to specific DNA sites and thereby activate or repress the transcription machinery at that point. Thus binding of a particular protein to a specific site is needed. In practice it is possible to obtain quite high specific binding energies, but not arbitrarily high: if specific binding is very high, then often the protein will also bind substantially to non-specific DNA sites.

The non-specific binding can be taken into account by using Eq. (7.33) for $Z(s)$, but with the free dimer concentration [CI] given by a modified conservation equation. In the limit where we ignore depletion due to the few dimers bound to the operators, the equation reads:

$$N = n_M + 2n_D + 2n_D \cdot L_{\text{DNA}} \cdot \mathrm{e}^{-\Delta G_u/k_B T} / V \quad (7.38)$$

$$Z(s) = \left(\frac{n_D}{V}\right)^{i(s)} \mathrm{e}^{-\Delta G(s)/k_B T} \quad (7.39)$$

where $i(s)$ is the number of dimers bound to operators in occupancy state $s$, and $V$ is the volume measured in $\text{M}^{-1}$. Here $n_D$ is the number of free dimers in the cell volume $V$. In the above equation $L_{\text{DNA}}\mathrm{e}^{-\Delta G_u/k_B T}$ is the total contribution from non-specific binding in terms of the length of DNA in the *E. coli*, $L_{\text{DNA}} \sim 5 \times 10^6$ base pairs (bp), and the typical non-specific binding per base pair, $\Delta G_u$. $\Delta G_u$, is the standard free energy corresponding to 1 M concentration, and the factor $1/V \sim 1$ nM converts to the corresponding concentrations in Eq. (7.38). The non-specific binding becomes significant when

$$\mathrm{e}^{-\Delta G_u/k_B T} \cdot L_{\text{DNA}}/V \sim 1 \quad (7.40)$$

which with $L_{\rm DNA} \sim 10^7$ happens when $\Delta G_{\rm u} \sim -3$ kcal/mol. Thus when a DNA binding protein binds stronger than 3 kcal/mol to non-specific sites, then the $10^7$ binding sites on the DNA win over the about $10^9$ unbound states in the cell. To be more detailed, ignoring the specific binding the $N$ CI molecules would be partitioned into

$$N = n_{\rm M} + 2n_{\rm D} \quad \text{without} \quad \text{DNA} \tag{7.41}$$

$$N = n_{\rm M} + 2n_{\rm D}\left(\frac{V + L_{\rm DNA}{\rm e}^{-\Delta G_{\rm u}/k_{\rm B}T}}{V}\right) \quad \text{with} \quad \text{DNA} \tag{7.42}$$

corresponding to having an additional statistical weight of $(L_{\rm DNA}/V){\rm e}^{-\Delta G_{\rm u}/k_{\rm B}T}$ for dimers sequestered by non-specific binding to the DNA. Thus we can simply replace

$$V \rightarrow V + L_{\rm DNA}{\rm e}^{-\Delta G_{\rm u}/k_{\rm B}T} \tag{7.43}$$

in Eq. (7.38) and subsequently ignore non-specific binding. The overall effect of non-specific binding to DNA is to provide an additional volume for dilution of the molecules of size $L_{\rm DNA}{\rm e}^{-\Delta G_{\rm u}/k_{\rm B}T}$, where $L_{\rm DNA}$ is the number of binding sites on the DNA that should be compared with the volume $V \sim 6 \times 10^8 \times (\text{cell volume}/\mu\text{m}^3)$. The effect of non-specific CI–DNA binding on PRM and PR activity is illustrated in Fig. 7.16.

Finally we would like to stress that non-specific binding is not only a cost that has to be accounted for. It is usefull because it adds robustness to the system. In

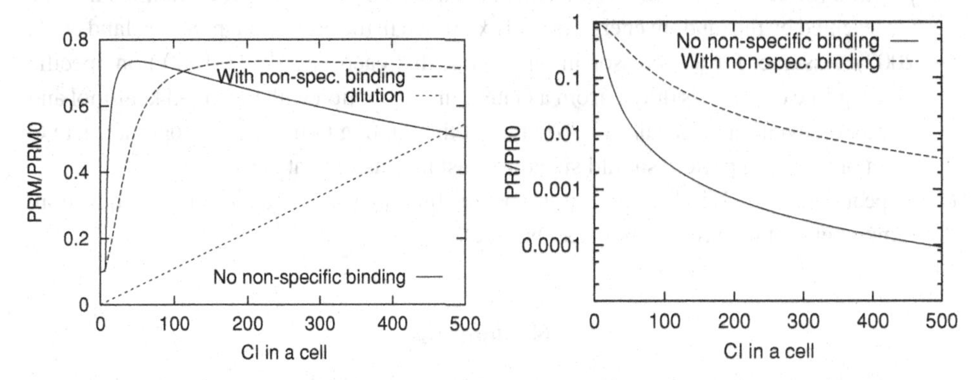


Figure 7.16. PRM and PR activity as functions of numbers of CI proteins in a bacterial cell that is assumed to have zero Cro molecules. The figure illustrates the effect of a non-specific CI–DNA binding of $\Delta G_{\rm u} = -4$ kcal/(mol · bp), by comparison with behavior in Fig. 7.15 where we ignored non-specific binding. Non-specific binding is found to weaken the response to, and eventual changes in, CI concentration. The dilution line shows the rate of CI depletion due to *E. coli* growth and division.

fact, if nearly all transcription regulators are anyway bound non-specifically to the DNA, then an overall change of protein–DNA binding constant by, for example, a factor 1000 will not change the relative strength of the operator to non-specific binding. Thus the non-specific binding effectively buffers the specific bindings against changes in protein–DNA binding constants. In contrast if there were no non-specific binding, then the occupancy of the operator state would be solely given by a free energy of specific binding, which would greatly increase if salt concentration decreased. As most transcription factors are bound non-specifically, a change in salt concentration makes a proportional change in both specific and non-specific binding, and the occupancy of operator sites becomes less salt dependent. For *E. coli* changes in intrinsic salt concentration are commonly induced as a response to a changed external osmotic pressure on the cell.

### *Questions*

(1) The linear dimensions of a human cell are about 10 times those of a bacterium. The human DNA consists of $3 \times 10^9$ base pairs. If one (wrongly!) assumes that the human cell can be viewed as one big bag of proteins and DNA, what would the non-specific binding be that makes it equally likely for a protein to be found on the DNA as in the cell volume?

(2) Consider a DNA binding protein in an *E. coli* cell that binds to 90% of its DNA with $\Delta G = -3$ kcal/mol and to 10% of the DNA with $\Delta G = -5$ kcal/mol. What is the probability that such a protein will be free in the cell?

(3) The interaction between a DNA binding protein and the DNA can be written as a sum of individual interactions between amino acids and the base pairs at the corresponding position (Stormo & Fields, 1998). Thus the ensemble of non-specific bindings may be represented by the random energy model examined in the protein chapter (Gerland *et al.*, 2002). Assume, for a repressor in an *E. coli*, that each of the 5 000 000 non-specific binding free energies is drawn from a Gaussian distribution with mean $-3$ kcal/mol and standard deviation $-2$ kcal/mol. What must the binding to the specific operator site O be in order that a protein should spend at least half its time at O?

(4) Repeat Question (3) if the typical non-specific binding was $+3$ kcal/mol (trick question, remember that the protein may also be free).

## DNA looping

Given the huge number of non-specific DNA sites in a cell, they may often outcompete the specific DNA binding to a particular operator. The biological remedy for this is to increase cooperativity, i.e. to build operator sites that demand the simultaneous binding of several proteins, such that each protein is bound both to the DNA and other proteins. Formally, if we have two proteins that each bind with energy $\Delta G_s$ to a specific site, an energy $\Delta G_p$ for two proteins that bind to each other, and an energy $\Delta G_u$ for the proteins binding to a non-specific site, then the competition between specific and non-specific sites for two repressor dimers is governed by

$$\Delta G(\text{specific}) = 2\Delta G_s + \Delta G_p \tag{7.44}$$

$$\Delta G(\text{non-specific}) = 2\Delta G_u - 2k_B T \ln(L_{DNA}) \tag{7.45}$$

The ln term is the entropy associated with the number of states accessible to a non-specifically bound repressor ($= L_{DNA}$). In the above equation we have ignored the much smaller contribution from non-specific binding where both repressors bind to each other. This can be ignored because this binding implies an entropy loss corresponding to an additional $T \ln(L_{DNA}) \sim 8$ kcal/mol in translational free energy, which is sufficient to overrule the $\sim -3$ kcal/mol binding associated to two CI dimers forming a tetramer.

The difference between specific and non-specific binding is

$$\Delta G(\text{specific}) - \Delta G(\text{non-specific}) = 2\Delta G_s + \Delta G_p - 2\Delta G_u + 2k_B T \ln(L_{DNA}) \tag{7.46}$$

if there is cooperativity. If there is no cooperativity the difference is

$$\Delta G(\text{specific}) - \Delta G(\text{non-specific}) = 2\Delta G_s - 2\Delta G_u + 2k_B T \ln(L_{DNA}) \tag{7.47}$$

and thus any cooperative binding $\Delta G_p < 0$ indeed favors specificity (see Fig. 7.17).

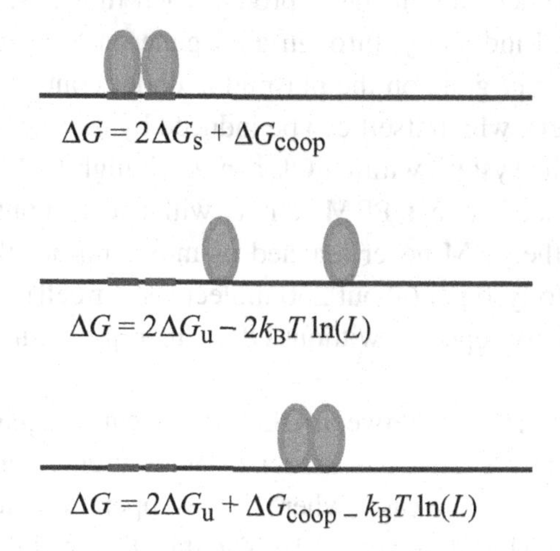


Figure 7.17. Cooperativity as a mechanism to beat non-specific binding. When cooperative binding between two transcription factors $\Delta G_{coop}$ contributes less than the free energy of a transcription factor on all possible non-specific DNA binding sites, then cooperative non-specific binding is not possible. The remaining competition between the upper two scenarios shows that cooperativity helps the specific binding sites. Notice that the same competition takes place when comparing with other non-specific effects, for example proteins in the cell volume, or proteins sequestered by the cytoplasm. In all cases a small cooperative energy favors specific bindings.

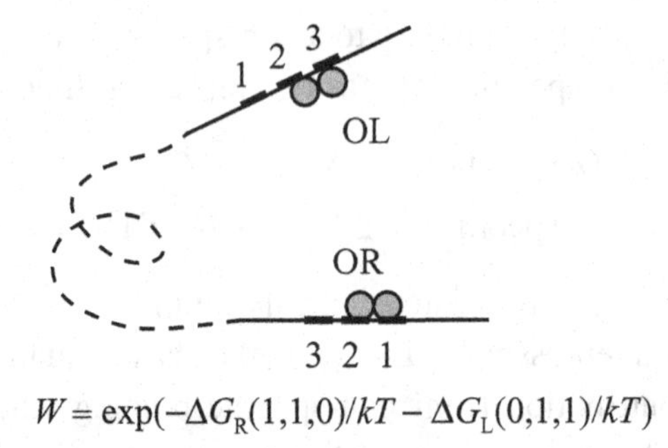


Figure 7.18. OL–OR association, separated by a large loop of DNA.

In the case of λ, we have already seen cooperativity between CI bound to OR1 and CI bound to OR2. It turns out that there are additional mechanisms that add to cooperativity. On the λ-DNA there is also an operator left (OL), ~2.4 kbase to the left of OR, and having almost identical sequences; see Fig. 7.5. OR and OL can interact through DNA looping; see Fig. 7.18. The effect of this loop is to repress PRM and PR activity as quantified by Dodd *et al.* (2001) and Révet *et al.* (2000), respectively.

In the experiment of Dodd *et al.*, the PRM activity is measured through a reporter gene that produces a detectable protein each time PRM is activated. The CI level is controlled indirectly, through a *cI* gene on a plasmid (see Glossary) in the bacterium. The *cI* gene on the plasmid is under control of the lacZ operon (operator + promoter), which itself can be induced by the chemical (IPTG). Dodd *et al.* reported that the system without OL can reach high PRM activity, reflecting the factor 10 difference between PRM activity with and without CI bound to OR2. However, with OL the PRM never reached its maximum level. At CI concentration corresponding to lysogen (about 250 molecules per cell), PRM with OL only had 40% of the activity reported without OL. This repression must involve DNA looping.

This additional interaction between OL and OR can be quantified in terms of some binding energies. That is, we associate binding energies with OR and OL occupation patterns in the two cases where the two operators interact (closed loop) or do not interact (open loop); see Fig. 7.19. The statistical weight of a given binding pattern $s$ is then either

$$Z(\text{open}) = f(C) \cdot e^{-\Delta G_R(s)/k_B T} \cdot e^{-\Delta G_L(s)/k_B T} \tag{7.48}$$

or

$$Z(\text{closed}) = f(C) e^{-\Delta G_R(s)/k_B T} \cdot e^{-\Delta G_L(s)/T} \cdot e^{-\Delta G(\text{complex}, s)/T} \tag{7.49}$$

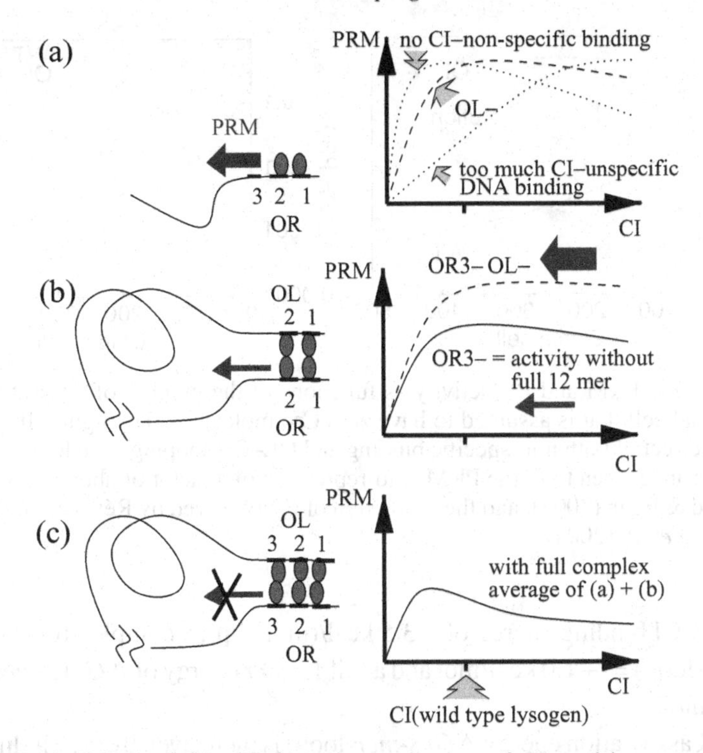


Figure 7.19. Effect of OL–OR loop on production of CI, and how to estimate involved binding affinities by fitting PRM activity. (a) PRM activity without OL–OR complex. This is experimentally observed in an OL-mutant. Given CI–OR bindings, the CI concentration where PRM becomes active is determined by CI non-specific binding $\Delta G_u = -3.5$ kcal/mol. (b) PRM activity without OR3 is experimentally investigated through a mutant where CI does not bind OR3; neither does it bind OR3 and OL. It teaches us that the middle configuration reduces PRM by about 35%, reflecting that RNA polymerase has a smaller "on" rate to PRM when OL–OR 8-*mer* is formed. (c) The full effect of averaging (a) + (b) in normal $\lambda$ (wild type). The decline of PRM with CI is increasingly dominated by the complete repression due to formation of the 12-*mer* complex. The overall decline at large [CI] is fitted by $\Delta G$(12-*mer*-loop) $= -3.0$ kcal/mol. These were analyzed experimentally by Dodd *et al.* (2001), where CI was controlled externally.

depending on whether OL–OR interact (Eq. (7.49)) or not (Eq. (7.48)). Here $f(C)$ is the concentration-dependent factor from Eq. (7.34). $\Delta G$ (complex, $s$) is the net free energy associated to OL–OR bound to each other. It involves both the direct binding energy and the counteracting contribution from entropy cost of bringing OL and OR together. The binding energies depend on the state $s$, which now includes both the OR and the OL occupation pattern.

In Fig. 7.20 we show a calculated profile for PRM and PR that fits the known repression of CI at the normal lysogenic concentration of about 250 molecules per cell. In Fig. 7.19 we illustrated the effect of the key parameters in such a fit: a

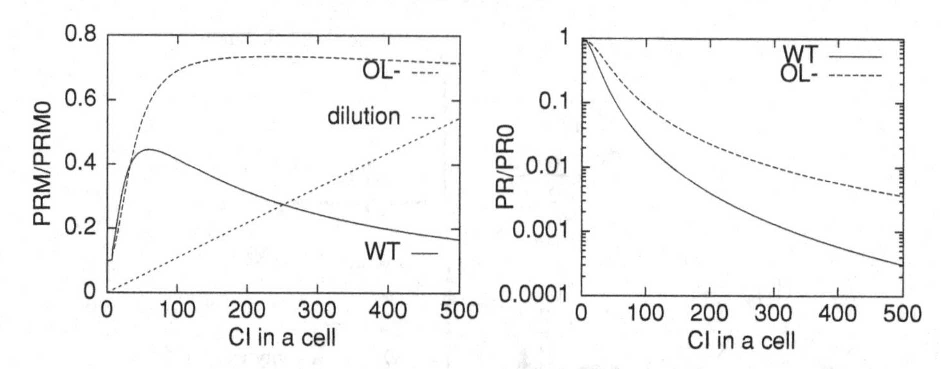


Figure 7.20. PRM and PR activity as functions of the number of CI proteins in a bacterial cell that is assumed to have zero Cro molecules. The figure illustrates the full effect of both non-specific binding and OL–OR looping. The loop binding energies are chosen to fit the PRM auto-repression of a factor of about 3 observed by Dodd & Egan (2002), and the repression of PR observed by Révet *et al.* (2000) and Dodd *et al.* (2004).

non-specific CI binding energy of $-3.5$ kcal/(mol · bp), a complex forming energy $\Delta G(8\text{-}mer\text{-loop}) = -1.0$ kcal/mol and a full 12-*mer* energy of $\Delta G(12\text{-}mer\text{-loop}) = -3.0$ kcal/mol.

The L–R association energy $\Delta G(8\text{-}mer\text{-loop})$ is then given by the binding energy and the entropy associated with closure of the DNA loop:

$$\Delta G(8\text{-}mer\text{-loop}) = \Delta G_8 - T\Delta S(\text{loop}) \tag{7.50}$$

Here $\Delta G_8$ is the energy of two CI dimers on OL interacting with two CI dimers on OR, and $\Delta S$(loop) is the entropic part of the free energy associated with the DNA loop of length given by distance between OL and OR, $l \sim 3$ kb.

The loop entropy $\Delta S$(loop) can be estimated as the ratio of the number of accessible states for a closed loop and a random coil: $v/V_{(\text{OL}-\text{OR})}$. Here $V_{(\text{OL}-\text{OR})}$ is the random coil volume and $v$ is an "interaction volume", the latter is of the order of the volume of an occupied operator site, say $v \sim (5\text{ nm})^3$. In principle there will also be an entropy cost due to rotational confinement from alignment of CI dimers at OL and OR. Also the comparison to a random coil DNA is a simplification; the DNA in the *E. coli* chromosome is in fact supercoiled. However, in the simplest estimate, the entropy cost of loop closure $\Delta S(\text{loop}) = S(\text{bound}) - S(\text{open})$ is given by

$$e^{\Delta S(\text{loop})/k_B} = \frac{v}{V_{(\text{OL}-\text{OR})}} = \frac{v}{\left(\left(\frac{l}{l_k}\right)^{0.6} \cdot l_k\right)^3} \tag{7.51}$$

where $l_k \sim 300$ bp is the Kuhn length of the DNA, $v$ is the volume that OL–OR gets confined to when binding to each other, and $V_{(OL-OR)}$ was the volume spanned by the DNA between OL and OR, ie. $V_{(OL-OR)} \sim R^3$ in terms of its radius of gyration $R \sim l_k(l/l_k)^{3/5}$. As a result

$$\frac{\Delta S}{k_B} \sim \ln\left(\frac{v}{V_{(OL-OR)}}\right) = \ln\frac{v}{l_k^3} - \frac{9}{5}\ln\frac{l}{l_k} \tag{7.52}$$

where a reasonable value for the length-independent term would be $\ln(l_k^3/v) \sim 3\ln(100/5) \sim 8$, because $l_k = 300$ bp $\sim 100$ nm. Thus $\Delta S \sim 13$ and

$$\Delta G(8\text{-}mer\text{-loop}) = \Delta G_8 - T\Delta S(\text{loop}) \tag{7.53}$$

Therefore, using the fitted $\Delta G(8\text{-}mer\text{-loop}) = -1$ kcal/mol and $T\Delta S(\text{loop}) = -8$ kcal/mol, we estimate $\Delta G_8 \approx -9$ kcal/mol. Assuming that each OR–CI–CI–OL binding contributes with equal energy, this tetramerization energy for each pair of CI dimers would then be of the order of $-4.5$ kcal/mol. Thus we would predict that the 12-*mer* complex has an additional $\Delta G$ of $-4.5$ kcal/mol, and $\Delta G(12\text{-}mer\text{-complex}) = -5.5$ kcal/mol, whereas it was found to be only $-3.0$ kcal/mol. The discrepancy in part results from a loop entropy that is too big, as DNA is probably supercoiled and thus the entropy cost in making the loop in one dimension along the supercoil is smaller than closing a random loop in three dimensions.

To summarize this section, we have seen that true transcription factors often exhibit significant non-specific binding to DNA. To fight this non-specificity, transcription factors use cooperative binding to each other. DNA looping can add new levels of cooperativity to genetic controls and thereby help to increase specificity.

## *Questions*

(1) Write $P(s)$ in terms of the number of non-specific bound CI dimers, instead of as a function of free dimer concentration.

(2) Correct the estimate for $\Delta G(12\text{-}mer\text{-loop})$ using the fact that the 8-*mer* association has some additional entropy due to the two different orientations of OL relative to OR. Estimate $\Delta G(12\text{-}mer\text{-loop})$ if one takes entropy reduction due to alignment of DNA strands with OL–OR binding. Assume that OL–OR DNA align to within 30° of each other.

(3) The lacZ repressor dimer binds with $K = 10^{-13}$M binding, whereas the tetramer binds with about $K_t = 0.3 \cdot 10^{-13}$M. The part of the DNA which regulates Lac forms a closed loop of length 26 nm, and the persistence length of the DNA may be set to $l_0 = 50$ nm. Estimate the elastic energy of the loop (use Chapter 3); see Balaef *et al.* (1999).

(4) Non-specific DNA bindings: an *E. coli* has $5 \times 10^6$ base pairs. Dimer Cro binds specifically to the operator OR with a binding energy of about $\Delta G_s = -13$ kcal/mol and non-specifically to a random DNA site with about $\Delta G_u = -4.5$ kcal/mol. Calculate

the fraction of free dimer Cro, specifically bound Cro and non-specifically bound Cro as function of numbers of Cro in the cell. The dimerization energy of Cro is $-8.7$ kcal/mol; calculate everything as functions of the numbers of monomer Cro in the cell.

(5) Consider a piece of double-stranded DNA with "sticky ends" (ending in single-stranded segments that are complementary). Assume that these end segments are five bases long, and have a binding energy of $\Delta G = -2$ kcal/mol per bp. What is the maximum length of the DNA if you want it to form a stable loop? What is the minimum length?

## Combinatorial transcription regulation

One aspect of looping is that it makes it possible to influence a given promoter from many different operators. This opens extensive options for combinatorial regulation, which indeed seems typical in eukaryotes (see also Davidson *et al.* (2002)). In this regard it is interesting to explore the possibilities for simple combinatorial control (Buchler *et al.*, 2003). In Fig. 7.21 we show some simple ways to make a simple logical output as a function of two proteins A and B, as suggested by Buchler *et al.* In all cases the output in the form of RNAp binding to the promoter, or not, is monitored. If it binds, it may form an open complex and later initiate transcription of the gene. If the RNAp does not bind, the gene is effective turned off. In practice the Boolean nature of the logic is limited by both the sigmoidal shape (see right panel of Fig. 7.10) of any binding curve, and by the basal activity of the promoter. Whereas the sigmoidal binding curve refers to the non-discrete transition from off to on as a function of input concentration, the basal activity refers to the finite activity of even the non-activated promoter.

### *Questions*

(1) Calculate promoter activity as a function of A and B concentrations in the four cases shown in Fig. 7.21. Assume that A or B bind to their operators with binding constants 1, and that the heterodimer AB binds with binding constant 0.01. RNAp binds perfectly when either A or B is present (binding probability 1), except in Fig. 7.21c, where we assume that RNAp binds to the promoter with probability 1 if there is no heterodimer present. Assume that a bound RNAp corresponds to a promoter activity of 1. Calculate (plot as a two-dimensional surface in three dimensions) promoter activity in all four cases of Fig. 7.21 as function of A and B concentrations between 0.01 and 100. Set AB dimerization constant to be equal 1.

(2) Repeat the above plots, assuming that A and B are both dimers, with a dimerization constant of 10 (and AB is then a tetramer).

## Timescales for target location in a cell

To discuss the dynamics associated with transcriptional regulation we introduce a few basic equations related to the diffusive motion of particles inside the closed

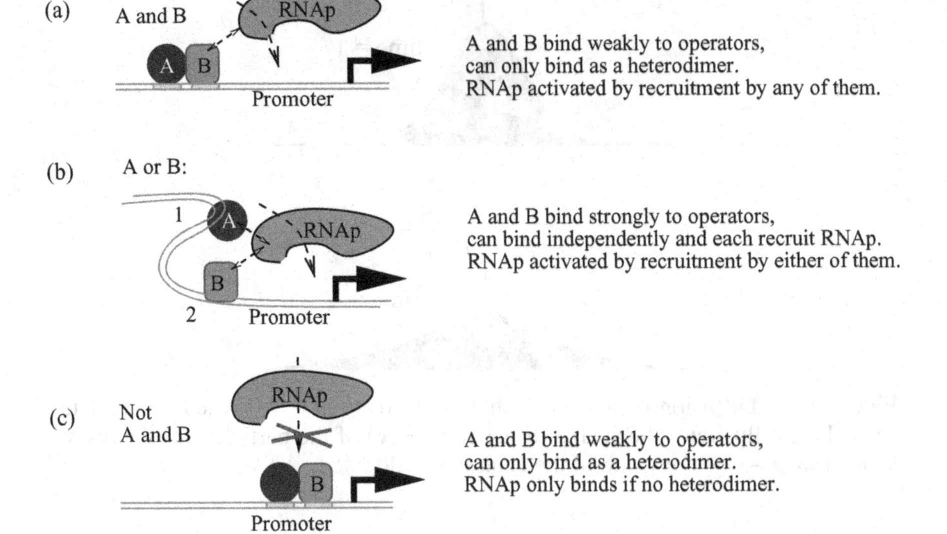


(d) Combinatorical regulation: XOR(A,B) = (A or B) and Not(A and B):

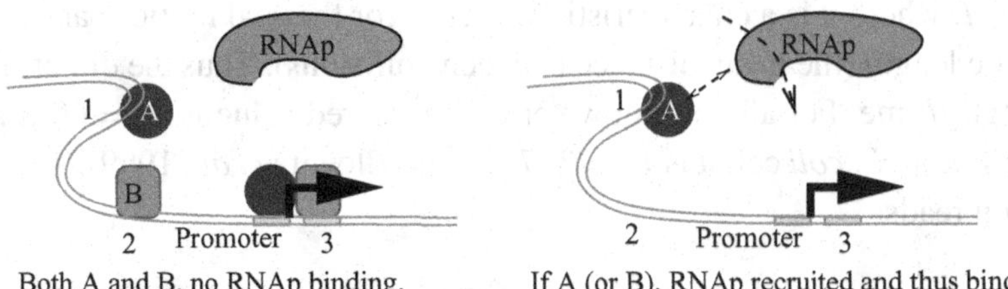


Figure 7.21. Combinatorical transcription logic, with protein A and protein B determining activation of a gene downstream of the promoter. The RNAp binding is regulated either by recruiting it to the promoter, or by preventing it from binding to the promoter. Thus in (a) and (b) RNAp needs either A or B in order to bind to the promoter. In (c) RNAp can bind without help, provided there is no AB complex on the promoter. In (d) A, B bind strongly to operators at positions 1 and 2, respectively. Only the heterodimer AB binds substantially to the operator at position 3 and can thereby repress the promotor. In (a), (b) and (d) the RNAp can bind only by recruitment.

bacterial volume. The inside of a bacterium is a very crowded solution of macromolecules. We will assume that protein motion inside the *E. coli* is diffusive and can be characterized by a diffusion constant $D$ that can be calculated from the mobility $\mu$ using the fluctuation dissipation equation $D = k_B T \mu$ (see Appendix). For a spherical protein of radius $r$, $\mu$ is given by the Stokes relation $\mu = 1/(6\pi r \eta)$ and its diffusion constant

$$D = \frac{k_B T}{6\pi \eta r} \tag{7.54}$$

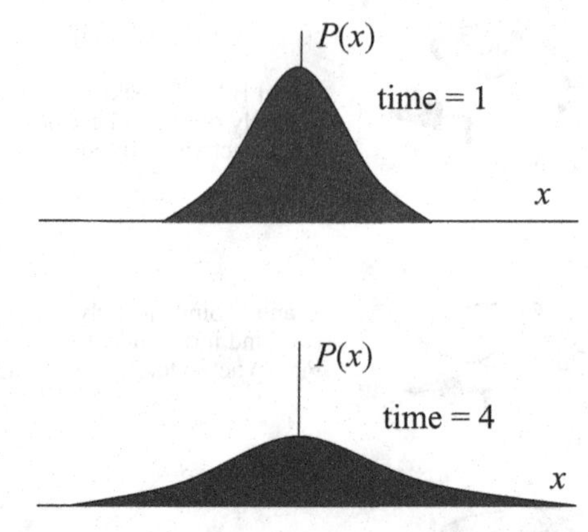


Figure 7.22. Diffusion of a particle that is at position $x = 0$ when $t = 0$. The shaded area illustrates the spatial distribution $P(x)$ of the particles at two times. When time $t \to 4t$ the width of the Gaussian is doubled.

where $\eta$ is the viscosity of the medium ($= 0.001$ kg/(m · s) for water). In the Appendix we will discuss diffusion in more detail, including the diffusion constant $D = v \cdot l$, where $v$ is a characteristic velocity (for thermal motion) and $l$ is a characteristic length (the mean distance between collisions). Thus the dimensions of $D$ are length$^2$/time. For albumin in water the measured value is $D = 60\ \mu\text{m}^2/\text{s}$. For GFP inside an *E. coli* cell it is $D = 3\text{–}7\ \mu\text{m}^2/\text{s}$ (Elowitz *et al.*, 1999). The diffusion equation reads

$$\frac{\mathrm{d}}{\mathrm{d}t} P(\mathbf{r}, t) = D \left( \frac{\mathrm{d}^2 P}{\mathrm{d}x^2} + \frac{\mathrm{d}^2 P}{\mathrm{d}y^2} + \frac{\mathrm{d}^2 P}{\mathrm{d}z^2} \right) \tag{7.55}$$

where $P(\mathbf{r}, t)$ is the probability of finding a particle at position $\mathbf{r}$ at time $t$. A particle that starts at $\mathbf{r} = \mathbf{0}$ at time $t = 0$, will at time $t$ be found at position $\mathbf{r}$ with probability

$$P(\mathbf{r}, t) = \frac{1}{\sqrt{4\pi t D}} \exp\left( -\frac{\mathbf{r}^2}{4Dt} \right) \tag{7.56}$$

Technically, this equation is the solution of Eq. (7.55) with a Gaussian distribution as initial condition for $P$. In other words the diffusion equation has the property that a Gaussian evolves into a Gaussian, illustrated in Fig. 7.22. Also note that Eqs. (7.55) and (7.56) hold equally well for the concentration $C(\mathbf{r}, t)$ of many independently diffusing particles.

From Eq. (7.56) the typical time it takes a particle to diffuse through a distance $d$ is

$$t_{\text{diffusion}} = d^2/2D \tag{7.57}$$

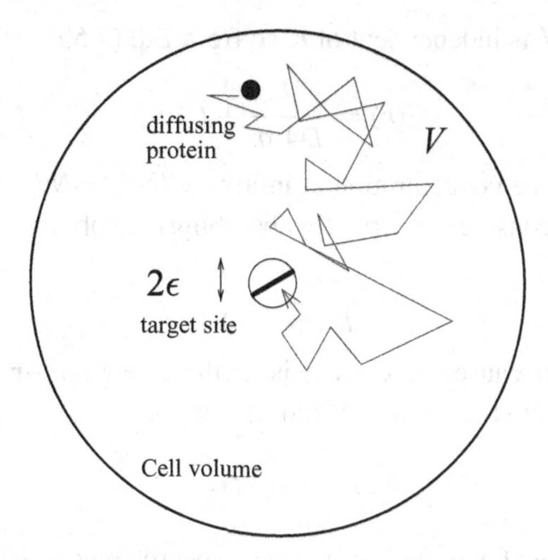


Figure 7.23. A diffusing particle in a volume $V$ will at some time enter a smaller sub-volume of radius $\epsilon$. In the text we estimate the time it takes for the first encounter with this smaller volume. This time sets the diffusion-limited time for a reaction between the particles in a cell volume, where $\epsilon$ is the radius of the reaction volume.

Thus for a bacterial cell of diameter $d = 1\ \mu\text{m}$, a protein will typically have reached the opposite end after a time

$$t_{\text{diffusion}} = d^2/2D = 0.1\ \text{s} \tag{7.58}$$

Thus, when considering timescales $\gg 0.1$ s, proteins that diffuse are homogeneously distributed in the *E. coli* cell.

We now consider the time it takes a molecule to find a binding site if the only process is diffusion. This is the time it takes the molecule to visit a specific reaction center, which we model as a spherical volume of radius $\epsilon \ll d/2$. For the sake of this argument, envision the following: far away from the reaction volume there is a concentration of particles $N/V$, which is held constant. The reaction volume is a sphere of radius $\epsilon$ that is perfectly absorbing for the particles (see Fig. 7.23). For any distance $r > \epsilon$ this gives rise to a steady state flux into the absorbing volume:

$$J = -D\, 4\pi r^2 \frac{dC}{dr} \tag{7.59}$$

$J$ is the current density integrated over a spherical shell of radius $r$, hence the factor $4\pi r^2$; the units of $J$ are particles/s; more formally, Eq. (7.59) follows from writing the diffusion Eq. (7.56) in spherical coordinates:

$$\frac{dC}{dt} = \frac{1}{r^2} D \frac{d}{dr}\left(4\pi r^2 \frac{dC}{dr}\right) \tag{7.60}$$

In the steady state $J$ is independent of $r$, so from Eq. (7.59) we have

$$C(r) = \frac{J}{D4\pi r} + C(\infty) \tag{7.61}$$

Using the fact that the concentration at infinity $C(\infty) = N/V$ and $C(\epsilon) = 0$ (the reaction volume is considered perfectly absorbing) we obtain

$$\frac{-J}{D4\pi r} = \frac{N}{V} \tag{7.62}$$

for $r > \epsilon$. ($J < 0$ because the current is in the direction $-\mathbf{r}$.) Thus the rate of molecules coming into the volume of radius $\epsilon$ is

$$|J| = 4\pi\epsilon D\frac{N}{V} \tag{7.63}$$

The time it takes one of $N$ molecules to find a specific reaction center of radius $\epsilon$ is then

$$\tau_{\text{on}} = 1/|J| = \frac{V}{4\pi D\epsilon N} \tag{7.64}$$

Expressed in terms of density $N/V$, this is called the Smoluchowski equation. For a cell of volume $V = 1\ \mu\text{m}^3$, the diffusion-limited encounter time with a region of radius $\epsilon \sim 1$ nm is about

$$\tau_{\text{on}} \sim 20\ \text{s}/N \tag{7.65}$$

where the number $N$ of a given protein type in an *E. coli* is between 1 and $10^4$.

We can now address the off rate associated to establishing an equilibrium distribution. Assuming that the on rate is diffusion limited, i.e. given by $\tau_{\text{on}}$ in the above equation, the off rate must be

$$\frac{1}{\tau_{\text{off}}} = \frac{1}{\tau_{\text{on}}} V \exp(\Delta G/k_{\text{B}}T) = 4\pi D\epsilon[M]\exp(\Delta G/k_{\text{B}}T) \tag{7.66}$$

where $\Delta G$ is the binding free energy for, say, a CI molecule to the operator under consideration. We also introduced $[M]$ to emphasize that the $D\epsilon$ with unit length$^3$/s should be measured in a volume corresponding to $1\ M^{-1} = 1.6\ \text{nm}^3$. For a binding energy $\Delta G = -12.6$ kcal/mol and $\epsilon = 1$ nm the off time is

$$\tau_{\text{off}} \sim 20\ \text{s} \tag{7.67}$$

This is a situation where 1 molecule in 1 $\mu\text{m}^3$ will make one operator half occupied.

## *Questions*

(1) Derive the Smoluchowski equation in two dimensions.

(2) Derive the Smoluchowski equation in three dimensions by dimensional analysis, using the condition that the capture rate must be proportional to the concentration.

(3) Estimate the typical rate of *cro* translation in the lysogenic state, assuming (i) that RNAp initiates transcription immediately upon binding to the promoters, and (ii) that PR is about as strong as PRM when CI is bound to OR2.
(4) Plot $\tau_{off}$ as a function of binding energy $\Delta G$, and find the $\Delta G$ value where $t_{off}$ equals the *E. coli* cell generation time (say, 1 h).

## Facilitated target location

The picture of how a protein finds its specific binding site on the DNA is more complicated if one takes into account non-specific binding, which allows the protein to diffuse *along* the DNA for some time. If non-specific binding is, say, a factor of $10^5$ weaker compared with the typical specific binding of say $-12.6$ kcal/mol, then the average time the protein spends on the DNA at a non-specific location is $t_{walk} = \tau_{off} \approx 10^{-4}$ s instead of the 20 s at the specific site. If diffusion along the DNA is comparable with free diffusion, this results in a walk of about $\Delta x \approx \sqrt{ZDt_{walk}} \approx 20$ nm. After a walk of this characteristic size, the protein leaves the DNA, to be recaptured later by another segment of the DNA. Thus the protein walks partly in three dimensions, partly in one dimension (see Figs. 7.24 and 7.25).

One can ponder about the time spent walking in one or three dimensions, respectively (Hippel & Berg, 1989). This depends strongly on (i) the binding affinity to DNA and (ii) the density of DNA. In vitro experiments focus on rather short DNA pieces with much lower density than inside an *E. coli*. Typical measured "on-rates" to specific sites on DNA in test tubes increase by up to a factor of 20 when the length of flanking DNA varies from, say, 20 to about $10^3$ base pairs (Surby & Reich, 1996). In contrast, in vivo conditions present the challenge of searching the very long DNA before a specific site is located. Thus what was facilitated in vitro, can easily be time consuming in vivo.

In this section we discuss the search for a specific site in the living *E. coli* cell. The time it takes takes to find a specific site of size $\epsilon$ depends on the time spent in one or three dimensions on the DNA. Let us call the typical time on one visit to the

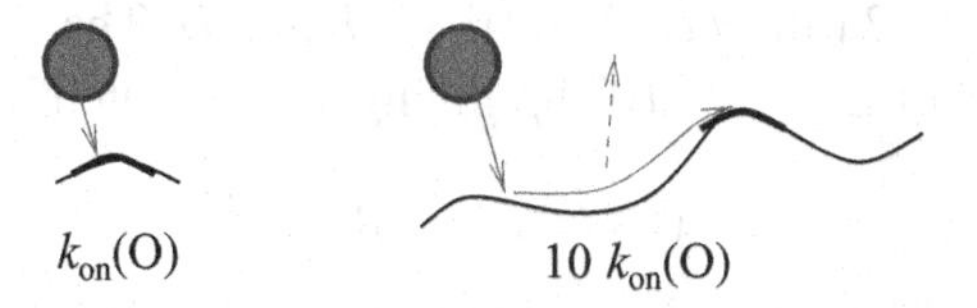


Figure 7.24. Facilitated target search in vitro. A protein searching for a specific operator site (O) is helped by flanking DNA, as the original three-dimensional search on the left is replaced by an easier search for the larger DNA, followed by a one-dimensional search along the flanking DNA. One typically observes increased on rates, $k_{on}$ of up to a factor 20. The upper limit is set in part by the non-specific binding strength that naturally limits the time the protein can stay on the flanking DNA.

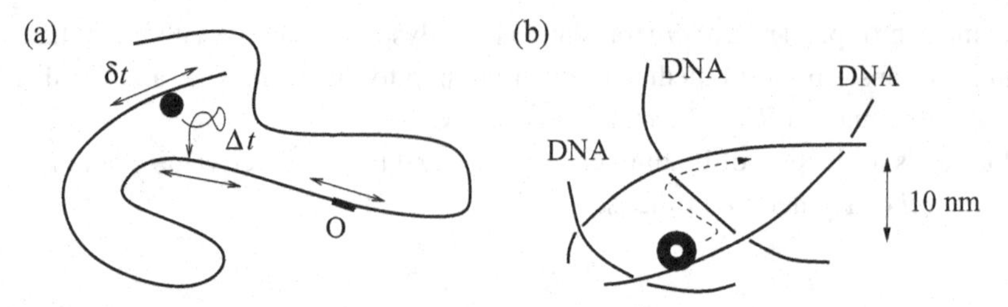


Figure 7.25. A protein searching for a specific operator site inside a cell. (a) The protein diffuses along the DNA, and diffuses in three dimensions. The ratio between the time spent in these two states is set by the non-specific binding. When this binding is very small, the overall search takes a long time because the target is difficult to locate; when the non-specific binding is very large the search is slowed down because most time is spent on repeatedly searching along the same segment of DNA. (b) In vivo facilitated search, where no time is spent in three dimensions, and the one-dimensional search involves only small segments between interacting DNA pieces. With a typical in vivo density, parts of the DNA that are separated by millions of base pairs along the DNA are separated by only 20 nm in three dimensions, thus making the jumping DNA search fairly fast.

DNA $\Delta t$. For $\Delta t \to \infty$ the protein is always bound to the DNA and the search is purely one dimensional. Then the search time is

$$\tau = \frac{L^2}{2D_1} \tag{7.68}$$

where $D_1$ is the diffusion constant for the protein along the DNA. One expects $D_1 < D \sim 5\ \mu\text{m}^2/\text{s}$, but even when they are similar, the time to locate the specific site (in *E. coli* with 1.5 mm of DNA) will be of order $\tau = (1500\ \mu\text{m})^2/(2.5\ \mu\text{m}^2/\text{s}) = 200\,000\ \text{s} \sim 2$ days. This is a very large time, and thus there must be a jump on shorter timescales.

If one assumes in vivo binding resembles the one from in vitro measurements, the measured increase of "on-rates" to specific sites on DNA in test tubes when the length of flanking DNA varies up to about $10^3$ base pairs implies that the protein will search at least $l = 500$ bp $\sim 0.2\ \mu$m on each encounter. Thus one encounter takes the time $\Delta t \sim (0.2\ \mu\text{m})^2/D_1 \sim 0.01$ s if $D_1 = D$. The number of encounters should cover in total of $L = 5 \times 10^6$ bp giving a "facilitated" search time of

$$(L/l)\Delta t \sim 100\ \text{s} \tag{7.69}$$

plus the (insignificant) time spent in jumping between the $L/l$ different segments. This facilitated time from Eq. (7.69) is slower than the three-dimensional search without non-specific binding, but much faster than a pure one-dimensional search. We will now see that (i) the time spent in three dimensions does not contribute, and (ii) the actual facilitated search time would be even shorter because in vivo conditions presumably allow proteins to jump faster.

The time to jump from one part of the DNA to another part depends on the density of DNA. When a protein is unbound, any part of the DNA is a target, and the rate to reach the total length $L$ of DNA can thus be estimated by summing up the rates to reach all parts. If we assume that DNA is everywhere in the cell, the rate of binding to any part $\mathrm{d}L$ of the DNA is

$$\text{rate} \sim \frac{4\pi D \mathrm{d}L}{V} \tag{7.70}$$

Summing all contributions, the time spent for a three-dimensional search for a site on the whole DNA is therefore

$$\delta t \sim \frac{V}{4\pi DL} \sim 10^{-5}\ \text{s} \quad \text{for} \quad L = 1.5\ \text{mm},\ V = 1\ \mu\text{m}^3 \tag{7.71}$$

The real time may be even shorter, because DNA is located in an *E. coli* cell. Thus, compared with typical residence times on DNA (in vitro estimate $\Delta t \sim 0.01$ s), the time spent in three dimensions is presumably insignificant.

The walk length $l$ and thus the $\Delta t$ estimate from above is an upper estimate, and may be shorter because the protein does not need to leave the DNA entirely when jumping to another segment of DNA. The typical distance "dist" between nearby DNA in *E. coli* can be estimated from

$$\pi \left(\frac{\text{dist}}{2}\right)^2 \cdot \frac{L}{\text{length of cell}} = \text{area of cross section of cell} \tag{7.72}$$

where the cross section of the cell is divided into a number of areas given by the number of times the DNA crosses the available cell volume $V =$ (length of cell) $\cdot$ (area of cell). Thus if the $L = 1500\ \mu$m *E. coli* DNA is everywhere within a cell volume of 1 $\mu$m$^3$, the average distance between neighbor DNA is dist $= 30$ nm. If, on the other hand, one takes into account that the DNA is typically confined to about a tenth of the cell volume, then dist $\sim 10$ nm.

We now want to estimate the typical distance $l$ along one DNA section, between subsequent intersections of neighboring DNA. We define an intersection as being where the distance is small enough to allow a protein to jump between the DNA sections, while remaining in contact with DNA all the time. If the typical protein diameter is $b = 4$ nm, the distance $l$ can be estimated by considering the intersections as a mean free path problem. In this case the total length of the DNA, $L = 1.5$ mm $= 1.5 \times 10^9$ nm is subdivided into $L/(2b)$ balls that for now we assume to be randomly distributed in the volume $V$. This gives a density $\rho = L/(2bV)$. The intersection cross section of a piece of DNA with a protein with a crossing DNA section is that of a particle with radius $b$: $\sigma = \pi b^2$. Thus the length between intersections of this particle with one of the DNA sections is

$$l = \frac{1}{\rho\sigma} = \frac{2V}{\pi L b} \approx 150\ \text{nm} \tag{7.73}$$

This distance should be compared with the experimentally measured walking distance of about 500 bp $\sim$ 150 nm, teaching us that non-specific binding may be of a strength where the protein jumps between DNA strands instead of falling off. If this is the case the search time becomes

$$\tau \sim \left(\frac{l^2}{D_1}\right) \cdot \frac{L}{l} = \frac{Ll}{D_1} \tag{7.74}$$

This is an equation that expresses the fact that when one doesn't know where to go, the goal is reached fastest by making big steps (by letting the local walk length $l \to 0$ and thus making frequent big jumps).

Inserting $L = 1.5$ mm as the length of *E. coli* DNA, $l \sim 150$ nm and assuming that the diffusion along DNA is similar to that in the bulk: $\tau \approx 50$ s. This is slower than, but comparable to, the original three-dimensional search. For more densely packed DNA, the search is even faster because very frequent jumps between different segments eliminate the costly repetitions of one-dimensional random walkers ($V \to V/10$ makes $\tau = 50$ s $\to \tau = 5$ s). Maybe this is one additional reason for maintaining the *E. coli* DNA in a small fraction of the cell volume. Another, probably more important, reason is that localization simplifies DNA partitioning when the *E. coli* cells divide.

## *Questions*

(1) Consider the intermediate case where the protein may be both on and off the DNA. Argue that the rate of facilitated target location into binding site of size $\epsilon$ behaves as

$$\text{rate} = \frac{1}{\tau} \approx \frac{4\pi}{V} \frac{D\delta t}{\Delta t + \delta t} \left(\sqrt{2D_1 \Delta t} + \epsilon\right) \tag{7.75}$$

where the non-monotonic behavior reflects the gain in the search by increasing effective binding sites ($\epsilon \to \sqrt{2D_1\Delta t} + \epsilon$), as opposed to the penalty by binding to non-specific sites ($\delta t + \Delta t$). Assume that the diffusion constant $D$ in the medium is the same as $D_1$ along the DNA. What is $\Delta t$ for the maximum rate of location of the target (in units of $\delta t$ when $\epsilon$ is small)? Schurr (1979) estimates $D_1 \sim D/100$ assuming that the protein spirals around the DNA as it diffuses along the backbone. How would that change the optimal $\Delta t$?

(2) Argue that the rate of escaping to distance $R$ away from a one-dimensional DNA strand before recapturing scale as $r \propto (2\pi Dl)/(\ln(R/b))$, where $b$ is the diameter of DNA, $l$ is the length of DNA and where one assumes that touching distance $= b$ always leads to absorption (this is the diffusion limited case, reaction is instant when possible).

(3) For a finite piece $l$ of DNA, argue for the capturing rate as a function of its length, taking into consideration first that it is a rod (of radius $b$), then that it is a random coil (with persistence length $l_p$).

## Traffic on DNA

DNA is not only the object for binding/unbinding events and simple diffusion. It is also a one-dimensional highway with substantial directed traffic. In fast-growing *E. coli* the DNA polymerase protein (DNAp) that polymerize the new chromosome and thus passes any point every 25 min, disrupting any protein–DNA complex, and splitting the DNA into separated strands. Other molecules are in constant action, including gyrases and topo-isomerases that maintain the topological properties of functional DNA. Finally DNA is constantly transcribed by RNAp, which moves along the DNA while separating the DNA strands locally and transcribing them into mRNA. This last transcription activity opens up a new level of regulation (Adhya & Gottesman, 1982; Callen *et al.*, 2004), where a promoter on one strand fires RNAp into the RNAp bound to a promoter on the opposite strand (see Fig. 7.26). In this way, opposing promoters can repress each other's activity and one can build a new intricate way of regulating genes.

To model the promoter interference consider first an isolated promoter, with properties determined through the "on rate", $k_{\text{on}}$, and the "firing rate", $k_{\text{f}}$, from a so-called "sitting duck" complex (see Sneppen *et al.*, 2005). The sitting duck complex is similar to the open complex; it represents a simple description of a state

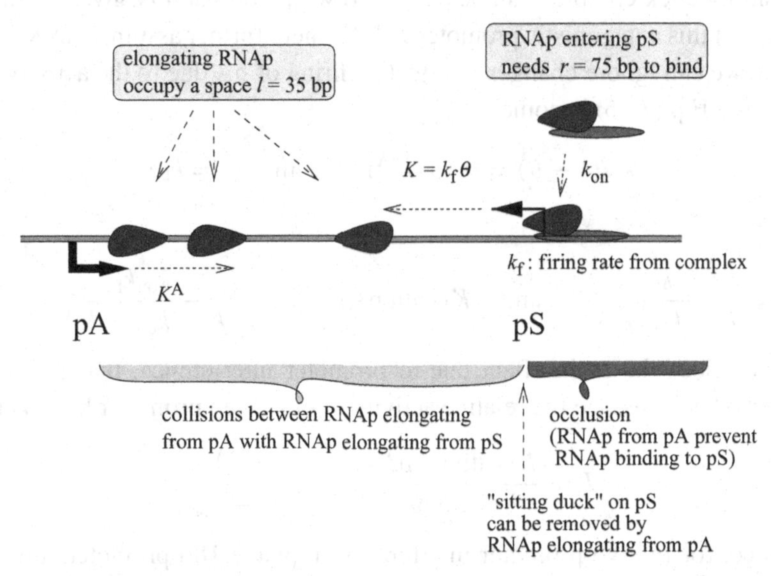


Figure 7.26. Geometry of promoter interference, where an aggressive promoter pA influences the activity of a sensitive promoter pS by firing RNAp through it. There are several mechanisms for such interference: occlusion, collisions between moving RNAps and collisions between RNA from pA with RNAp sitting on pS. In practice, for small distances $N$ between the convergent promoters, the last interference mechanism, the "sitting duck" interference, is the most important.

of the RNAp that can leave the promoter only through transcription initiation (or by other non-equilibrium interventions). Thus we simply describe the transcription initiation by a two-step process: a one-way formation of a sitting duck complex with rate $k_{\text{on}}$, and a one-way initiation of elongation from the sitting duck complex with rate $k_{\text{f}}$. The average occupancy of this complex $\theta$ and the total activity $K$ of the undisturbed promoter is given by

$$k_{\text{on}}(1-\theta) = k_{\text{f}}\theta \quad \text{with} \quad K = k_{\text{f}}\theta \tag{7.76}$$

This balance equation states that RNAp can enter pS only if there is no sitting duck complex. Thus the average occupancy of an undisturbed promoter

$$\theta = \frac{k_{\text{on}}}{k_{\text{f}} + k_{\text{on}}} \quad \text{and} \quad K = k_{\text{f}}\theta = \frac{k_{\text{on}}k_{\text{f}}}{k_{\text{f}} + k_{\text{on}}} \tag{7.77}$$

teaches us that a strong promoter needs to have both a large on rate, $k_{\text{on}}$, and a large firing rate, $k_{\text{f}}$, once RNAp is on the promoter. In the above equations we have obviously ignored the fact that RNAps have a finite length, and thus that they also occupy a promoter for some time ($\sim Kl/v$, $l \approx 35$ bp is the length of elongating complex, $v \sim 40$ bp/s is its velocity) after they have left.

Now with an antagonistic promoter pA firing into the above promoter, the so-called sitting duck complex can be destroyed with a probability given by the total firing rate of this antagonistic promoter $K^{\text{A}}$. We here further assume that $K^{\text{A}} \gg K$, such that we can ignore changes in effective firing of pA due to the activity of pS. In that case, Eq. (7.76) becomes

$$k_{\text{on}}(1-\theta) = (k_{\text{f}} + K^{\text{A}})\theta \quad \text{with} \quad K = k_{\text{f}}\theta \tag{7.78}$$

giving

$$\theta = \frac{k_{\text{on}}}{k_{\text{f}} + k_{\text{on}} + K^{\text{A}}} \quad \text{and} \quad K(\text{with pA}) = k_{\text{f}}\theta = \frac{k_{\text{on}}k_{\text{f}}}{k_{\text{f}} + k_{\text{on}} + K^{\text{A}}} \tag{7.79}$$

which describes the main effect due to promoter interference, that the stronger promoter pA will reduce the relative activity of our given promoter by a factor

$$\mathcal{I} = \frac{K(\text{without pA})}{K(\text{with pA})} = 1 + \frac{K^{\text{A}}}{k_{\text{f}} + k_{\text{on}}} \tag{7.80}$$

In practice, for in vivo promoter interference in phage 186 promoters this factor is found (Callen *et al.*, 2004) to be 5.6, for a pair of promoters where the weak promoter is a factor 10 weaker that the aggressive promoter pA, $K = K^{\text{A}}/10$. Equation (7.80) predicts somewhat less interference, but it also ignores effects associated with occlusion), as well as to collisions of RNAp from pA with RNAp that has left the sitting duck complex (see Questions or p. 189) and also corrections associated with the correlations between subsequent firings of RNAp from pA (see Sneppen *et al.*, 2005).

Promoter interference is in fact documented in the core regulation in P2-like phages, but could also be of relevance in establishment of lysogens in $\lambda$ through PR–PRE interference. In *E. coli* about 100 promoters are known to be placed face to face at fairly close distance, making promoter interference part of the regulation of $\sim$5% of the promoters. Furthermore, promoter interference may also act when promoters fire in parallel, as was seen in the original demonstration of promoter interference by Adhya & Gottesman (1982).

## *Questions*

(1) Reconsider Eq. (7.77) when taking into account that RNAp needs time $Kl/v$ to leave the promoter before a new RNAp can bind to it. Here $l = 35$ is the length of elongating RNAp, $v \sim 40$ bp/s is its velocity. At what promoter firing strength does this correction become more than a factor 2?

(2) Re-express $\mathcal{I}$ in terms of the relative promoter strength $K^{\mathrm{A}}/K$ and the so-called aspect ratio $\alpha = k_{\mathrm{on}}/k_{\mathrm{f}}$. Which value of $\alpha$ gives maximal interference? Discuss why interference decreases for both very small and very large $\alpha$.

(3) Consider occlusion, the fact that an entering RNAp from pA prevents an RNAp from binding to pS for a time given by $l + r = 35 + 75$ base pairs (see Fig. 7.26). (The $r = 75$ bp is the length an RNAp occupies when bound to a promoter.) What is the interference factor $\mathcal{I}$ if one includes this occlusion effect? Calculate $\mathcal{I}$ for $K = k_{\mathrm{on}}/2 = k_{\mathrm{f}}/2 = 0.01/\mathrm{s}$, $K^{\mathrm{A}} = 0.1/\mathrm{s}$ and $v = 40$ bp/s.

(4) Assume that RNAp from pS (see Fig. 7.26) has to travel a distance $N - 40$ before it has escaped possible collision with RNAp from PA. Here $N$ is the distance between promoters, the $N - 40$ takes into account that each RNAp on a promoter occupies 20 bp ahead of the promoter start position. How does $\mathcal{I}$ change with increasing $N$?

(5) Implement a stochastic model for promoter interference on a computer. Use the values from Question (3) and set $N = 100$ and compare results with Eq. (7.79).

## Stability and robustness

Upon infection of an *E. coli* cell, the $\lambda$-phage enters either a pathway leading to lysis, or it enters lysogeny, in which it can be passively replicated for very long times. Indeed, the wild-type rate of spontaneous loss of $\lambda$ lysogeny is only about $10^{-5}$ per cell per generation, a life-time of order 5 years. Moreover, this number is mainly the result of random activation of another part of the genetic system (the SOS response involving RecA), whereas the intrinsic loss rate has in several independent experiments been found to be less than $10^{-7}$ per cell per generation (Little *et al.*, 1999), and possibly as low as $\sim 10^{-9}$ if one considers only switching due to spontaneous fluctuations in the finite number of CI in a cell, excluding events where mutations have changed either CI or PRM (J. W. Little, personal communication).

In Fig. 7.27 we illustrate the result of an experiment by Toman *et al.* (1985), where the two possible states are recorded from the color of the bacteria that host the phage. This is accomplished by using a "defective" λ-phage that cannot lyse, and by adding a reporter gene *gal* downstream of *cro*, which signals when *cro* is being transcribed. Namely, if *cro* is transcribed then *gal* is also transcribed, and the enzyme it produces, β-galactoxidase, catalyzes a reaction that turns a substrate (which is added to the bacterial culture) into a red dye. Thus colonies in the lysogenic state are white, while colonies in the lytic state are red.

The experiment illustrates not only that the lysogen is stable, but also that the lytic state would be metastable, if it didn't normally lead to cell death. Fig. 7.27 illustrates another important feature of cell control, namely the possibility of having two states with the same genome. This is called epigenetics, and is a property that cells in our own body utilize massively. That is, we have at least 250 different cell types, all with the same genetic material. Thus, stability of cell differentiation is obviously important for multicellular organization.

We now describe a quantitative model for the stability of the λ-phage switch. The stability of the switch depends on the CI level, which will force the lytic state back

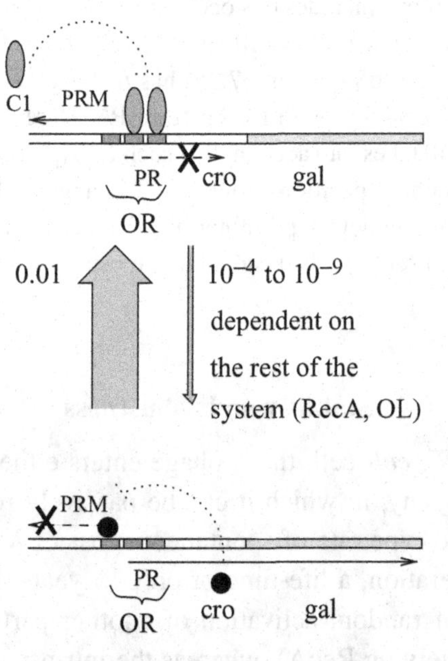


Figure 7.27. Dynamics of switching from lysis to lysogenic state in a defective λ-phage that cannot escape the *E. coli* chromosome (Toman *et al.*, 1985). The state of the phage was recorded through the color exhibited when the gene *gal* was expressed. Lysis gave red colonies, while the lysogenic state gave white colonies.

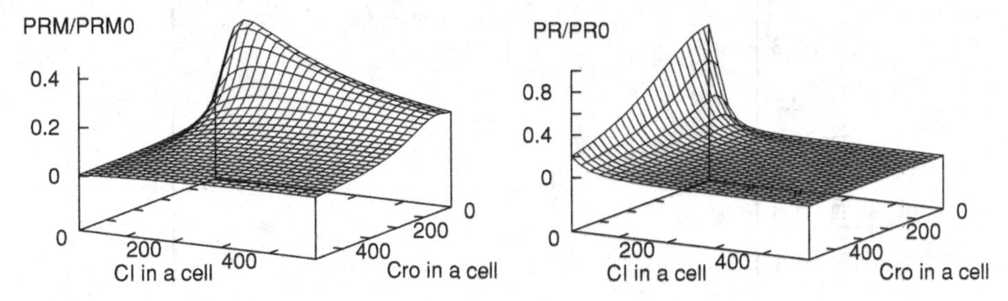


Figure 7.28. PRM and PR activity as functions of both CI and Cro with same parameters as used in Fig. 7.20, and a Cro non-specific binding of −3.5 kcal/(mol·bp). In the absence of CI and Cro the firing rate of PR is 16 times the firing rate of PRM (Dodd *et al.*, 2004). The two plots define the productions term in Eq. (7.81) for CI and Cro respectively. In the left-hand plot we see that, for any CI level, the PRM activity decreases monotonically with [Cro], which reflects the fact that Cro represses production of CI. The right-hand plot, on the other hand, demonstrates that PR activity decreases quickly especially with [CI], which reflects the cooperative repression of PR. In both cases the autorepression by the protein of its own promoter is much weaker that its repression of the opponent's promoter.

into the lysogenic state, and can be quantified by the spontaneous rate of escape from lysogeny to the lytic state. The dynamics of the lysogen–lytic states, for a cell with a given number of CI and Cro molecules ($N_{\mathrm{CI}}$, $N_{\mathrm{Cro}}$), can be modeled with Langevin equations of the type (Aurell *et al.*, 2002)

$$\frac{\mathrm{d}N}{\mathrm{d}t} = \text{production} - \text{decay} + \text{noise} \tag{7.81}$$

for both CI and Cro, with production given as a function of CI and Cro concentrations through the chemistry determined earlier in this chapter, and shown in Fig. 7.28. The decay term is the sum of dilution due to cell division and degradation. The noise is caused by random events and is both production and decay. The strength of the noise term can be quantified by its $N$-dependent variance, $\sigma^2(N)$, which can be calculated as the sum of the variance of production and the variance of decay

$$\sigma^2(N) = \sigma^2_{\mathrm{production}}(N) + \sigma^2_{\mathrm{decay}}(N) \tag{7.82}$$

as the variance of any sum of independent processes is given by the sum of variances of each process. In lysogen conditions the noise term can alternatively be measured by single-cell analysis of a suitable placed reporter gene (see Fig. 7.29).

Equation (7.81) can be dynamically simulated; see Aurell *et al.* (2002). At each time interval one updates $N_{\mathrm{CI}}$ and $N_{\mathrm{Cro}}$ by a change that consists of a deterministic part and a stochastic part. Such a simulation can easily take into account discrete events such as cell divisions or genome duplications, and can, in principle, be done on a very detailed level with single molecule resolution. The result of such a simulation is shown in Fig. 7.30. In Fig. 7.31 we illustrate trajectories in the state space for the *dynamics* of the switch. There are two stable fixed points (corresponding to

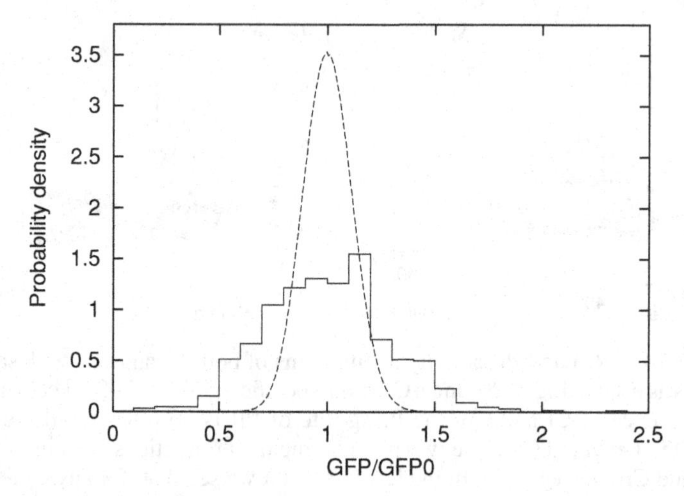


Figure 7.29. Histogram shows data for size of noise term in Eq. (7.81) (from Bæk *et al.*, 2003). The data are indirect, showing cell-to-cell distribution of the reporter gene GFP, placed under control of PRM such that it resembles the CI level in the individual *E. coli* cell. For comparison we show a theoretical estimate (broken line) of the size of fluctuations from the same construct. Here it is assumed that fluctuations are associated with the finite number of molecules, in particular the finite number of mRNA transcripts, $N_{\mathrm{mRNA}}$. Also the random division of molecules of the two daughter cells will in principle contribute to fluctuations. In any case, simulations show that the spread divided by mean of number of molecules in a cell is given by $\sigma/\mathrm{mean} = 1/\sqrt{N_{\mathrm{mRNA}}} \approx 0.1$. With our current knowledge of mRNA transcription of CI (1–5 CI per mRNA transcripts) and CI number in a cell ($\sim$250) the expected fluctuations are smaller than the experimentally observed fluctuations are $\sigma/\mathrm{mean} \approx 0.25$ (see Bæk *et al.*, 2003).

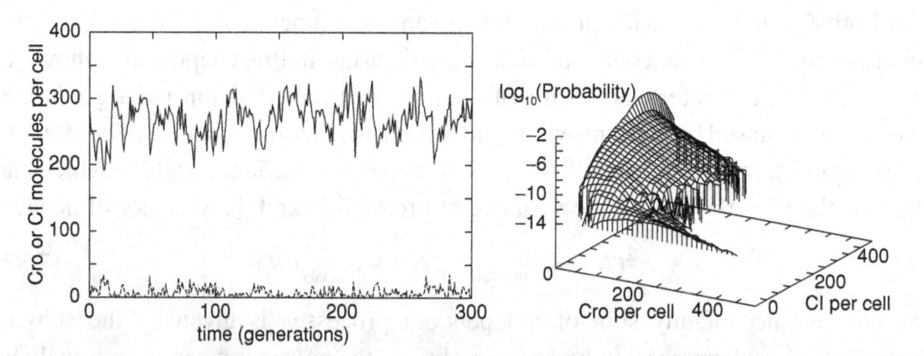


Figure 7.30. The left-hand plot shows a stochastic simulation of the number of CI and Cro inside an *E. coli* lineage (from Aurell *et al.*, 2002). The upper curve shows CI level from generation to generation, and the lower curve the corresponding Cro level. Notice that CI and Cro are anti-correlated: the presence of Cro represses the CI, and decreases the CI amount in a given cell. The right-hand plot shows a histogram of simulated probability for visiting various CI and Cro states (from Aurell *et al.*, 2002). The tail at low CI concentrations corresponds to the spontaneous transitions to lysis. These are transition events that happen very seldom, as is reflected in their very low probability (notice that $z$-axis is logarithmic).

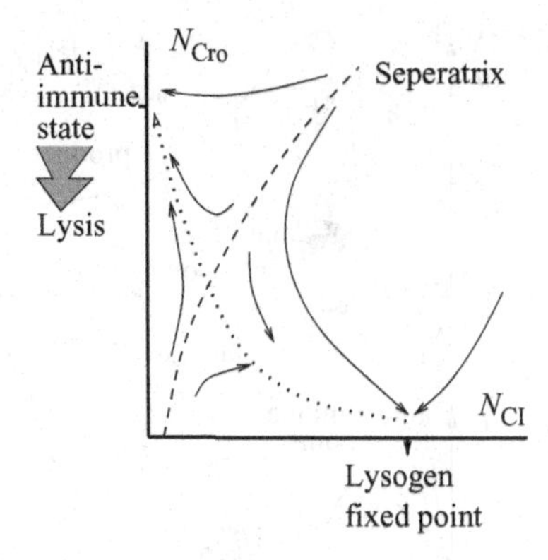


Figure 7.31. Dynamics of the lysogen–lysis switch. At high CI (~250 per cell) and low Cro we find the lysogenic state. At low CI and high Cro (~300 per cell) we finds the lysis, which can be realized as a steady state by a defect λ-phage that cannot lyse see Fig. 7.27. Without fluctuations, all trajectories would be deterministic, and no random switching would occur.

solutions of Eq. (7.81) such that $\mathrm{d}N/\mathrm{d}t = 0$), namely the lysogen state (large $N_{\mathrm{CI}}$, small $N_{\mathrm{Cro}}$) and the lytic state (small $N_{\mathrm{CI}}$, large $N_{\mathrm{Cro}}$). The arrows in the diagram show how one flows into these fixed points from given initial conditions.

For illustrative purposes, and in order to discuss why the transition in fact happens so rarely, we consider a one-dimensional graphical illustration of the stability issue (Fig. 7.32). This corresponds to the CI production rate along the most-probable trajectory that connects lysogeny to lysis, illustrated by the dotted arrow in Fig. 7.31. We then consider only CI fluctuations along this idealized path. That is, we here simplify the two variables $N_{\mathrm{CI}}$ and $N_{\mathrm{cro}}$ into a single effective coordinate describing the state of the cell. For a full discussion using the Friedlin–Wentzell formalism (1984), see Aurell & Sneppen (2002).

In Fig. 7.32 we show the production rate of CI. The straight dotted line is the "decay" of CI in Eq. (7.81) and it represents a decay that is proportional to CI. The points where the curves cross are the fixed points ($\mathrm{d}N/\mathrm{d}t = 0$), which can be either stable or unstable. As long as CI remains above the unstable fixed point $N_{\mathrm{u}}$, lysogeny will typically be restored. The position of the unstable fixed point is believed to be at $N_{\mathrm{u}} \sim (0.1 \to 0.2) \cdot N_{\mathrm{lysogeny}}$. This provides an intrinsic stability for the lysogenic state.

Both the questions of stability and the response to active CI degradation can also be illustrated in terms of a potential as seen Fig. 7.32b. The "effective potential" is obtained by integrating the drift term $f$ = production − decay, which acts as a "force" on the variable $N$

$$V(N) = -\int_0^N f(N')\mathrm{d}N' \tag{7.83}$$

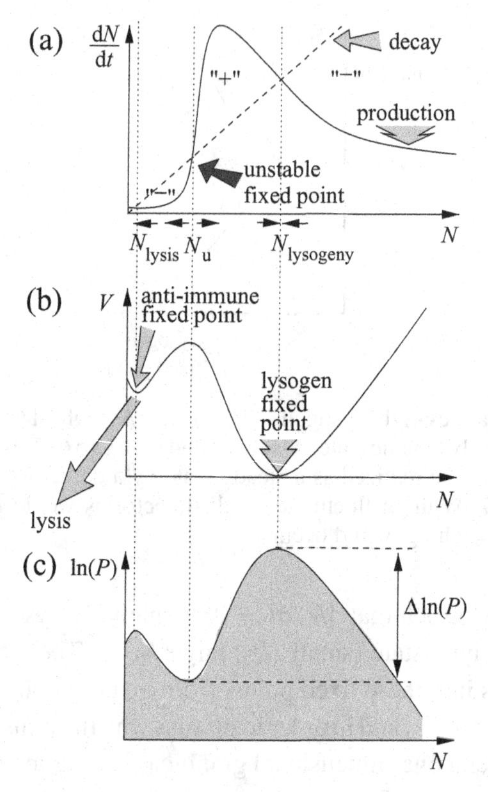


Figure 7.32. (a) The solid line shows production of CI due to activation and repression of PRM along an idealized one-dimensional lysis–lysogeny trajectory. The dashed line shows the dilution due to the cell growth. The intersections between the production and dilution define fixed points that are either stable on unstable. Part (b) shows the effective potential obtained by integrating the production from the top curve. The transition from lysogeny to the lysis can be treated as a Kramer's escape problem with $N$ as the coordinate. Part (c) illustrates that the potential has a concrete meaning, in terms of probability distribution for CI number inside one *E. coli* cell.

with the $N$-dependent noise $\sigma(N)$, representing the amplitude of the noise in Eq. (7.81).

The transition from the state at $N_{\text{lysogeny}}$ over the unstable (saddle) point $N_{\text{u}}$ to the lysis state at $N_{\text{lysis}} \sim 0$ is a first exit problem, and can be treated with standard techniques (see Kramer's escape problem in the Appendix). The system is unstable at the saddle point and thus the probability of staying there is minimal, as illustrated in Fig. 7.32c. Following Kramer, the transition rate from lysogeny to lysis over the saddle point is:

$$r \propto r_0 \cdot \exp\left(-\int_{N_s}^{N_u} \frac{f(N)\mathrm{d}N}{\sigma^2(N)/2}\right) \tag{7.84}$$

where the prefactor $r_0$ is about one event per bacterial generation (see Aurell & Sneppen, 2002). Thus the noise quantified in terms of the denominator $\sigma^2/2$ plays the role of an effective temperature, whereas the averaged production–decay defines the mean potential. The stability at lysogen is essentially given by $\exp(\Delta \ln(P))$ shown in Fig. 7.32c.

The overall lesson from the stability analysis shown above is that this problem is similar to escape from a potential well, with coordinates being chemical concentrations, and potential walls given by CI levels and chemical affinities. As a result the stability strongly depends on parameters that decide CI and Cro levels and affinities in the living cell, roughly in the form:

$$\text{rate(lysogeny} \rightarrow \text{lysis)} \propto \exp(\Delta\exp(\Delta G/T)) \tag{7.85}$$

where $\Delta\exp(\Delta G)$ is a symbolic notation for the difference between lysogenic CI numbers and destabilizing levels of CI and Cro at the saddle towards lysis. It is possible to obtain the observed stabilities within the constraint set by present day experiments in biochemistry and known PRM, PR activity as function of Cro and CI; see Aurell *et al.* (2002) and Sneppen (2003). However, as the stability is the exponential of an exponential, it is not robust. Even a very small change in $\Delta G$ for any of the interactions typically makes a dramatic change in the stability of the system. This suggests that we do not really understand what drives the switch, especially as Little *et al.* (1999) reported large robustness with respect to CI and Cro bindings to the OR1 and OR3 positions, respectively. In fact the mutants considered were one with OR=OR3–OR2–OR3 and one with OR=OR1–OR2–OR1. The fact that these mutants have substantial stability thus challenges the current view of how the switch works.

In particular, the measured stability of the OR3–OR2–OR3 mutant is surprising in view of the fact that because OR3 binds CI about a factor 50 less than OR1, PR is less repressed and this mutant is thus expected to have about 50 times larger amounts of Cro in the cell. This suggests that the current view of stability, based on the balance between a good molecule (CI) and a bad molecule (Cro), should be exchanged for a stability based on a balance where CI and Cro are intermixed in a more subtle way: maybe some moderate amount of Cro stabilizes the switch. A stochastic model that reproduces all of the observed mutant stabilities can be constructed, when we use the fact that PRM is repressed by about a factor of 2.5 by the OL–CI–OR complex, and we assume (see Fig. 7.33):

(a) that a Cro bound to any operator site on OL or OR makes formation of closed loop impossible, and thus de-represses PRM;
(b) that there is some modification in free energy bindings, for example about $-1.5$ kcal/mol cooperative binding between Cro and CI would stabilize the 323 mutant.

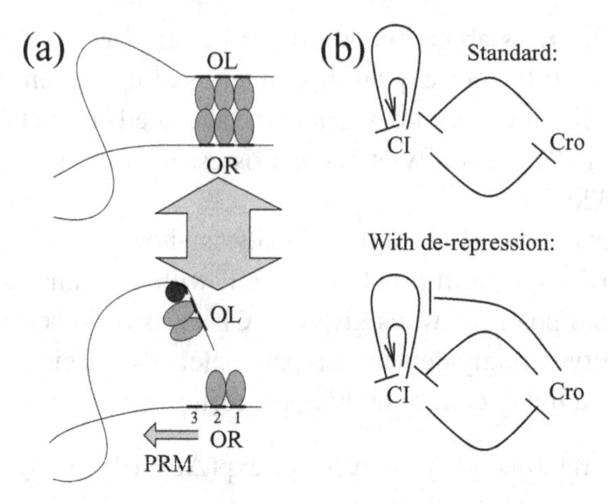


Figure 7.33. (a) Suggested role for Cro in lysogen stabilization acting with CI to enhance CI production (Sneppen, 2003). The little round molecules represent a dimer Cro. They may disrupt the OL–OR complex that represses PRM, and thereby open the way for additional production of CI. In WT Cro often binds to OR3 and represses PRM, but in the mutants it at least as often binds to other operator positions and thus de-represses PRM. (b) Network illustration of Cro's suggested role, illustrated in terms of activation (arrow) or repression (line with vertical bar on end). By de-repressing CI repression of itself, a moderate amount of Cro together with CI may increase CI production. This predicts a switch that is robust towards changes in operator design. As we will see later, robustness is believed to be a main evolutionary constraint for design of biological systems.

The first assumption is plausible in view of the fact that Cro bound to DNA is known to bend it by about 40° (see Albright & Matthews, 1998), and thereby Cro will force the negatively charged DNAs towards each other and disrupt the complex. The second assumption has not, at present, been tested experimentally. In any case, the fact that λ-phage switch is not truly understood is also conveyed by the strange properties of the *hyp* mutant[1] (see Eisen *et al.*, 1982). This particular mutant, with a lysogen that contains a substantial amount of Cro, has properties that demonstrates that Cro also can act in helping CI to maintain lysogen.

Finally we return to the relation between stability of the lysogenic state and the function of the switch: to make the phage lyse correctly when it is right to do so. An illustrative connection between stability and switch function can be found in the

[1] The *hyp* mutant of λ is a challenge to the standard model of λ-phage in this chapter. Note that *hyp* is normal λ except that PRE is replaced with a constitutive promoter that is weaker by a factor of ~30 compared fully activated PRE, but retains this activity even without CII. The properties of *hyp* are: (i) *hyp* is immune to normal λ and λvir (normal λ is not immune to λvir as this virulent phage cannot repress PRM and PL); (ii) *hyp* with CI replaced by temperature-sensitive CI will not induce PL even when the temperature is raised to a level where the repressor cannot dimerize; (iii) *hyp* shows no plaques, and thus does not infect well; (iv) the double mutant *hyp* cro- behaves as cro- *hyp*cro- behaves as λcor-. Thus in *hyp*, Cro works on the lysogenic side of the switch, in direct contrast with the standard model of the λ-phage switch (Harwey Eisen, personal communication); *hyp* has 60% more CI in lysogen than wild-type λ.

Table 7.1. *Properties of OR mutants as given by Little* et al. *(1999), with stability (= lysis frequency in lysis per cell per generation) measured for the RecA strain of* E. coli

| Phage | CI in lysogen | Lysis frequency | Burst size if lysis upon infection | UV dose to induce lysis |
|---|---|---|---|---|
| $\lambda^+$ | 100% | $4 \times 10^{-7}$ to $10^{-9}$ | 56 | high |
| $\lambda$121 | 25–30% | $3 \times 10^{-6}$ | 38 | medium |
| $\lambda$323 | 60–75% | $2 \times 10^{-5}$ | 26 | small |

The robustness is manifested by the fact that the 121 and 323 mutants form fairly stable lysogens. This is surprising, especially because 323 has very weak repression of Cro production.

paper by Little *et al.* (1999) (see Table 7.1), where not only stability but also lysis properties and stability against UV induction are recorded for a number of different mutants. From Table 7.1 it is seen that these properties correlate with stability (lysis frequency), and not with (for example) the CI level in lysogens. Thus stability seems to reflect the overall working condition of the phage. In Fig. 7.34 we summarize the present established knowledge of the λ-phage core decision proteins, including some key players from *E. coli*.

## *Questions*

(1) A very simplified model of the OR switch is obtained by assuming that PRM activity demands CI at OR2 but not at OR3. Thus production effectively takes place in state (011). The probability of this state is $\propto N_D^2$, where $N_D$ is number of dimers. When $N_D$ is large, (111) begins to dominate, thus

$$\text{production} \sim P_R(011) \propto N_D^2 / \sum_{i=0,3} \alpha_i N_D^i \sim N_D^2 - \alpha N_D^5 \tag{7.86}$$

where the sum in the denominator takes into account statistical weights of all combinations of CI bound to the three operator sites. Consider the full $dN/dt$ equation with production, decay $N/\tau$ and assume noise $= \sqrt{\Gamma N}\eta$, where $\eta$ has mean 0 and variance 1 (that is $g^2(N) = \Gamma N$). Discuss the stable and unstable fixed points, and subsequently develop a Kramers' formula for escape from lysogen to lysis.

(2) Consider a simplified model for CI production where its rate $dN/dt = R_{\text{free}} = \text{rate} \cdot N^4/(N^4 + N_c^4)$ with $N_c = 50$. If cell generation is 1 h and average CI level is 300, what is the value of "rate"?

(3) Build a computer model for development of CI level in an *E. coli* lineage that takes into account fluctuations in CI numbers on cell division. Hint: assume that each RNA transcript gives one CI, and selects sufficiently short time intervals to have at most one

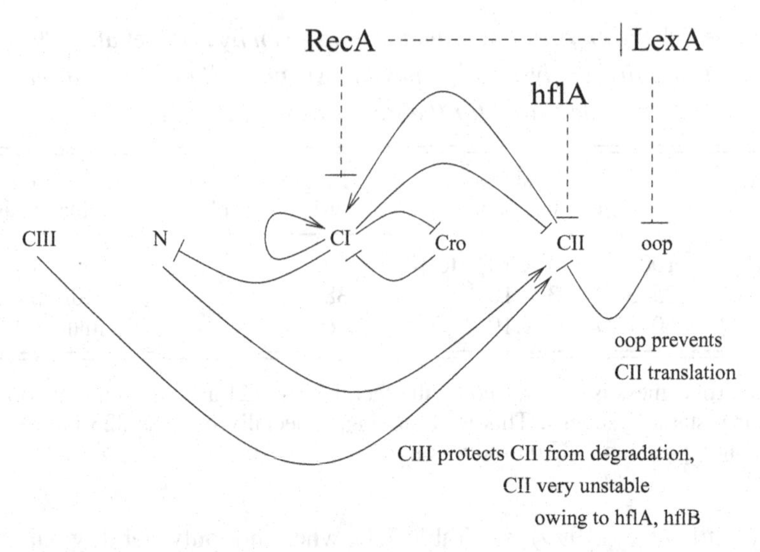


Figure 7.34. Part of the regulatory network of the λ-phage, with positive regulators indicated by arrows, and negative regulators indicated by arrows with perpendicular lines at the end. Part of the control is done through degradation of already-produced proteins, and thus could easily be missed if we examine only transcriptional control mechanisms. The core switch is decided by CI–Cro mutual repression, which secures domination by only one of them. We have added the CII, CIII regulatory system, which further interacts with the host metabolic state through hflA and hflB host proteases. These proteases degrade CII quickly. CIII protects CII from degradation. The dashed lines indicate regulation from host proteins, with RecA being the key player in lysis induction. Notice also that not only proteins, but also the small (77 base pair) mRNA piece oop, are part of the regulatory system. The oop mRNA binds to complementary CII mRNA and the double-stranded mRNA so formed is cleaved quickly by the RNAaseIII from *E. coli* that cleaves double-stranded mRNA, which is more than 20 base pairs long. This is regulated down by a factor 4 when LexA binds to its promoter. If repression by LexA is removed, the lysis emits only half as many phages (Krinkle *et al.*, 1991).

transcript each time. At cell divisions, place each available CI in one of the daughter cells randomly.

(4) Extend the model by assuming a negative feedback from CI on its own production parametrized by $dN/dt = R_{\text{total}} = R_{\text{free}} \cdot (1 - N^2/(N^2 + N_{\text{d}}^2))$, with $N_{\text{d}} = 100$. This extension roughly approximates the average effect of OL on OR through OL–CI–OR binding. Simulate on a computer the dynamical changes in CI with this additional negative (stabilizing) feedback on CI production.
(5) Green fluorescent molecules (GFP) are reporter molecules that can be used to monitor selected genes in a cell. If GFP is linked to the CI gene, then on average one GFP protein will be produced for each CI molecule. However, GFP is not negatively autoregulated, and thus may fluctuate more than CI. Simulate the fluctuations of GFP in the model from Question (4).

(6) Consider Little's 121 and 323 mutants. Explain why, in an OL-strain, 121 has less CI in its lysogen than 323, and that this again has less than the wild type. Is this argument robust for the inclusion of OL?

(7) LexA repression of RecA, and subsequent cleavage of LexA by activated RecA, is at the center of the SOS response system. LexA is a dimer, to which monomers bind with $K_d \sim 1\,\mu$M on the RecA operator, whereas the dimers bind with $10^{-9}$ M. Show that a LexA dimerization constant of $\sim 1\,\mu$M is consistent with this picture. Notice that there is about 1 μM LexA in the *E. coli* cell, whereas RecA repressed is about 2 μM, and RecA is activated up to 50 μM.

## The 186 phage

Given the huge interest and impact the study of λ-phage has had, it may be worthwhile to ponder about the universality of the functions of this phage. Do all temperate phages use the same design as λ, or is the regulation of λ just one of many possible ways in which one may obtain a switch? The answer to this question is that there indeed exist temperate phages with completely different ways of organizing the positive feedback associated with a switch. One example is the P2 phage family, in which the mutual inhibition of the two key proteins involves promoters that fire into each other (see Fig. 7.35) instead of firing away from the same operator complex. The best characterized phage in this family is the 186 phage. B. Egan, I.

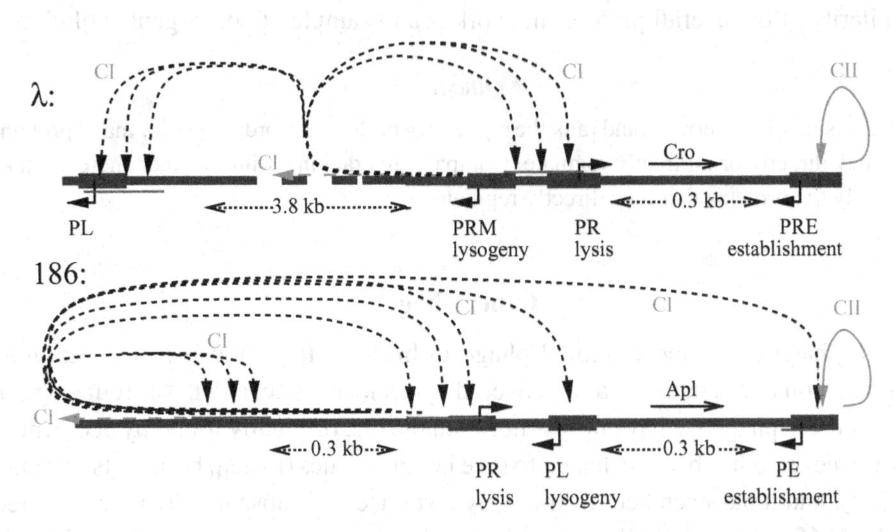


Figure 7.35. Comparison of core control of λ-phage with 186 phage. The geometry of the λ-phage regulation is found in all lambdoid phages, including, for example, HK22. The 186 on the other hand represents the P2 phage family. Only the binding of the lysogen maintaining gene CI is shown in both phages; the action of the antagonistic protein Apl in 186 is not fully understood. It is emphasized that CI(186) has no homology with CI(λ).

Dodd, B. Callen, K. E. Shearwin and collaborators have characterized 186 to a level of detail that compares with that obtained for the λ-phage. In Fig. 7.35 we show the core of the regulatory network for the two phages, for comparison, including the location of involved promoters on the DNA.

The overall regulation of λ and 186 is very similar (see Fig. 7.36), in particular on the level of the protein regulatory network. Both phages have essentially the same regulatory paths, with a few differences associated with replacing a direct regulation with an intermediate protein. Thus one could believe that the two phages were closely related. This, however, is not the case. The only homologous protein between the two phages is the integrase protein Int, which is known to be common for most phages. However, one does not need to consider that level of detail in order to find fundamental differences. When considering the molecular network, where promoters and RNA regulatory elements are included, the two phages differ. Thus the CI(λ) self-activation through binding to OR2 in λ is replaced by CI repression of promoter interference; see the promoter location on Fig. 7.35. Further, CI(186) does not fully repress all promoters leading to Int in 186, and thus Int is present in lysogens in 186. Finally, λ uses anti-sense RNA in three instances (anti-Q, oop and sib) for CII repression of Q, for repression of CII and for repression of Int. In 186 these functions are either done directly by CI(186) or not done at all.

Presumably the evolutionary origin of the two networks is different, and thus the similarity of the overall proteins network is an example of convergent evolution.

*Question*

(1) Consider the λ network and rank the regulatory proteins according to how many proteins they directly or indirectly regulate. Compare this ranking with the one where we take only the proteins that they directly regulate.

## Other phages

The λ-phage is the most studied phage in biology. It is, however, not the most abundant on our planet. It was discovered by accident once, and never found again. In fact the λ-phage we have now is not even the one originally found by accident in some freezer. It has been reshaped to give larger plaques (to help biologists working with it), and it has even been completely lost once and subsequently reconstructed in the 1960s (A. Campbell, personal communication). Then why bother with this very peculiar organism?

One reason is that close relatives of λ-phages exist: the ϕ-phage and the 434 phage, both differing from the λ-phage only by having different immunity regions (from OL to and including the *cro* region). On a wider scale, the λ represents a

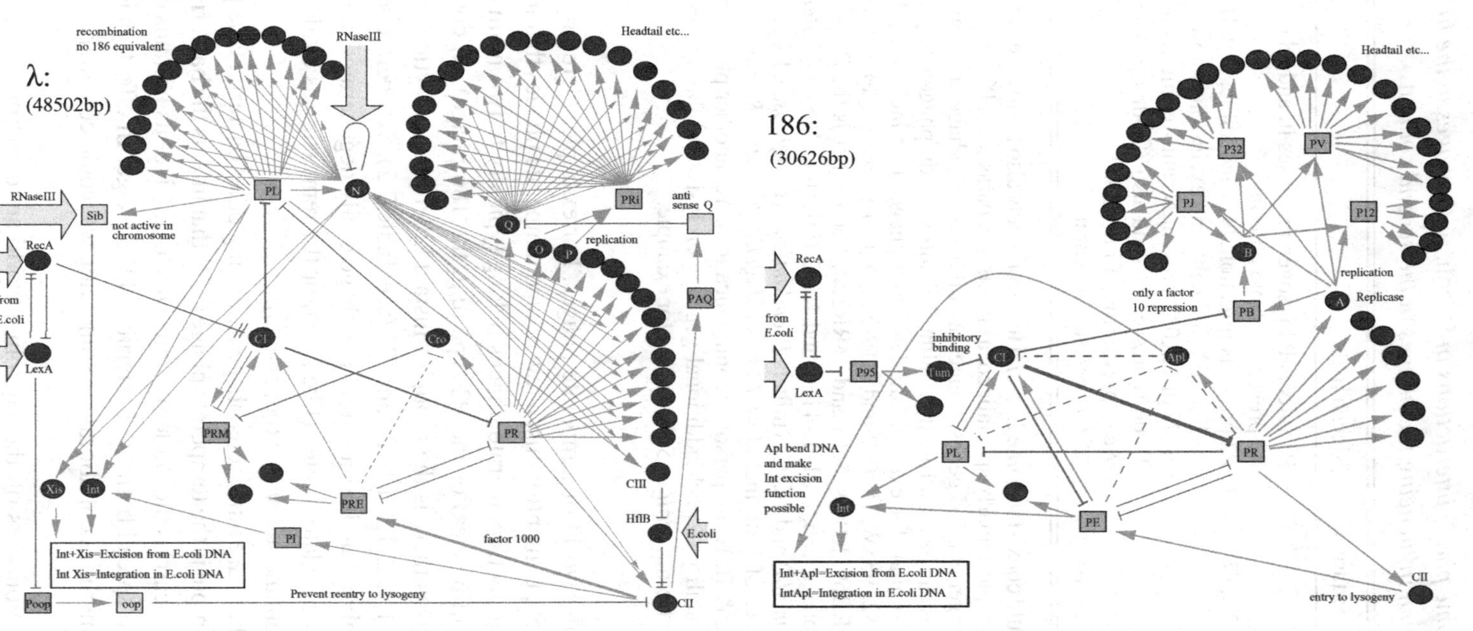


Figure 7.36. Comparison of full protein + promoter networks of the λ-phage with the 186 phage. Notice that the 186 phage genome is smaller. Further, what is present in both networks are similarly regulated. However, the regulatory mechanism differs, and in fact there are essentially no protein homologies between the two phages.

Table 7.2. *Some phage interactions in* E. coli, *with phages in the first column influencing the phages in the second column*

| Phage | Phage target | Mechanism |
|---|---|---|
| P4 | P2 | P4 infects P2 lysogen and lyses it, and takes all |
| P2 | λ | P2 prevents λ from entering |
| P2 | T4,T5,T7 | P2 prevents infection. Non-inducible, no CII |
| 186 | | P2-like, UV inducible, no exclusions known |
| HK022 | λ | NUN of HK022 prevent Ns anti-termination in λ |
| T4 | T6 | T4 de-represses T6 (and also 30 other phages) |
| T6 | T2 | Weak prevention of infection? |
| RB69 | T2,T4,T6 | RB69 prevents infection |

Many phages could coexist in some *E. coli*. However, many also try to take control of the *E. coli*. P4 is particularly specialized, as it acts as a parasite on P2, and relies on many of the larger phages' genes. Phages are often related; phage taxonomy of 144 known phages can be found at http://salmonella.utmem.edu/phage/tree. Some phages are very similar, for example the lambdoid phages that include λ, φ80, 434, HK97, 21, 82, 933W, HK022 and H19-B. Another phage group consist of the T4 relatives RB69, RB49, Aeh1 and 44RR2.8t. Yet another is the P2 group, which includes phage 186, HP1, L-413C and K139 (infecting *Vibrio Cholerae*) and PSP3 (*Salmonella*) and φCTX (infects *Pseudomonas Aeruginosa*). P2 phages infect several types of host, including *E. coli*, *Shigella*, *Serratia*, *Kelbsiella* and *Yersinia* (I. Dodd & B. Egan, personal communication). Yet another phage, p22, is known for its ability to pack additional genes upon lysis, and through this it may move genes from *E. coli* to *Salmonella*, for example.

substantial part of the world of temperate phages. For example, there are φ, 434, P2, HK022 and RB69, to mention just a few related temperate phages that infect (in particular) *E. coli*. In fact, temperate phages belong typically to two main classes: the λ-type (lambdoid phages) or the P2 type (e.g. the 186 phage). In contrast to these temperate phages there is the group of instant killer phages (lytic phages) of *E. coli*, including T1, T2, T3, T4, T5 and T7 (see Table 7.2). Temperate and lytic phages can be closely related, as a temperate phage easily degenerates into a lytic one by losing a few regulatory sites (like λvir), or the repressor (as evidenced by clear plaques in spontaneous lysis experiments).

Beyond *E. coli* there are numerous phages infecting other bacteria. One particularly prominent example is the temperate phage CTX that infects the bacteria *Vibrio Cholerae*, and in fact carries the cholera toxin. This toxin gives us humans diarrhea and thereby helps to spread both the bacteria and the phage. In effect phages and bacteria thereby sometimes collaborate in something we would consider as a "war" against us.

In fact, collaboration/wars on the microbial world are common. Many phages collaborate, or compete with each other for control or spreading of their host. For example, the CTX story above includes collaboration by the phage KSF and phage CTX that helps phage RS1 enter *Vibrio Cholerae* (Davis *et al.*, 2002). Subsequently, with a lysogenic CTX in the host, RS1 induces lysis of CTX and thereby kills *V. Cholerae* but at same time opens the way for spreading CTX to other *V. Cholerae*. As mentioned above, without CTX *V. Cholerae* is harmless. We as humans are exposed to a network of phage–phage and phage–bacteria interactions, a network that may result in epidemics of lethal diseases.

## References

Ackers, G., Johnson, A. & Shea, M. (1982). Quantitative model for gene regulation by λ phage repressor. *Proc. Natl Acad. Sci. USA* **79**, 1129–1133.

Adhya, S. & Gottesman, M. (1982). Promoter occlusion: transcription through a promoter may inhibit its activity. *Cell* **29**, 939.

Albright, R. A. & Matthews, B. W. (1998). How Cro and λ-repressor distinguish between operators: The structural basis underlying a genetic switch. *Proc. Natl Acad. Sci. USA* **95**, 3431–3436.

Aurell, E., Brown, S. & Sneppen, K. (2002). Stability puzzles in phage λ. *Phys. Rev.* E **65**, 051914.

Aurell, E. & Sneppen, K. (2002). Epigenetics as a first exit problem. *Phys. Rev. Lett.* **88**, 048101-1.

Bæk, K. T., Svenningsen, S., Eisen, H., Sneppen, K. & Brown, S. (2003). Single-cell analysis of lambda immunity regulation. *J. Mol. Biol.* **334**, 363.

Balaef, A., Mahadevan, L. & Schulten, K. (1999). Elastic rod model of a DNA loop in the lac operon. *Phys. Rev. Lett.* **83**, 4900.

Bernstein, J. A., Khodursky, A. B., Lin, P.-H. & Cohen, S. N. (2002). Global analysis of mRNA decay and abundance in *Escherichia coli* at single-gene resolution using two-color fluorescent DNA microarrays. *Proc. Natl Acad. Sci. USA* **99**, 9697–9702.

Bestor, T. H., Chandler, V. L. & Feinberg, A. P. (1994). Epigenetic effects in eucaryotic gene expression. *Dev. Genet.* **15**, 458.

Buchler, N. E., Gerland, U. & Hwa, T. (2003). On schemes of combinatorial transcription logic. *Proc. Natl Acad. Sci. USA* **100**, 5136.

Callen, B., Dodd, I., Shearwin, K. E. & Egan, B. (2004). Transcriptional interference between convergent promoters caused by elongation over the promoter. *Molecular Cell* **14**, 647.

Casadesus, J. & D'ari, R. (2002). Memory in bacteria and phage. *BioEssays* **24**, 512–518.

Darling, P. J., Holt, J. M. & Ackers, G. K. (2000). Energetics of l cro repressor self-assembly and site-specific DNA operator binding II: cooperative interactions of cro dimers. *J. Mol. Biol.* **302**, 625–638.

Davidson, E. H. *et al.* (2002). A genomic regulatory network for development. *Science* **295**(5560), 1669–2002.

Davis, B. M., Kimsey, H. H., Kane, A. V. & Waldor, M. K. (2002). A satellite phage-encoded antirepressor induces repressor aggregation and cholera toxin gene transfer. *Embo J.* **21**, 4240–4249.

Dodd, I. B., Perkins, A. J., Tsemitsidis, D. & Egan, B. J. (2001). Octamerization of l CI repressor is needed for effective repression of PRM and efficient switching from lysogeny. *Genes and Development* **15**, 3013–3022.

Dodd, I. B. & Egan, J. B. (2002). Action at a distance in CI repressor regulation of the bacteriophage 186 genetic switch. *Molec. Microbiol.* **45**, 697–710.

Dodd, I. B., Shearwin, K. E., Perkins, A. J., Burr, T., Hochschild, A. & Egan, J. B. (2004). Cooperativity in long-range gene regulation by the lambda CI repressor. *Genes Dev.* **18**, 344–354.

Eisen, H. *et al.* (1982). Mutants in the y region of bacteriophage λ constitutive for repressor synthesis: their isolation and the characterization of the *Hyp* phenotype. *Gene* **20**, 71–81; 83–90.

Elowitz, M. B., Surette, M. G., Wolf, P. E., Stock, J. B. & Leibler, S. (1999). Protein mobility in the cytoplasm of *Escherichia coli. J. Bacteriol.* **181**, 197–203.

Elowitz, M. B. & Leibler, S. (2000). Synthetic oscillatory network of transcriptional regulators. *Nature* **403**, 335.

Friedman, D. I. (1992). Interaction between bacteriophage λ and its *Escherichia coli* host. *Curr. Opin. Genet. Dev.* **2**, 727–738.

Gerland, U., Moroz, J. D. & Hwa, T. (2002). Physical constraints and functional characteristics of transcription factor–DNA interactions. *Proc. Natl Acad. Sci. USA* **99**, 12 015.

Friedlin, M. & Wentzell, A. (1984). *Random Perturbations of Dynamical Systems*. New York: Springer Verlag.

Hawley, D. K. & McClure, W. R. (1982). Mechanism of activation of transcription initiation from the lambda PRM promoter. *J. Mol. Biol.* **157**, 493–525.

Hoyt, M. A., Knight, D. M., Das, A., Miller, H. I. & Echols, H. (1982). Control of phage λ development by stability and synthesis of CII protein. Role of the viral CIII and host *hflA*, *himA* and *himD* genes. *Cell* **31**, 565–573.

Johnson, A. D., Poteete, A. R., Lauer, G., Sauer, R. T., Ackers, G. K. & Ptashne, M. (1981). Lambda repressor and cro–components of an efficient molecular switch. *Nature* **294**, 217–223.

Krinkle, L., Mahoney, M. & Lewit, D. L. (1991). The role of oop antisense RNA in coliphage λ development. *Molec. Microbiol.* **5**, 1265–1272.

Kourilsky, P. (1973). Lysogenization by bacteriophage lambda. I. Multiple infection and the lysogenic response. *Mol. Gen. Genet.* **122**, 183–195.

Liang, S.-T., Bipatnath, M., Xu, Y.-C., Chen, S.-L., Dennis, P., Ehrenberg, M. & Bremer, H. (1999). Activities of constitutive promoters in *Escherichia coli. J. Mol. Biol.* **292**, 19–37.

Little, J. W., Shepley, D. P. & Wert, D. W. (1999). Robustness of a gene regulatory circuit. *EMBO J.* **18**, 4299–4307.

Obuchowski, M., Shotland, Y., Simi, K., Giladi, H., Gabig, M., Wegrzyn, G. & Oppenheim, A. B. (1997). Stability of CII is a key element in cold stress response of Bacteriophage λ infection. *J. Bacteriol.* **179**, 5987–5991.

Pakula, A. A., Young, V. B. & Sauer, R. T. (1986). Bacteriophage λ *cro* mutations: effects on activity and intracellular degradation. *Proc. Natl Acad. Sci. USA* **83**, 8829–8833.

Pedersen, S., Bloch, P. L., Reeh, S. & Neidhardt, F. C. (1978). Patterns of protein synthesis in *E. coli*: a catalog of the amount of 140 individual proteins at different growth rates. *Cell* **14**, 179–190.

Rauht, R. & Klug, G. (1999). mRNA degradation in bacteria. *FEMS Microbiol. Rev.* **23**, 353.

Reichardt, L. & Kaiser, A. D. (1971). Control of λ repressor synthesis. *Proc. Natl Acad. Sci. USA* **68**, 2185.

Reinitz, J. & Vaisnys, J. R. (1990). Theoretical and experimental analysis of the phage lambda genetic switch implies missing levels of cooperativity. *J. Theor. Biol.* **145**, 295.

Révet, B., von Wilcken-Bergmann, B., Bessert, H., Barker, A. & Müller-Hill, B. (2000). Four dimers of λ repressor bound tp two suitably spaced pairs of λ operators form octamers and DNA loops over large distances. *Curr. Biol.* **9**, 151–154.
Ringquist, S., Shinedling, S., Barrick, D., Green, L., Binkley, J., Stormo, G. D. & Gold, L. (1992). Translation initiation in *Escherichia coli*: sequences within the ribosome-binding site. *Mol. Microbiol.* **6**, 1219–1229.
Schurr, J. M. (1979). The one-dimensional diffusion coefficient of proteins absorbed on DNA. Hydrodynamic considerations. *Biophys. Chem.* **9**, 413.
Shea, M. A. & Ackers, G. K. (1985). The OR control system of bacteriophage lambda – a physical-chemical model for gene regulation. *J. Mol. Biol.* **181**, 211–230.
Sneppen, K. (2003). Robustness of the λ phage switch: a new role for Cro (preprint).
Sneppen, K., Dodd, I., Shearwin, K. E., *et al.* (2005). A mathematical model for transcriptional interference by RNA polymerase traffic in *Escherichia coli*. Journal of Molecular Biology **346**, 339–409.
Stormo, G. D. & Fields, D. S. (1998). Specificity, free energy and information contest in protein–DNA interactions. *Trends Biochem. Sci.* **23**, 109–113.
Surby, M. & Reich, N. O. (1996). Facilitated diffusion of the EcoRI methyl-transferase is described by a novel mechanism. *Biochemistry* **35**, 2210.
Thomas, R. (1998). Laws for the dynamics of regulatory networks. *Int. J. Dev. Biol.* **42**, 479–485.
Toman, Z., Dambly–Chaudiere, C., Tenenbaum, L. & Radman, M. (1985). A system for detection of genetic and epigenetic alterations in *Escherichia coli* induced by DNA-damaging agents. *J. Mol. Biol.* **186**, 97–105.
Vind, J., Sorensen, M. A., Rasmussen, M. D. & Pedersen, S. (1993). Synthesis of proteins in *Escherichia coli* is limited by the concentration of free ribosomes. *J. Mol. Biol.* **231**, 678–688.
Vogel, U., Sørensen, M., Pedersen, S., Jensen, K. F. & Kilstrup M. (1992). Decreasing transcription elongation rate in *Escherichia coli* exposed to amino acid starvation. *Mol. Microbiol.* **6**, 2191–2000.
Vogel, U. & Jensen, K. F. (1994). Effects of guanosine 3′,5′-bisdiphosphate (ppGpp) on rate of transcription elongation in isoleucine-starved *Escherichia coli. J. Biochem. Chem.* **269**, 16 236–16 241.
von Hippel, P. H. & Berg, O. G. (1989). Facilitated target location is biological systems. *J. Biol. Chem.* **264**, 675–678.
Voulgaris, J., French, S., Gourse, R. L., Squires, C. & Squires, C. L. (1999a). Increased *rrn* gene dosage causes intermittent transcrition of rRNA in *E. coli. J. Bacteriol.* **181**, 4170–4175.
Voulgaris, J., Pokholok, D., Holmes, W. M., Squires, C. & Squires, C. L. (1999b). The feedback response of *E. coli* rRNA synthesis is not identical to the mechanism of growth rate-dependent control. *J. Bacteriol.* **182**, 536–539.

## Further reading

Arkin, A., Ross, J. & McAdams, H. H. (1998) Stochastic kinetic analysis of developmental pathway bifurcation in phage lambda-infected *Escherichia coli* cells. *Genetics* **149**, 1633–1648.
Asai, T., Zaporojets, D., Squires, C. & Squires, C. L. (1999). An *E. coli* strain with all chromosomal rRNA operons inactivated: complete exchange of rRNA genes between bacteria. *Proc. Natl Acad. Sci. USA* **96**, 1971–1976.

Berg, O. G., Winter, R. B. & von Hippel, P. H. (1982). How do genome-regulatory proteins locate their DNA target sites? *Trends Biochem. Sci.* **7**, 52–55.

Bremer, H., Dennis, P. & Ehrenberg, M. (2003). Free RNA polymerase and modeling global transcription in *Escherichia coli. Biochimie* **85**, 597–609.

Bremmer, H. & Dennis, P. P. (1996). In *Modulation of Chemical Composition and other Parameters of the Cell by Growth Rate, Escherichia coli and Salmonella*, ed. F. C. Neidhardt. ASM Press, pp. 1553–1569.

Capp, M. W. *et al.* (1996). Compensating effects of opposing changes in putrescine (2+) and $K^+$ concentrations on lac repressor–lac operator binding: in vitro thermodynamic analysis and in vivo relevance. *J. Mol. Biol.* **258**, 25–36.

Casjens, S. & Hendrix, R. (1988). Control mechanisms in dsDNA bacteriophage assembly. In *The Bacteriophages*, vol. 1, ed. R. Calendar. New York: Plenum Press, pp. 15–91.

Csonka, L. N. & Clark, A. J. (1979). Deletions generated by the transposon Tn10 in the srl recA region of the *Escherichia coli* K-12 chromosome. *Genetics* **93**, 321–343.

Fong, R. S.-C., Woody, S. & Gussin, G. N. (1993). Modulation of PRM activity by the lambda $p_R$ promoter in both the presence and absence of repressor. *J. Mol. Biol.* **232**, 792–804.

(1994). Direct and indirect effects of mutations in $\lambda$ $P_{RM}$ on open complex formation at the divergent $P_R$ promoter. *J. Mol. Biol.* **240**, 119–126.

Freidlin, M. & Wentzell, A. (1984). *Random Perturbations of Dynamical Systems*. New York/Berlin: Springer-Verlag.

Friedberg, E. C., Walker, G. C. & Siede, W. (1995). *DNA Repair and Mutagenesis*. Washington, DC: ASM Press.

Gardner, T. S., Cantor, C. R. & Collins, J. J. (2000). Construction of a genetic toggle switch in *Escherichia coli. Nature* **403**(6767), 339–342.

Glass, L. & Kauffman, S. A. (1973). The logical analysis of continuous, non-linear biochemical control networks. *J. Theor. Biol.* **39**, 103–129.

Hänggi, P., Talkner, P. & Borkevic, M. (1990). Reaction-rate theory: Fifty years after Kramers. *Rev. Mod. Phys.* **62**, 251–341.

Hawley, D. K. & McClure, W. R. (1983). The effect of a lambda repressor mutation on the activation of transcription initiation from the lambda $P_{RM}$ promoter. *Cell* **32**, 327–333.

Hendrix, R. W. & Duda, R. L. (1992). Lambda PaPa: not the mother of all lambda phages. *Science* **258**(5085), 1145–1148.

Herman, C., Ogura, T., Tomoyasu, T. *et al.* (1993). Cell growth and lambda phage development controlled by the same essential *Escherichia coli* gene, ftsH/hflB. *Proc. Natl Acad. Sci. USA* **90**, 10 861–10 865.

Herman, C., Thénevet, D., D'Ari, R. & Bouloc, P. (1995). The HFlB protease of *E. coli* degrades its inhibitor $\lambda$cIII. *J. Bacteriol.* **179**, 358–363.

Hoopes, B. C. & McClure, W. R. (1985). A cII-dependent promoter is located within the Q gene of bacteriophage $\lambda$. *Proc. Natl Acad. Sci. USA* **82**, 3134–3138.

Jana, R., Hazbun, T. R., Mollah, A. K. M. M. & Mossing, M. C. (1997). A folded monomeric intermediate in the formation of lambda Cro dimer–DNA complexes. *J. Mol. Biol.* **273**, 402–416.

Jana, R., Hazbun, T. R., Fields, J. D. & Mossing, M. C. (1998). Single-chain lambda Cro repressors confirm high intrinsic dimer–DNA affinity. *Biochemistry* **37**, 6446–6455.

Jensen, K. F. (1993). The *Escherichia coli* K-12 "wild-types" W3110 and MG1655 have an rph frameshift mutation that leads to pyrimidine starvation due to low pyrE expression levels. *J. Bacteriol.* **175**, 3401–3407.

Johnson, A. D., Meyer, B. J. & Ptashne, M. (1979). Interactions between DNA-bound repressors govern regulation by the lambda phage repressor. *Proc. Natl Acad. Sci. USA* **76**, 5061–5065.

Kennell, D. & Riezman, H. (1977). Transcription and translation initiation frequencies of the *Escherichia coli* lac operon. *J. Mol. Biol.* **114**, 1–21.

Kim, J.G., Takeda, Y., Matthews, B. W. & Anderson, W. F. (1987). Kinetic studies on Cro repressor–operator DNA interaction. *J. Mol. Biol.* **196**, 149–158.

Koblan, S. K. & Ackers, G. K. (1991). Energetics of subunit dimerization in bacteriophage lambda cI repressor: linkage to protons, temperature and KCl. *Biochemistry* **30**, 7817–7821.

(1992). Site-specific enthalpic regulation of DNA transcription at bacteriophage $\lambda$ $O_R$. *Biochemistry* **31**, 57–65.

Kramers, H. A. (1940). Brownian motion in a field of force and the diffusion model of chemical reactions. *Physica* **7**, 284–304.

Lederberg, E. M. (1951). Lysogenicity in *E. coli* K12. *Genetics* **36**, 560.

Little, J. W. (1993). LexA cleavage and other self-processing reactions. *J. Bacteriol.* **175**, 4943–4950.

Little, J. W. & Mount, D. W. (1982). The SOS-regulatory system of *Escherichia coli. Cell* **29**, 11–22.

Lobban, P. & Kaiser, A. D. (1973). Enzymatic end-to-end joining of DNA molecules. *J. Mol. Biol.* **781**, 453–471.

Maier, R. S. & Stein, D. S. (1997). Limiting exit location distribution in the stochastic exit problem. *SIAM J. Appl. Math.* **57**, 752–790.

Maurer, R., Meyer, B. J. & Ptashne, M. (1980). Gene regulation at the right operator (OR) of bacteriophage $\lambda$. *J. Mol. Biol.* **139**, 147–161.

McAdams, H. H. & Arkin, A. (1997). Stochastic mechanisms in gene expression. *Proc. Natl Acad. Sci. USA* **94**, 814–819.

Miller, J. H. (1972). *Experiments in Molecular Genetics*. Cold Spring Harbor, New York: Cold Spring Harbor Laboratory Press.

Monot, J. & Jacob, F. (1961). General conclusions: teleonomic mechanisms in cellular metabolism, growth and differentiation. *Cold Spring Harbor Symp. Quant. Biology* **26**, 389–401.

Murakami, K. S. *et al.* (2002). Structural basis of transcription initiation: an RNA polymera holoenzyme-DNA complex. *Science* **296**, 1285–1290.

Mustard, J. A. & Little, J. W. (2000). Analysis of *Escherichia coli* RecA interactions with LexA, CI, and UmuD by site-directed mutagenesis of recA. *J. Bacteriol.* **182**, 1659–1670.

Nelson, H. M. & Sauer, R.T. (1985). Lambda repressor mutations that increase the affinity and specificity of operator binding. *Cell* **42**, 549–558.

Neufing, P. J., Shearwin, K. E. & Egan, J. B. (2001). Establishing lysogenic transcription in the temperate coliphage 186. *J. Bacteriol.* **183**, 2376–2379.

Powell, B. S. *et al.* (1994). Rapid confirmation of single copy lambda prophage integration by PCR. *Nucl. Acids Res.* **22**, 5765–5766.

Pray, T. R., Burz, D. S. & Ackers, G. K. (1999). Cooperative non-specific DAN binding by octamerizing $\lambda$ CI repressors: a site-specific thermodynamic analysis. *J. Mol. Biol.* **282**, 947–958.

Ptashne, M. (1992). *A Genetic Switch; Phage $\lambda$ and Higher Organisms*. Blackwell Scientific Publications & Cell Press.

Ptashne, M. & Gann, A. (1997). Transcriptional activation by recruitment. *Nature* **386**, 569.

Record, M. T., Courtenay, E. S., Cayley, D. S. & Guttman, H. G. (1998). Biophysical compensation mechanisms buffering *E. coli* protein–nucleic acid interactions against changing environments. *Trends Biochem. Sci.* **23**(5), 190–194.

Reed, M. R., Shearwin, K. E., Pell, L. M. & Egan, J. B. (1997). The dual role of Apl in prophage induction of coliphage 186. *Molec. Microbiol.* **23**, 669–681.

Rgnier, P. & Grunberg-Manago, M. (1989). Cleavage by RNase III in the transcripts of the metY-nusA-infB operon of *E. coli* releases the tRNA and initiates the decay of the downstream mRNA. *J. Mol. Biol.* **210**, 293.

Roberts, J. W. & Devoret, R. (1983). Lysogenic induction. In *Lambda II*, ed. R. W. Hendrix, J. W. Roberts, F. W. Stahl & R. A. Weisberg. Cold Spring Harbor Laboratory, New York: Cold Spring Harbor, pp. 123–144.

Ross, W., Ernst, A. & Gourse, R. L. (2001). Fine structure of *E. coli* RNA polymerase-promoter interactions: alpha subunit binding to the UP element minor groove. *Genes Dev.* **15**, 491–506.

Rozanov, D. V., D'Ari, R. & Sineoky, S. P. (1998). RecA-independent pathways of lambdoid prophage induction in *Escherichia coli*. *J. Bacteriol.* **180**, 6306–6315.

Shean, C. S. & Gottesman, M. E. (1992). Translation of the prophage lambda cI transcript. *Cell* **70**, 513–522.

Shearwin, K. E., Brumby, A. M. & Egan, J. B. (1998). The tum protein of coliphage 186 Is and antirepressor. *J. Biol. Chem.* **273**, 5708–5715.

Shearwin, K. E. & Egan, J. B., (2000). Establishment of lysogeny in bacteriophage 186. *J. Biol. Chem.* **275**, 29 113–29 122.

Takeda, Y., Ross, P. D. & Mudd, C. P. (1992). Thermodynamics of Cro protein–DNA interactions. *Proc. Natl Acad. Sci. USA* **89**, 8180–8184.

Takeda, Y., Sarai, A. & Rivera, V. M. (1989). Analysis of the sequence-specific interactions between Cro repressor and operator DNA by systematic base substitution experiments. *Proc. Natl Acad. Sci. USA* **86**, 439–443.

Zinder, N. & Lederberg, J. (1952). Genetic exchange in *Salmonella*. *J. Bacteriol.* **64**, 679–699.

# 8

# Molecular networks

Kim Sneppen

Cells are controlled by the action of molecules upon molecules. Receptor proteins in the outer cell membrane sense the environment and may subsequently induce changes in the states of specific proteins inside the cell. These proteins then again interact and convey the signal further to other proteins, and so forth, until some appropriate action is taken. The states of a protein may, for example, be methylation status, phosphorylation or allosteric conformation as well as sub-cellular localization. The final action may be transcription regulation, thereby making more of some kinds of proteins, it may be chemical, or it may be dynamical. A chemical response would be to change the free concentration of a particular protein by binding it to other proteins. A dynamical response could be the activation of some motor, as in the chemotaxis of *E. coli*.

The presently known regulatory network of yeast is shown in Fig. 8.1. The action of proteins in this network is to control the production of other proteins. The control is done through genetic regulation discussed previously, through control of mRNA degradation, or possibly through the active degradation of the proteins.

Regulatory genetic networks are essential for epigenetics and thus for multicellular life, but are not essential for life. In fact, there exist prokaryotes with nearly no genetic regulation. Figure 8.2 shows the number of regulators as a function of genome size for a number of prokaryotic organisms. One notices that those with a very small genome hardly use transcriptional regulation. More strikingly, it appears that the number of regulators, $N_{\text{reg}}$, grows much faster than the number of genes, $N$, it regulates. If life was just a bunch of independent $\lambda$ switches, this would not be the case. That is, if living cells could be understood as composed of a number of modules (genes regulated together) each, for example, associated with a response to a corresponding external situation, then the fraction of regulators would be independent of the number of genes $N$. Also, if life was simply hierarchical with each gene controlling a certain number of downstream genes the number of

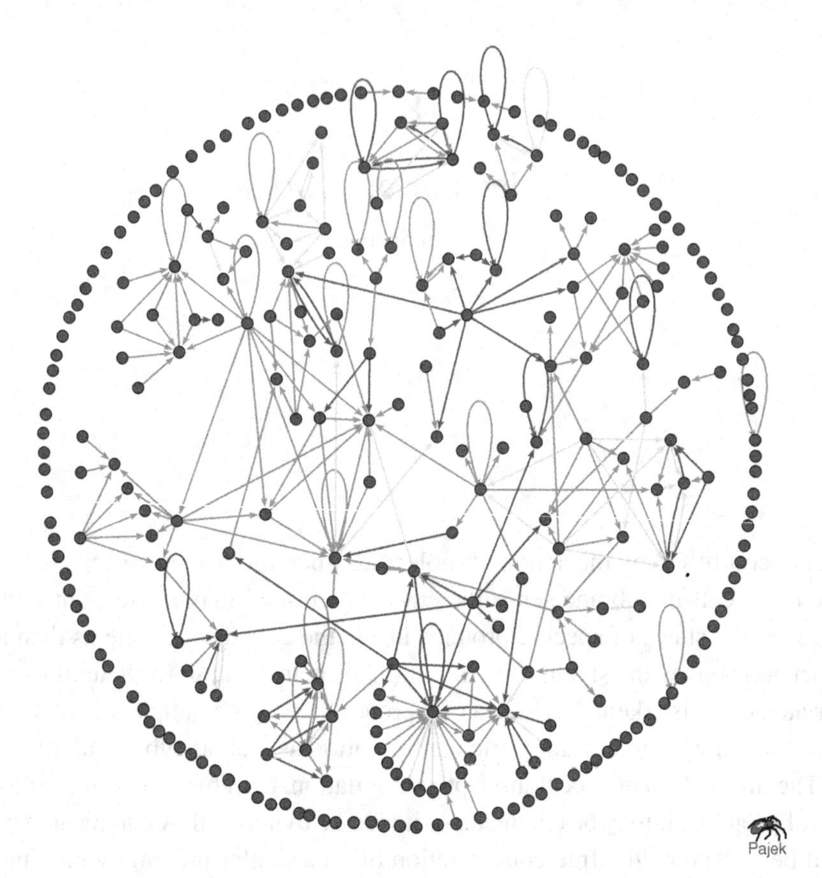


Figure 8.1. Networks of transcriptional regulatory proteins in yeast, showing all proteins that are known to regulate at least one other protein. Arrows indicate the direction of control, which may be either positive or negative. Functionally the network is roughly divided into an upper half that regulate metabolism, and a lower half that regulate cell growth and division. In addition there are a few cell stress response systems at the intersection between these two halves.

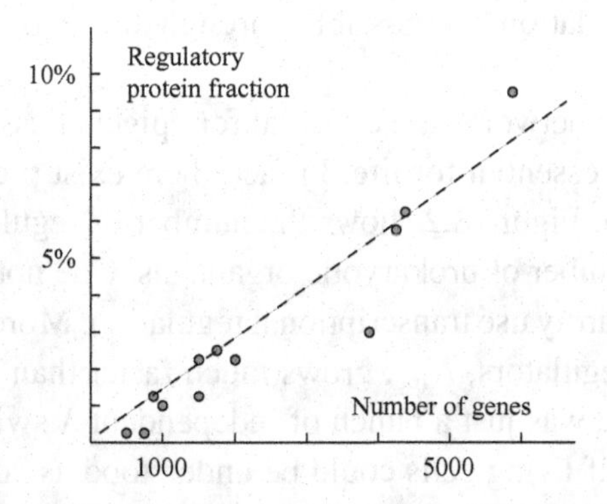


Figure 8.2. Fraction of proteins that regulate other proteins, as a function of size of the organisms gene pool (Stover *et al.*, 2000). The smallest genome is *M. Genitalium* (480 genes); the largest genome is *P. Aeruginosa* (5570 genes). The linear relation demonstrates that each added gene should be regulated with respect to all previously added genes.

regulators would grow linearly with $N$. Further, if life would have been controlled using the maximum capacity for combinatoric control, one could manage with even fewer transcription factors. In fact if the regulation of each gene would include all transcription regulators one could, in principle, specify the total state of all genes with only $\log_2(N)$ regulators.

To see this, imagine that one had 10 regulators that can be on or off. Then one can, in principle, specify $2^{10}$ different states, and thus specify all possible states for up to $N = 2^{10}$ genes.

To summarize:

$$N_{\text{reg}} \propto \log_2(N) \quad \text{if a combinatorial hierarchy} \tag{8.1}$$

$$N_{\text{reg}} \propto N \quad \text{if a simple hierarchy} \tag{8.2}$$

$$N_{\text{reg}} \propto N \quad \text{if independent modules} \tag{8.3}$$

where independent modules refer to the particularly simple regulation where genes are grouped into clusters, each regulated by one or a few regulators.

When sampling all prokaryotes (see Fig. 8.2) one finds

$$N_{\text{reg}} \propto N \cdot N \quad \text{for prokaryotes} \tag{8.4}$$

The fact that $N_{\text{reg}}/N$ grows linearly with $N$ indicates that each added gene or module should be regulated with respect to all other gene modules. Thus, prokaryotic organisms show features of a highly integrated computational machine. The above scaling was also reported, and in fact extended, by Nimwegen (2003), who demonstrated that, for eukaryotes, $N_{\text{reg}} \propto N \cdot N^{0.3}$. In any case the mere fact that the fraction of regulators increases with genome size shows that networks are indeed important. Networks are not just modular, they show strong features of an integrated circuitry, even on the largest scale.

In addition to the control of production of proteins, there is a huge number of protein–protein mediated interactions, where one protein changes the status of another. This may be through changes of its three-dimensional structure (allosteric modification), or by methylation or phosphorylation or binding other molecules to it. A diagram of such a modification is outlined in Fig. 8.3 where protein A catalyzes protein B's transition to state B*. For example, the * could refer to addition of a phosphate group from the ATP freely floating in the cell. A is then an enzyme, which in this case increases one particular reaction rate by diminishing its barrier; see the right-hand panel in Fig. 8.3. There are different types of enzymes, classified into six major groups that we list in order to convey an impression of the possible types of signaling in a cell.

- Oxidoreductases (EC class 1), transfer electrons (three-body).
- Transferases (EC class 2), transfer functional groups between molecules.
- Hydrolases (EC class 3), break bonds by adding $H_2O$.

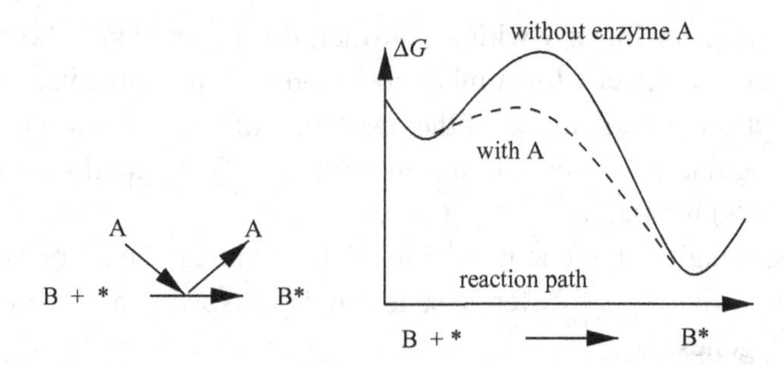


Figure 8.3. A catalyzes B's transition to state B*. In principle, A would then also catalyze the opposite process, as the barrier in both directions becomes smaller. However, if B* is consumed by other processes, the process becomes directed.

- Lyases (EC class 4), form double bonds (adding water or ammonia).
- Isomerases (EC class 5), intramolecular arrangements.
- Ligases (EC class 6), join molecules with new bonds (three-body).

In the yeast *Saccromyces Cerevisiae* there are about 6000 proteins, of which ~3000 have known functions. Of these 3000 proteins about 1570 are categorized in one of the six enzyme classes above. As enzymes often work through binding to another protein, knowledge of the enzymatic networks can be obtained by screening all possible protein–protein bindings in the cell. One option for doing this is the two hybrid method. An example of a network of protein–protein binding partners is shown later in this chapter. Signal transmission with protein binding and enzymatic reactions will be fast compared with timescales associated to protein production.

In summary, the total molecular network in a cell is the combination of fast signaling by enzymes, the slower adaptation associated with regulated changes in protein concentrations, and these two combined with the metabolic networks. This last network is responsible for production of amino acids, ATP, GTP and other basic building blocks.

In the next three sections we will analyze regulatory and signaling networks. We start microscopically by quantifying local properties as defined by degree distribution, and then continue with fairly local topological patterns in forms of local correlations. Finally we will analyze the global topology of a network in the perspective of its communication ability. After these three different levels of topological analysis, we will return to modeling the elementary parts of a molecular network. This modeling may also be seen as a natural continuation and extension of the concepts introduced in Chapter 7.

## *Questions*

(1) Consider an army with hierarchical signal transmission. Assume that an officer of rank $j$ always has two officers of rank $j-1$ to command. Further assume that each

officer/soldier is, at most, directly commanded by one above himself. What is the fraction of soldiers at the bottom of the hierarchy? Assume that one eliminates a fraction $p$ of the soldiers independent of their rank. What is the average size of connected groups of soldiers? Estimate the probability that a bit of information can be transmitted between two random soldiers.

(2) Consider a bureaucracy where everyone has to spend 10 s on any paper that is produced. If it takes 1 h for one employee to produce one paper, and the working week is 40 hours, calculate the production (of papers) per week as a function of number of employees. When does it not pay to hire more bureaucrats? In a molecular network analogy, the 10 s may represent non-specific bindings, and the 1 h may represent the specific/functional bindings.

## Broad degree distributions in molecular networks

A common feature of molecular networks is the wide distribution of directed links from individual proteins. There are many proteins that control only a few other proteins, but also there exist some proteins that control the expression level of many other proteins. It is not only proteins in the regulatory networks (see Fig. 8.4) that have this wide variety of connectivities (see Fig. 8.5). Metabolic networks and protein signaling networks also have a large variety of connectivities.

The distribution of proteins with a given number of neighbors (connectivity) $K$ may (very crudely) be approximated by a power law

$$N(K) \propto 1/K^{\gamma} \tag{8.5}$$

with exponent $\gamma \sim 2.5 \pm 0.5$ (Jeong *et al.*, 2001) for protein–protein binding networks, and exponent $\gamma \sim 1.5 \pm 0.5$ for "out-degree" distribution of transcription regulators (see Fig. 8.5). Notice that the broad distribution of the number of proteins regulated by a given protein, the "out-degree", differs from the much narrower distribution of "in-degrees". We now discuss features and possible reasons for why life may have chosen to organize its signaling in this way.

One aspect of a broad distribution of connectivity in a network is the possible amplification of signals. Consider a signal that enters a node, and make the extreme assumption that it is transmitted along all exit links (unspecific broadcasting). Thus it is amplified by a factor $K_{\text{out}}$. However, not all nodes have equal chances to amplify signals. The probability of entering a node is proportional to $K_{\text{in}}$. Thus, on average, one visits nodes with probability $\propto K_{\text{in}}$ and the weighted average amplification factor in a *directed* network (Newman *et al.*, 2001):

$$\mathcal{A} = \frac{\langle K_{\text{in}} K_{\text{out}}(\text{given } K_{\text{in}}) \rangle}{\langle K_{\text{in}} \rangle} = \frac{\langle K_{\text{in}} K_{\text{out}} \rangle}{\langle K_{\text{in}} \rangle} \tag{8.6}$$

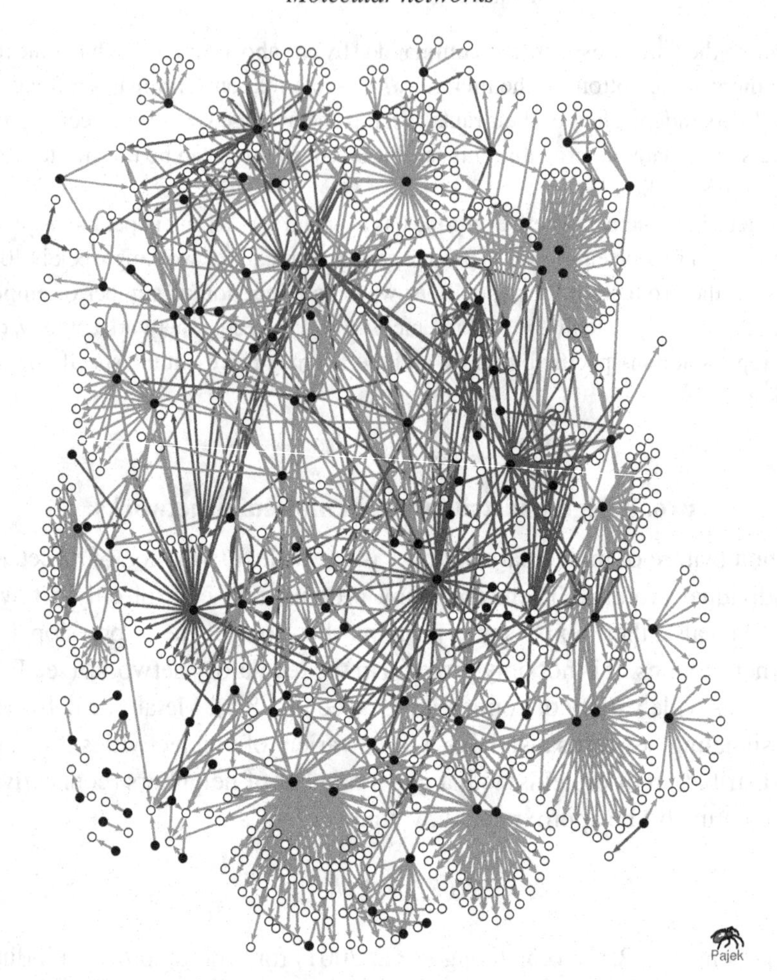


Figure 8.4. Presently known transcription regulations in yeast (*Saccromyces Cerevisiae*). Compared with Fig. 8.1, this figure also includes those that were regulated, but on the other hand only the one that regulates/is regulated through the transcription stage. Notice that yeast has many more positive regulators than *E. coli*. May be the histone–DNA complexes in eukaryotes naturally suppress transcription, and thereby non-regulated genes automatically become silenced.

The first equality assumes that there are no correlations between the degree of a node and the degree of its neighbors. The second equality assumes that there are no correlations between a given protein's "in" and "out" degrees. For undirected random network the potential amplification would be

$$\mathcal{A} = \frac{\langle K(K-1)\rangle}{\langle K\rangle} \tag{8.7}$$

If all nodes have about the same connectivity, we recover the simple result that when $\langle K\rangle = 2$ then $\mathcal{A} = 1$. Thus to have marginal transmission, one input signal,

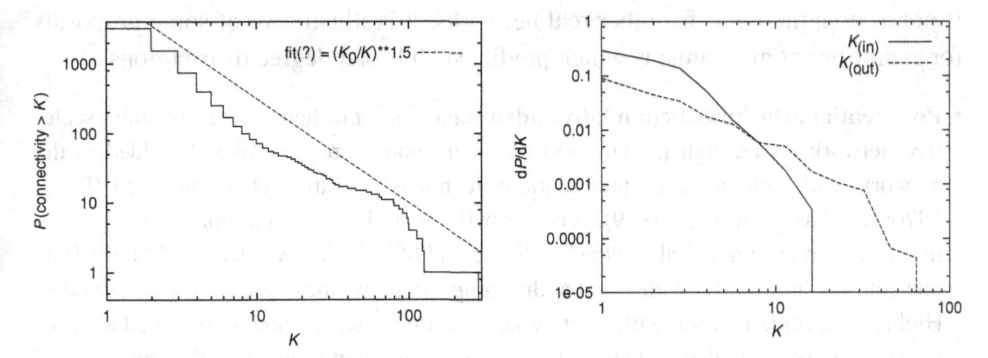


Figure 8.5. Left: overall distribution of edge degree in the two hybrid measurements of Ito *et al.* (2001). We show the cumulative distributions $P(> K)$, as these allow better judgment of suggested fits. Notice $P(> K) \propto 1/K^{1.5}$ corresponds to $N(K) = \mathrm{d}P/\mathrm{d}K \propto 1/K^{2.5}$. Right: $N(K) = \mathrm{d}P/\mathrm{d}K$ for a regulatory network in yeast, separated into the in-link and the out-link connectivity distributions, respectively. Notice that out-links are broader, not far from $N(K) \propto 1/K^{1.5}$.

on average, should lead to one output signal through a new exit. When $\mathcal{A} > 1$ signals can be exponentially amplified, and thus most signals can influence signaling over the entire network. For broad connectivity distributions $\mathcal{A}$ typically depends on the node with highest connectivity. To see this, assume that the number of neighbors is power-law distributed (Eq. (8.5)). Then

$$\mathcal{A} = \frac{\int_1^N (K^2 \mathrm{d}K)/K^\gamma}{\int_1^N (K \mathrm{d}K) K^\gamma} - 1 \sim N^{3-\gamma} \tag{8.8}$$

for $\gamma > 2$. That is, in case of $\gamma > 2$ the denominator becomes independent of $N$ in the limit of large $N$. One notices that, for $\gamma < 3$, $\mathcal{A}$ depends on the upper cut off in the integral, which represents the node (protein) with highest connectivity. Most real-world networks are fitted with exponents between 2 and 3. We stress that the estimate in Eq. (8.8) is valid only when the network consist of nodes that are randomly connected to each other. Further, it assumes that signals always are going anywhere where there are connections. For molecular networks these two conditions are not fulfilled.

The power-law distribution in Eq. (8.5) is often referred to as scale-free, because it signals that the change in behavior is independent of scale:

$$\frac{N(aK)}{N(K)} = \frac{N(a)}{N(1)} \tag{8.9}$$

That is, each time we multiply $K$ by a certain factor $a$, the frequency $N$ decreases with a given other factor $(1/a^\gamma)$. Although scale-free nature of molecular network is not as convincing as for other real networks, it is illustrative of some proposals for evolutionary mechanisms which predict such broad degree distributions.

- **Preferential attachment/cumulative advantage.** One mechanism for obtaining scale-free networks is through growth models, where nodes are subsequently added to the network, with links attached preferentially to nodes that are highly connected (Price, 1976; Barabasi & Alberts, 1999). It is a growth model based on minimal information in the sense that each new link is attached to the end of a randomly selected old link. Thus one connects new nodes with a probability proportional to the degree of the older nodes. Highly connected nodes therefore grow faster. After $t$ steps, $t$ nodes are added and, for the simplest version, also $t$ edges. Let $n(k, t)$ be the number of nodes with connectivity $k$ at time $t$. The evolution of $n$ is given by (Bornholdt & Ebel, 2001)

$$n(k, t+1) - n(k, t) = \frac{(k-1) \cdot n(k-1, t) - k \cdot n(k, t)}{\sum kn(k)} \text{ for } k > 1 \tag{8.10}$$

because the probability to add a link to a specific node of connectivity $k$ is $k/\sum kn(k)$. Now each added node is associated with two edge ends, and thus $\sum_k kn(k, t) = 2t$. Accordingly, the continuous limit is

$$\frac{\mathrm{d}n}{\mathrm{d}t} = -\frac{1}{2t}\frac{\mathrm{d}(k \cdot n)}{\mathrm{d}k} \tag{8.11}$$

Now for large $t$ the relative frequency of most nodes will have a stationary distribution and $n(k)/t$ is constant. From this one obtains

$$\begin{aligned} \frac{\mathrm{d}n}{\mathrm{d}t} &= \frac{n(k)}{t} = -\frac{1}{2t}\frac{\mathrm{d}(k \cdot n)}{\mathrm{d}k} \Rightarrow \\ 2n &= -n - k \cdot \frac{\mathrm{d}n}{\mathrm{d}k} \Rightarrow n(k) \propto \frac{1}{k^3} \end{aligned} \tag{8.12}$$

In fact the obtained scaling behavior can be modulated by introducing further addition of links (see Fig. 8.6).

The preferential growth model was originally proposed in an entirely different context, relating to the modeling of human behavior quantified by the Zipf law (1949): that "law" states the empirical observation that incomes, or assets, or number of times particular words are used, all tend to be distributed with power laws of type $1/s^2$. Simon (1955) suggested that this reflected the human tendency to preferentially give to what already has. For networks, a feature of this history-dependent model is that the most connected nodes are also the oldest. Another feature is that steady-state preferential attachment and random elimination of nodes do not generate scale-free networks. Scale-free behavior relies on the ongoing growth process. This is an unrealistic restriction for molecular network for which current regulation presumably reflects a snapshot in some sort of steady-state evolutionary sampling. A third feature also at odds with protein networks in at least yeast, is that preferential attachment tends to link highly connected nodes with one another.

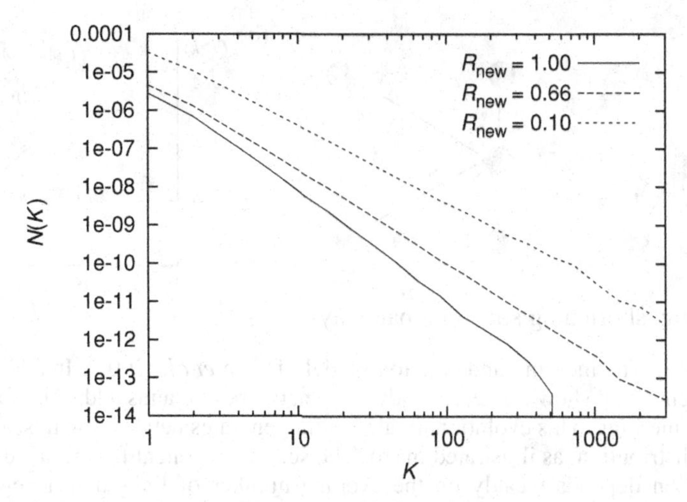


Figure 8.6. Connectivity distribution in a preferential attachment model for three different values of $R_{\text{new}}$. For all values of $R_{\text{new}}$ one obtains a power law. That is, at each timestep one node is added with probability $R_{\text{new}}$ and, if not, a link between two nodes is added. Ends of links are added preferentially, that is each end is linked to a node with a probability proportional to the connectivity of that node (see also Bornholdt & Ebel, 2001). The obtained power law depends on $R_{\text{new}}$ and approaches $1/K^3$ when $R_{\text{new}} = 1$. Thus only adding nodes develops a network with very steep $K$ distribution. Given that each newly added node has to come with at least one outgoing link it is impossible to obtain steeper distributions with growth guided by preferential attachment.

- **Threshold networks.** Another model for generating networks with power-law distributed connectivity is the threshold networks considered by Calderelli *et al.* (2002). In this model each protein $i$ is assigned an overall binding strength $G_i$ selected from an exponential distribution, $P(G) \propto \exp(-G)$. Then one assigns a link to all protein pairs $i, j$ where $G_i + G_j$ is larger than a fixed detection threshold $\Theta$. Thereby a network with scaling in some limited range is generated.

  Analytically the scaling comes about because a given protein is assigned a $G$ with probability $\exp(-G)$, and thereby the number of binding partners equals the number of proteins with $G' > \Theta - G$. This number is proportional to $\exp(+G)$. Therefore there is probability $P(> G) = \exp(-G)$ for having $N \propto \exp(+G)$ partners:

$$P(> N) = 1/N \Rightarrow \frac{\mathrm{d}P}{\mathrm{d}N} \propto 1/N^2 \tag{8.13}$$

Thus threshold networks generate scale-free networks (when threshold $\Theta$ is rather large). In contrast to the preferential attachment model, the threshold model is history independent. But in the above formulation it is very non-specific: good binders bind to all reasonably strong binders. Presumably real networks have a specific reason to

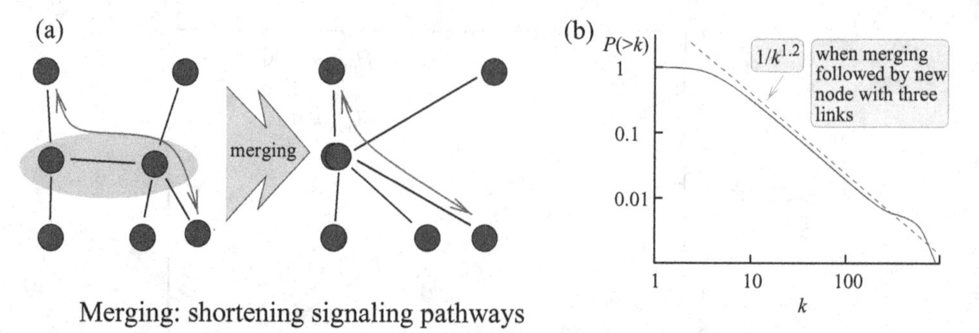


Figure 8.7. The merging and creation model of Kim *et al.* (2003). In addition to the merging step shown here, a steady-state network demands addition of a node for each merging. This evolutionary algorithm generates networks with scale-free degree distribution, as illustrated in (b). The scaling exponent for the steady-state distribution depends weakly on the average number of links that a new node attaches to the older ones.

have a broad connectivity distribution. Life would definitely not favor a prion-like mechanism where some proteins aggregate many other proteins into one giant connected clump.

- **Merging and creation (/duplication).** A third scenario for generating a scale-free network is the merging and creation model introduced by Kim *et al.* (2004). Here a scale-free network is generated by merging nodes and generating new nodes. In detail, one time step of this algorithm consists of selecting a random node, and one of its neighbors. These are merged into one node (see Fig. 8.7a). Subsequently one adds a new node to the network and links it to a few randomly selected nodes. As a result one may generate a nearly scale-free network with degree exponent $\gamma \sim 2.2$ (see Fig. 8.7b). The justification for this algorithm for protein networks would be merging in order to shorten pathways, and creation in order to generate new functions. In contrast to the preferential attachment model, the merging/creation scenario does not demand persistent growth. Instead it suggests an ongoing dynamics of an evolving network, which at any time has a very broad degree distribution.

The merging and creation model has its correspondence in physics, where it was originally suggested in the form of the aggregation and injection model for dust (Fields & Saslow, 1965). The mechanism is probably important in the creation of larger aggregates in the interstellar vacuum.

For protein interaction networks an ongoing merging process may seem hopeless. There is simply a limit to how large a protein can be. It is in any case interesting that the merging of proteins is a real phenomenon. Thus Mirny *et al.* (personal communication) analyzed various prokaryotes and concluded that proteins that are close along a metabolic pathway quite often merge. In living cells the creation, on the other hand, quite often comes from duplication and subsequent evolution of already existing proteins (see

Chapter 9). These duplication events can in themselves also contribute to a broad degree distribution (Pastor-Satorras *et al.*, 2003).

As a summary, we stress that any one of the above models presents only some possible evolutionary elements in obtaining a broad degree distribution. As we will subsequently see, the relative positioning of highly connected proteins relative to each other may provide us with a more functional view.

## *Questions*

(1) Write equations for protein production according to the feed-forward loop of the type shown in Fig. 8.10, that is A $\rightarrow$ B $\rightarrow$ C, while A $\rightarrow$ C. Let A be given by A $= \Theta(t)$ (step function), and simulate the amount of C when: (a) one assumes that either B or A have to be present in order for C to be produced, and (b) one assumes that both A and B have to be present. In all cases assume production from each arrow to be of the form $\mathrm{d}N_i/\mathrm{d}t = N_j/(N_j + 1)$ ($N_j =$ A or $N_j =$ B) and for each protein a spontaneous decay $\mathrm{d}N_i/\mathrm{d}t = -N_i$. If there are two input arrows their contribution should be added in (a), and multiplied in (b).

(2) Repeat the above simulation with the "and" gate when $\mathrm{A}(t) = \Theta(t) \cdot \Theta(\tau - t)$ with $\tau = 0.1$ and $\tau = 1.0$.

(3) Repeat (2) for a production where all Hill coefficients are 4 instead of 1.

(4) Simulate a network that grows with preferential attachment, say at each time step one node with one link is added with probability $R_{\mathrm{new}} = 0.1$, and if not, then only a link between two nodes is added (preferentially linked to nodes in both ends). Let the network reach a steady state, by removing nodes plus all their links with small probability $\epsilon \sim 0.001$. Quantify connectivity distribution plus nearest-neighbor correlations.

(5) Consider the model of (4), let $n(k, t)$ be the number of nodes with connectivity $k$ at time $t$. Find the analytical expression for the steady-state distribution of $n(k)$ for different values of $R_{\mathrm{new}}$.

(6) Simulate the threshold network model of Calderelli *et al.* (2002), using various thresholds. Also repeat the simulations for a Gaussian distribution, that is $P(G) \propto \exp(-G^2)$, instead of an exponential distribution.

(7) Consider the merging/creation model. Initially set all $k_i = 2$. Implement the merging model where one at each step $k_i, k_j \rightarrow k_i = k_i + k_j - \delta$ and $k_j = \delta$, where $\delta = 1, 2$ or 3 with equal probability. Accept moves only where all $k_i > 0$. What is the final distribution of $k$? What is the distribution if one allows negative $k$s?

(8) Prove that the scale-free condition in Eq. (8.9) implies a power-law connectivity distribution (Eq. 8.5).

(9) Simulate the merging/creation model in terms of a set of integers $k_i$, $i = 1, 2, \ldots n$, which are updated according to

$$k_i, k_j \rightarrow k_i = k_i + k_j - \delta, \quad \text{and } k_j = \delta \tag{8.14}$$

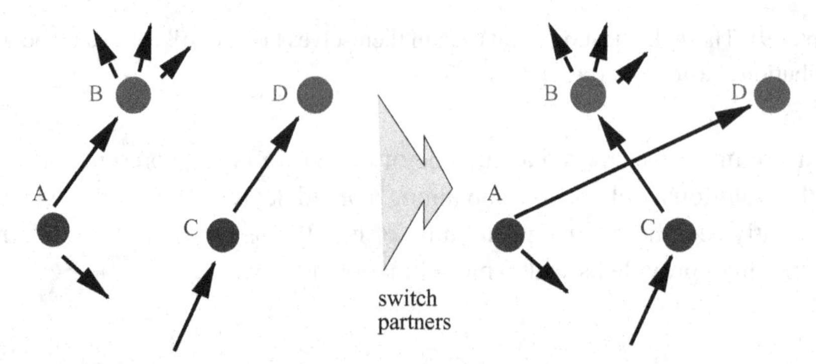


Figure 8.8. One step of the local rewiring algorithm (see Maslov & Sneppen, 2002a). We have a pair of directed edges A→B and C→D. The two edges then switch connections in such a way that A becomes linked to D, while C becomes linked to B, provided that none of these edges already exists in the network, in which case the move is aborted and a new pair of edges is selected. An independent random network is obtained when this is repeated a large number of times, exceeding the total number of edges in the system. This algorithm conserves both the in- and out-connectivity of each individual node.

where $\delta$ is a random number with mean equal to the average $\langle k \rangle$ in the total system. Show that this generates a steady-state distribution of the sizes of the numbers in the set $n(k) \propto (1/k^{\gamma})$ with $\gamma \sim 2.5$ when we allow updates only where all $k_i > 0$.

## Analysis of network topologies

We now want to discuss how to identify non-trivial topological features of networks. That is, we want to go beyond the single-node property defined by the degree distribution, and thus deal with the networks as objects that are indeed connected to each other. The hope is that, in the end, this may help us to understand the function–topology relationship of various types of network. The key idea in this analysis is to compare the network at hand with a properly randomized version of it. As we want to go beyond degree distributions we want to compare with random networks with exactly the same degree distribution as the real network we are analyzing. One way of generating such random networks is shown in Fig. 8.8. Technically the significance of any pattern is measured by its $Z$ score:

$$Z(\text{pattern}) = \frac{N(\text{pattern}) - \langle N_{\text{random}}(\text{pattern}) \rangle}{\sigma_{\text{random}}(\text{pattern})} \tag{8.15}$$

where $N_{\text{random}}(\text{pattern})$ is the number of times the pattern occurs in the randomized network. We can also say that

$$\sigma^2_{\text{random}}(\text{pattern}) = \langle N_{\text{random}}(\text{pattern})^2 \rangle \; - \; \langle N_{\text{random}}(\text{pattern}) \rangle^2 \tag{8.16}$$

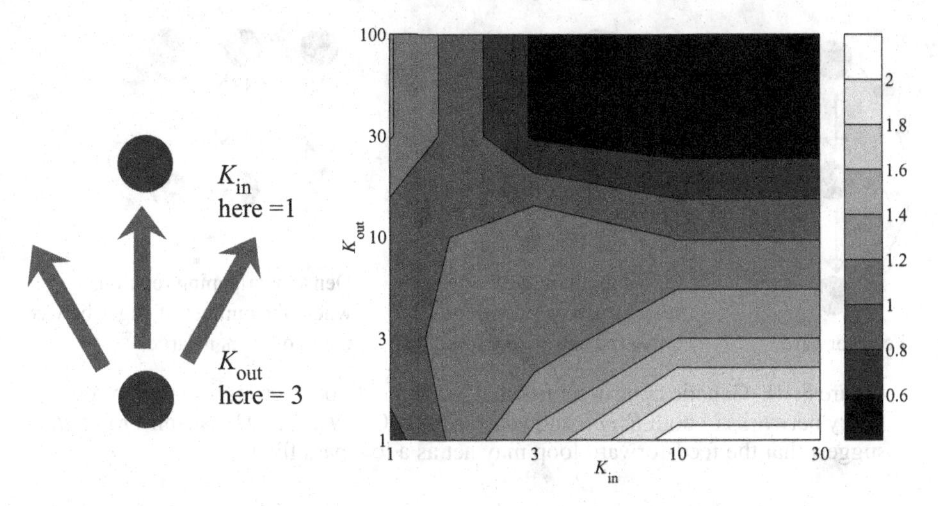


Figure 8.9. Correlation profile (Maslov & Sneppen, 2000a) showing the correlation between connected proteins in the regulatory network of yeast, quantified in terms of $Z$ scores. The single output modules are reflected in the abundance of high $K_{out}$ controlling single $K_{in}$ proteins. The dense overlapping regulons correspond to the abundance of connections between $K_{out} \sim 10$ with $K_{in} \sim 3$ proteins. Finally the suppression of connections between highly connected proteins shows that these tend to be on the periphery of the network.

is the variance of this number among the random networks. Considering patterns of links between proteins with various degrees, Maslov & Sneppen (2002a, b) reported a significant suppression of links between hubs, both for regulatory networks and for the protein–protein interaction networks in yeast (see Fig. 8.9). Similarly, by defining higher-order occurrences of various local patterns of control, Shen-Orr *et al.* (2002) found some very frequent motifs in gene regulation networks. These are illustrated in Fig. 8.10.

We stress that one should be careful when judging higher-order correlation patterns, because evaluation of these patterns is very sensitive to the null model against which they are judged. At least one should maintain the in and out degree of all nodes. But even when maintaining the degree distributions, short loops, for example, will easily appear to be hugely over-represented when compared with a randomly reshuffled network. This is because a simple randomization does not take into account the fact that proteins with similar functions tend to interact with each other (see Question (1) on p. 224). This locality in itself gives more loops. In fact one can often use the number of loops as a measure of locality, quantified in terms of the so-called cliquishness (Watts & Strogatz, 1998).

The tendency of highly connected proteins to be at the periphery of regulatory networks may teach us something about the origin of broad connectivity

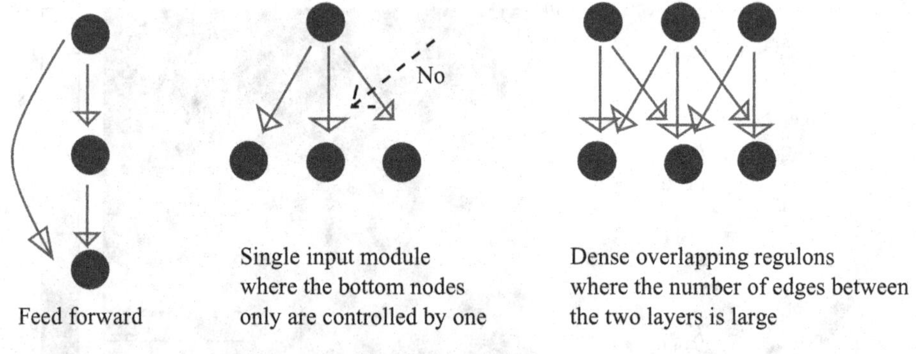


Figure 8.10. Genetic control motifs that are found to be over-represented in regulatory networks of both *E. coli* and yeast (Shenn-Orr *et al.*, 2002). Shenn-Orr *et al.* suggest that the feed-forward loop may act as a low pass filter.

distributions: maybe the hub proteins that "give orders" to many tend to give the same order to everyone below them. We found this feature in the phage networks discussed in Chapter 7. In λ the centrally placed CI is the major hub protein and it acts as a simple repressor for everything except itself. Similarly, the other hub proteins in λ, namely Cro, N and Q, all direct only one type of outcome. Thus for simple protein organisms, one may suggest: one regulatory protein, one commanding.

One may speculate that the broad-degree distribution of molecular networks is not an artifact of some particularly evolutionary dynamics (gene duplication, merging, etc.), but rather reflects the broad distribution of the number of proteins needed to do the different tasks required in a living system (Maslov & Sneppen, 2004). Some functions simply require many proteins, whereas many functions require only a few proteins. This scenario is not only supported by the tendency of highly connected proteins to sit on the periphery of the networks, but is also supported by the observation that there is essentially no correlation between connectivity of a regulatory protein and its importance as measured by its chance to be essential. That is, let us assume that highly connected proteins were involved in many functions, and also assume that each function had a certain likelihood of being essential. This would imply that the likelihood of being essential would grow linearly with the degree of a protein in the network. This is not the case, at least not for the yeast transcription network (Maslov & Sneppen, 2004). Thus the broad connectivity distributions in signaling and regulative molecular networks may reflect the widely different needs associated with the widely different functions that a living cell needs to cope with.

*Question*

(1) Construct a network of $N = 1000$ nodes subdivided into 10 different classes with 100 nodes in each. Generate a random network where each protein has about 3 links, and where each link has probability 0.75 to be between proteins of similar classes, and probability 0.25 to be between different classes. Calculate the number of loops, and compare this with the number of loops when all links are randomized.

## Communication ability of networks

A key feature of molecular, as well as most other, networks is that they define the channels along which information flows in a system. Thus, in a typical complex system one may say that the underlying network constrains the information horizon that each node in the network experiences (Rosvall & Sneppen, 2003). This view of networks can be formalized in terms of information measures that quantify how easy it would be for a node to send a signal to other specific nodes in the rest of the network (Sneppen *et al.*, 2004b). To do this one counts the number of bits of information required to *transmit* a message to a specific remote part of the network or, conversely, to *predict* from where a message is received (see Fig. 8.11).

In practice, imagine that you are at node $i$ and want to send a message to node $b$ in a given network (left panel in Fig. 8.11). Assume that the message follows the

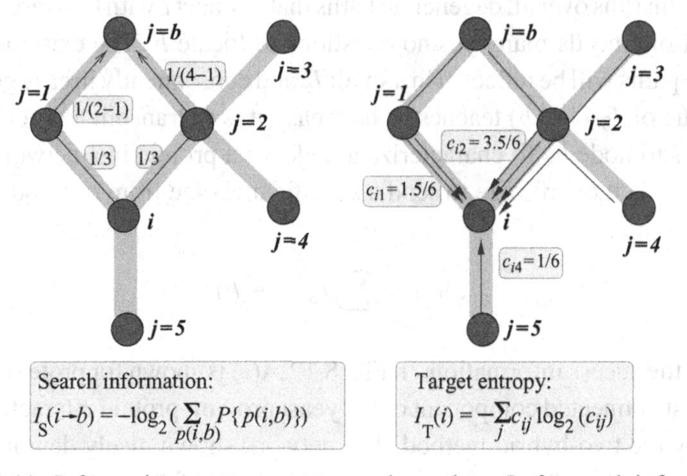


Figure 8.11. Information measures on network topology. Left: search information $S(i \to b)$ measures your ability to locate node $b$ from node $i$. $S(i \to b)$ is the number of yes/no questions needed to locate any of the shortest paths between node $i$ and node $b$. For each such path $P\{p(i, b)\} = (1/k_i), \prod_j 1/(k_j - 1)$, with $j$ counting nodes on the path $p(i, b)$ until the last node before $b$. Right: target entropy $T_i$ measures predictability of traffic to you located at node $i$. Here $c_{ij}$ is the fraction of the messages targeted to $i$ that passed through neighbor node $j$. Notice that a signal from $b$ in the figure can go two ways, each counting with weight 0.5.

shortest path. That is, as we are interested only in specific signals we limit ourselves to consider only this direct communication. If the signal deviates from the shortest path, it is assumed to be lost. If there are several degenerate shortest paths, the message can be sent along any of them. For each shortest path we calculate the probability to follow this path (see Fig. 8.11). Assume that without possessing information one would chose any new link at each node along the path with equal probability. Then

$$P\{p(i,b)\} = \frac{1}{k_i} \prod_{j \in p(i,b)} \frac{1}{k_j - 1} \tag{8.17}$$

where $j$ counts all nodes on the path from a node $i$ to the last node before the target node $b$ is reached. The factor $k_j - 1$ instead of $k_j$ takes into account the information we gain by following the path, and therefore reduces the number of exit links by one. In Fig. 8.11 we show the subsequent factors in going along any of the two shortest paths from node $i$ to node $b$. The total information needed to identify one of all the degenerate paths between $i$ and $b$ defines the "search information"

$$I_S(i \to b) = -\log_2 \left( \sum_{p(i,b)} P\{p(i,b)\} \right) \tag{8.18}$$

where the sum runs over all degenerate paths that connect $i$ with $b$. A large $I_S(i \to b)$ means that one needs many yes/no questions to locate $b$. The existence of many degenerate paths will be reflected in a small $I_S$ and consequently in easy goal finding.

The value of $I_S(i \to b)$ teaches us how easy it is to transmit a specific message from node $i$ to node $b$. To characterize a node, or a protein in a network, one may ask how easy is it on average to send a specific message from one node to another in the net:

$$\mathcal{A}(i) = \sum_b I_S(i \to b) \tag{8.19}$$

$\mathcal{A}$ is called the access information. In Fig. 8.12 $\mathcal{A}(i)$ is shown for proteins belonging to the largest connected component of the yeast protein–protein interaction network obtained by the two-hybrid method. The network shown nicely demonstrates that highly connected nodes are often on the periphery of the network, and thus do not provide particularly good access to the rest of the system. This is not what we see in a randomized version of the network, where all in and out degrees are maintained; the network is kept globally connected, but partners are reshuffled. In fact we quantify the overall ability for specific communication

$$\mathcal{I}_S = \sum_i \mathcal{A}(i) = \sum_{i,b} I_S(i \to b) \tag{8.20}$$

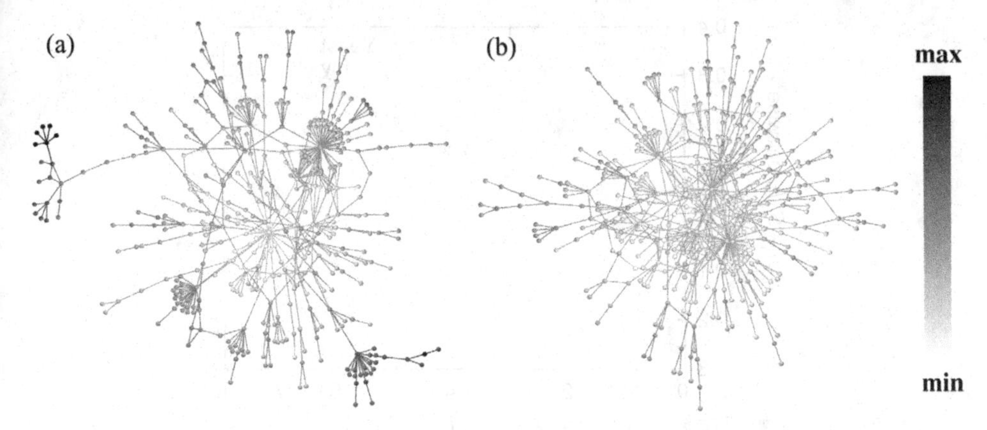


Figure 8.12. (a) Analysis of the protein–protein interaction network in yeast defined by the connected components of the most reliable data from the two-hybrid data of Ito *et al.* (2001). The value of the shown access information $\mathcal{A}_i$ increases from the light-colored area in the center to the darker area in the periphery. The dark colors mark nodes that have least access to the rest of the network, i.e. proteins that are best hidden from the rest. (b) A randomized version of the same network. One see that hubs are more interconnected and that typical $\mathcal{A}$ values are smaller (less dark).

and compare it with the value $\mathcal{I}_S(\text{random})$ obtained for a randomized network. In Fig. 8.14 we plot the $Z$ score defined as

$$Z = \frac{\mathcal{I}_S - \langle \mathcal{I}_S(\text{random})\rangle}{\sqrt{\langle \mathcal{I}_S(\text{random})^2\rangle - \langle \mathcal{I}_S(\text{random})\rangle^2}} \tag{8.21}$$

for the protein–protein network for both yeast (*Saccromyces Cerevisiae*) (Uetz *et al.*, 2000; Ito *et al.*, 2001) and fly (*Drosophila*) (Giot *et al.*, 2003), as well as for the hardwired Internet and a human network of governance (CEO) defined by company executives in USA where two CEOs were connected by a link if they are members of the same board. One sees that $\mathcal{I}_S > \mathcal{I}_S(\text{random})$ for most networks, except for the fly network. Thus most networks have a topology that tends to hide nodes.

The tendency to hide or communicate, respectively, can be quantified further by considering the average information $\langle S(l)\rangle$ needed to send a specific signal a distance $l$ inside the network (the average is over all nodes and all neighbors at distance $l$ to these nodes in the given network). This is done in Fig. 8.13. We see that $\langle S(l)\rangle - \langle S_{\text{random}}(l)\rangle$ has a minimum below zero for some rather short distance $l \sim 3$, whereas it becomes positive for large $l$. Thus the molecular signaling networks have relatively optimal topology for local specific communication, but at larger distances the proteins tend to hide.

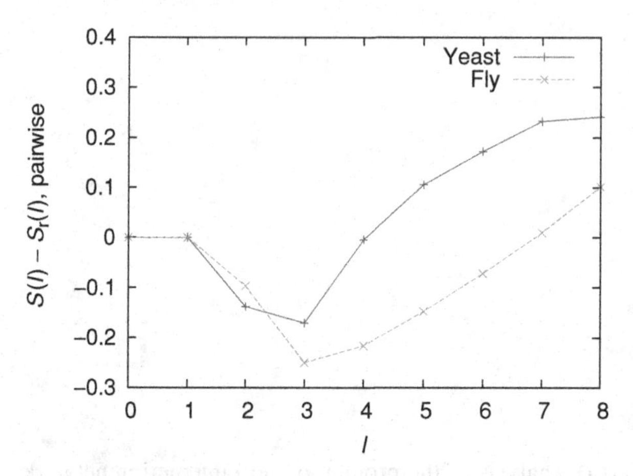


Figure 8.13. Average information needed to send a specific signal to a node at distance $l$, compared with the information needed if the network was random. Both cases refer to the protein–protein interaction networks, measured by the two-hybrid method. In both cases the random network is constructed such that, for each protein, its number of bait and its number of prey partners are conserved during the randomization.

In Fig. 8.14 we also show another quantity, the ability to predict from which of your neighbors the next message to you will arrive. This quantity measures predictability, or alternatively the order/disorder of the traffic around a given node $i$. The predictability based on the orders that are targeted *to* a given node $i$ is

$$I_T(i) = -\sum_{j=1}^{k_i} c_{ij} \log_2(c_{ij}), \tag{8.22}$$

where $j = 1, 2 \ldots, k_i$ denotes the links from node $i$ to its immediate neighbors $j$, and $c_{ij}$ is the fraction of the messages targeted to $i$ that passed through node $j$. As before our measure implicitly assumes that all pairs of nodes communicate equally with one another.

Notice that $I_T$ is an entropy measure, and as such is a measure of order in the network. In analogy with the global search information $I_S$ one may also define overall predictability of a network

$$\mathcal{I}_T = \sum_i I_T(i) \tag{8.23}$$

and compare it with its random counterparts. In general as the organization of a network appears more disorganized $\mathcal{I}_T$ increases and the number of alternative pathways increases.

In summary, networks are coupled to specific communication and their topology should reflect this. The optimal topology for information transfer relies on a

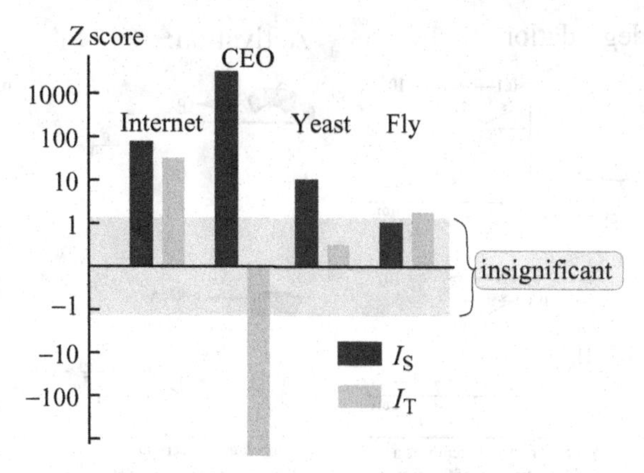


Figure 8.14. Measure of communication ability of various networks (Sneppen *et al.*, 2004b). A high $Z$ score implies relatively high entropy. In all cases we show $Z = (I - I_r)/\sigma_r$ for $I = I_S$ and $I_T$, by comparing with $I_r$ for randomized networks with preserved degree distribution. Here $\sigma_r$ is the standard deviation of the corresponding $I_r$, sampled over 100 realizations. Results within the shaded area of two standard deviations are insignificant. All networks have a relatively high search information $I_S$. The network of governance CEO shows a distinct communication structure characterized by local predictability, low $I_T$, and global inefficiency, high $I_S$.

system-specific balance between effective communication (*search*) and not having the individual parts being unnecessarily disturbed (*hide*). For molecular networks of both yeast and fly we observed that communication was good at short distances, whereas it became relatively bad at large distances. Thus local specificity is favored, whereas globally the proteins tend to hide. Presumably this reflects some hidden modularity of the protein interaction networks.

### *Questions*

(1) Calculate $\mathcal{A}_S(i)$ and $I_T(i)$ for all nodes $i$ in the network in Fig. 8.11.
(2) Rewire the network such that one degree of all nodes intact, and such that the network is connected. Calculate the total $I_S$ and $I_T$ for both the old network and the new network.

## Elementary dynamics of protein regulation

When constructing dynamical models of molecular networks it is important to know which types of interactions one is dealing with. A negative regulation may be merely inhibitive for the production, it may be blocking activity through binding, or it may direct the degradation of the regulated protein. Similarly, a positive regulation may activate production, it may change the protein property through, for example,

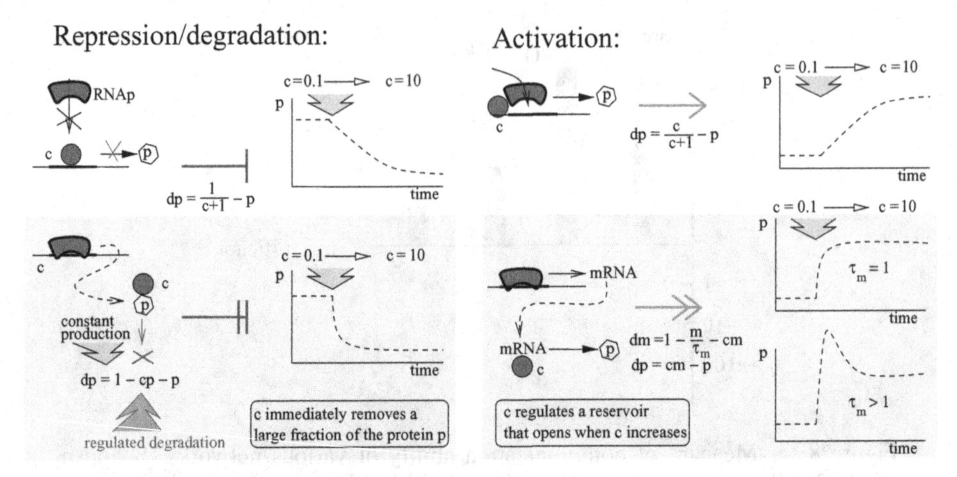


Figure 8.15. Ways to regulate the concentration of a protein (**p**), through negative and positive control, and through transcription regulation and something faster (from Bock & Sneppen, 2004). The upper panels are the simplest regulation option, where **c** simply regulates transcription of mRNA for **p**. In this case the time to obtain new steady state is set by the degradation rate of **p**. Fast supression or activation, respectively, are obtained when one use active degradation (lower left), and translational control (lower right). In lower right panel, **m** is inactive mRNA for protein **p** that accumulates in a reservoir for eventual conversion into a fast decaying active form.

phosphorylation and thus instantly activate a hidden reservoir of passive proteins. Also on fast timescales, it may activate a hidden reservoir of passive mRNA and thereby lead to a sudden burst in protein production. In any case, the method of regulation has a huge effect on the dynamics of the regulation.

Figure 8.15 shows different strategies for regulating the concentration of a protein. In all cases the external regulation takes place through a change in **c** that typically represents either a protein or a change in binding constant. The strategies shown may be combined; for example, a given protein may be both positively regulated by another protein, and negatively auto-regulated by itself. This is illustrated in Fig. 8.16.

Comparing the regulations in Fig. 8.15 one sees that protein degradation (for negative regulation), and translation control (for positive regulation) provide the most dramatic change in protein concentration. In accordance with this observation, it is typically these types of interaction that are found in relation to stress responses, for example the heat shock mentioned in the next section. A quantifiable example of translation control is found in the unfolded protein response in yeast (see Cox & Walter, 1996; Sidrauski *et al.*, 1997; and for the model see Bock & Sneppen, 2004).

In the next section we will describe in detail one particular network, where both transcriptional activation and protein degradation take place.

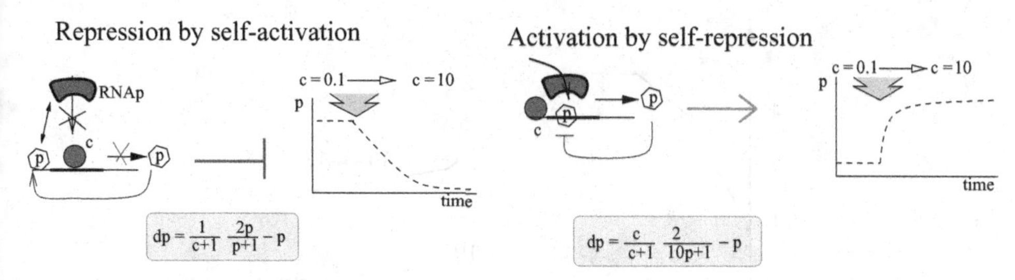


Figure 8.16. Illustration of the effect obtained when a protein activates and represses its own gene. Combined with an oppositely acting transcriptional regulator **c**, the self-regulation enhances the effect of the **c**. In the left-hand panel we see that the contrast between the repressed (**c** large) and non-repressed states is enhanced by self-activation (compare with Fig. 8.15). In the right-hand panel we seen that a positively acting regulator **c** may be helped to increase protein level faster, when the protein represses its own gene. This latter mechanism was suggested and studied experimentally by Rosenfeld *et al.* (2002). The suggested factors in the self-regulation are for Question (2).

*Questions*

(1) Simulate the different regulations in Fig. 8.15. Explain qualitatively why the translation response develops an increasing shock as $\tau$ becomes increasingly larger than 1 (= units of degradation time for the protein **p** in the figure).
(2) Simulate the different regulations in Fig. 8.16. Explain the meaning of the constants in the self-regulation.
(3) Consider the systems $\mathbf{dp/dt} = (\mathbf{c}/(\mathbf{c}+1)) \cdot (1/(1+10 \cdot \mathbf{p})) - \mathbf{p}$ versus $\mathbf{dp/dt} = (\mathbf{c}/(\mathbf{c}+1+10 \cdot \mathbf{p})) - \mathbf{p}$. Draw the genetic regulation corresponding to the two cases and investigate which gives the faster response under a $\mathbf{c} = 0.1 \rightarrow \mathbf{c} = 10$ input change.

## The heat shock network: an example of a stress response system

Protein stability and folding properties depend on temperature, as discussed in Chapter 5. To deal with temperature changes a living cell needs to have a mechanism to maintain all proteins folded even under extreme changes in external conditions. Such changes are common in nature. Think for example of an *E. coli* that occasionally moves from a cold water environment to the intestines of a mammal. Life has to deal with such shocks!

A number of proteins (about 5–10%) need chaperone proteins to fold. A chaperone is a molecular machine whose main mechanism is to bind to unfolded proteins, and facilitate their folding into their correct functional state. The expression level of these catalysts of protein folding changes in response to environmental stress. In particular, when a living cell is exposed to a temperature shock, the production of Chaperone proteins is transiently increased. This change in transcriptional activity

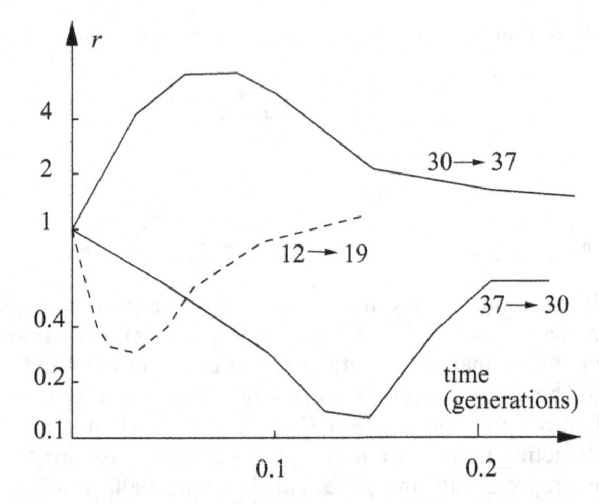


Figure 8.17. Heat shock response measured as the change in DnaK production rate as a function of time since a temperature shift. The production is normalized with overall protein production rate, as well as with its initial rate. In all cases the step $\Delta T = 7\,°\mathrm{C}$. The timescale is in units of one bacterial generation measured at the initial temperature. At $T = 37\,°\mathrm{C}$ the generation time is 50 min, at 30 °C it is 75 min and at 20 °C it is 4 h. Do not put too much emphasis on the flawed experiment for the late response to the temperature decrease experiment.

is called the heat shock. The response is seen for all organisms, and in fact involves related proteins across all biological kingdoms. The heat shock (HS) response in *E. coli* involves about 40 genes that are widely dispersed over its genome (Gross, 1996).

Following Arnvig *et al.* (2000) we will see that the heat shock indeed addresses protein stability and how this is communicated to a changed transcriptional activity. For *E. coli* the response is activated through the $\sigma^{32}$ protein. The $\sigma^{32}$ binds to RNA polymerase (RNAp) where it displaces the $\sigma^{70}$ sub-unit and thereby changes the affinity of RNAp to a number of promoters in the *E. coli* genome. This induces production of the heat shock proteins.

The heat shock is fast. In some cases it can be detected by a changed synthesis rate of, for example, the chaperone protein DnaK about 1 min after the temperature shift. Given that production of DnaK protein in itself takes this time, the fast change in DnaK production must be triggered by a mechanism that is at most a few chemical reactions away from the DnaK production itself. To quantify the heat shock, Arnvig *et al.* (2000) measured the dependence of $\sigma^{32}$ synthesis with initial temperature. In fact Arnvig *et al.* (2000) used the expression of protein DnaK because its promoter is activated only through $\sigma^{32}$.

In practice the heat shock shown in Fig. 8.17 is measured by counting the number of proteins produced during a short pulse of radioactive-labeled methionine.

Methionine is an amino acid that the bacteria absorb very rapidly, and then use in protein synthesis. After a large sample of bacteria have been exposed to a sudden temperature shift, small samples of the culture are extracted at subsequent times. Each sample is exposed to radioactive-labeled methionine for 30 s, after which non-radioactive methionine is added in huge excess ($10^5$ fold). During the 30 s of exposure, all proteins including DnaK will be produced with radioactive methionine. Protein DnaK can be separated from other proteins by 2-dim gel electrophoresis, a method that is briefly described later in this chapter. Finally the total amount of synthesis during the 30 s of labeled methionine exposure is counted through its radioactive activity.

The result is a count of the differential rate of DnaK production (i.e. the fraction DnaK constitutes of the total protein synthesis relative to the same fraction before the temperature shift). For the shift $T \to T + \Delta T$ at time $t = 0$

$$r(T, t) = \frac{\text{Rate of DnaK production at time } t}{\text{Rate of DnaK production at time } t = 0} \tag{8.24}$$

where the denominator counts steady-state production of DnaK at the old temperature $T$. Figure 8.17 displays three examples, all associated with temperature changes of absolute magnitude $\Delta T = 7\,°\text{C}$. When changing $T$ from 30 °C to 37 °C, $r$ increases to $\sim 6$ after a time of 0.07 generation. Later the expression rate relaxes to normal levels again, reflecting that other processes counteract the initial response. When reversing the jump, one observes the opposite effect, namely a transient decrease in expression rate. Thus the heat shock reflects some sort of equilibrium physics, where a reversed input gives reversed response.

Figure 8.17 also shows the effect of a temperature jump starting from a low temperature $T$. The shock is then reversed: a positive jump in $T$ causes a decreased expression $r$. Figure 8.18 summarizes the findings by plotting the value of $r = R$ where the deviation from $r = 1$ is largest, for a number of positive temperature quenches $T \to T + 7\,°\text{C}$. The dependence of $R$ on initial temperature $T$ is fitted by

$$\ln(R(T)) = (\alpha \Delta T)\,(T - T_s) \tag{8.25}$$

where $R(T = T_s = 19°) = 1$ and $\alpha \Delta T = \ln(R_1/R_2)/(T_1 - T_2) = 0.2 \cdot K^{-1}$ (i.e. $\alpha = 0.03 K^{-2}$). Notice that the $R = 1$ at 19°, corresponds to a $T$ shift from $T = T_s = 19$ to $T = T_s + \Delta T = 26°$. Thus a shift in temperature around $T_s + \Delta T/2 = 23$ will not give any heat shock. If we interpret $R$ as a ratio of a chemical binding constant $K$ at two different temperatures ($r$ does not change at time $= 0$), one may write

$$R = \frac{K(T + \Delta T)}{K(T)} \tag{8.26}$$

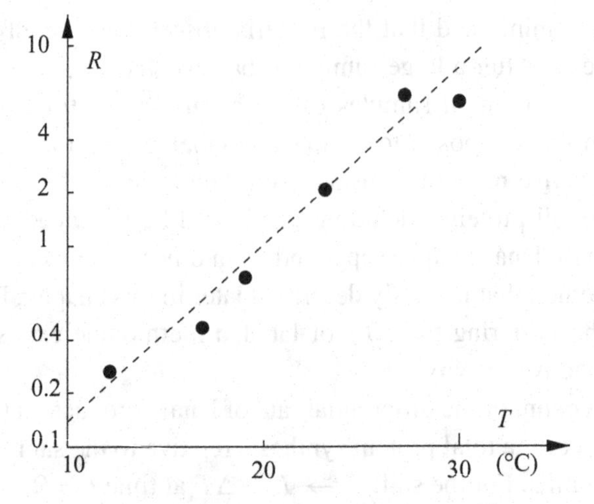


Figure 8.18. Induction fold $R$ (=extremum of $r$) for positive temperature jumps as a function of initial temperature. The straight line corresponds to the fit used.

which implies

$$\ln(R) = \ln(K(T + \Delta T)) - \ln(K(T)) = \frac{d\ln(K)}{dT}\Delta T \tag{8.27}$$

Inserting Eq. (8.25) into this expression we obtain

$$\ln(K) \approx \text{const} + \frac{\alpha}{2}(T - T_s)^2 \tag{8.28}$$

Setting $K = \exp(-\Delta G/k_B T)$ we obtain

$$\Delta G \approx \Delta G_s - \frac{\alpha k_B T}{2}(T - T_s)^2 \tag{8.29}$$

Thus one might expect $\Delta G$ to have a maximum at about $T = T_s + \Delta T/2 = T_s + 7°/2 \sim 23°$.

The association of the heat shock to a $\Delta G$ with a maximum at $T = T_s + \Delta T/2 \approx 23°$ presumably reflects the fact that many proteins exhibit a maximum stability at $T$ between 10°C and 30°C (see Fig. 8.19 and Chapter 5). Thus for a typical protein, $\Delta G = G(\text{unfolded}) - G(\text{folded})$ is maximum at 23°. In fact the size of the $\Delta G$ change inferred from the measured value of $\alpha = 0.03K^{-2}$ corresponds to the change in $\Delta G$ with temperature $T$ that one observes for typical proteins (see also Question (1) on p. 237).

To provide a positive response when one moves toward temperatures where proteins are less stable, one of the interactions along the signaling path from unfolded proteins to DnaK production should reverse the signal (be inhibitory). Figure 8.20 shows the known molecular network for regulation of the heat shock with the

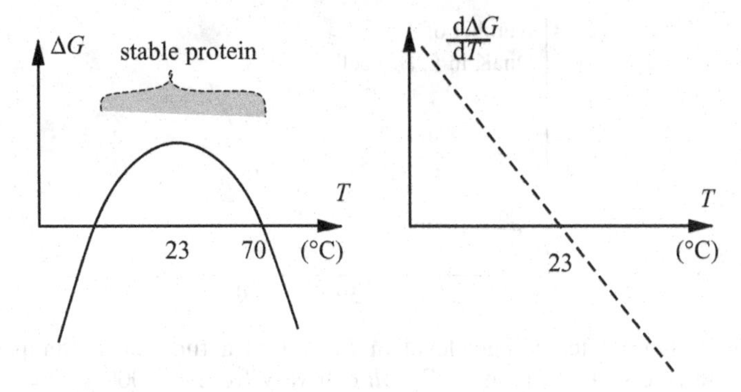


Figure 8.19. Typical stability of a protein, $\Delta G = G(\text{unfolded}) - G(\text{folded})$, and how fast this stability changes with temperature.

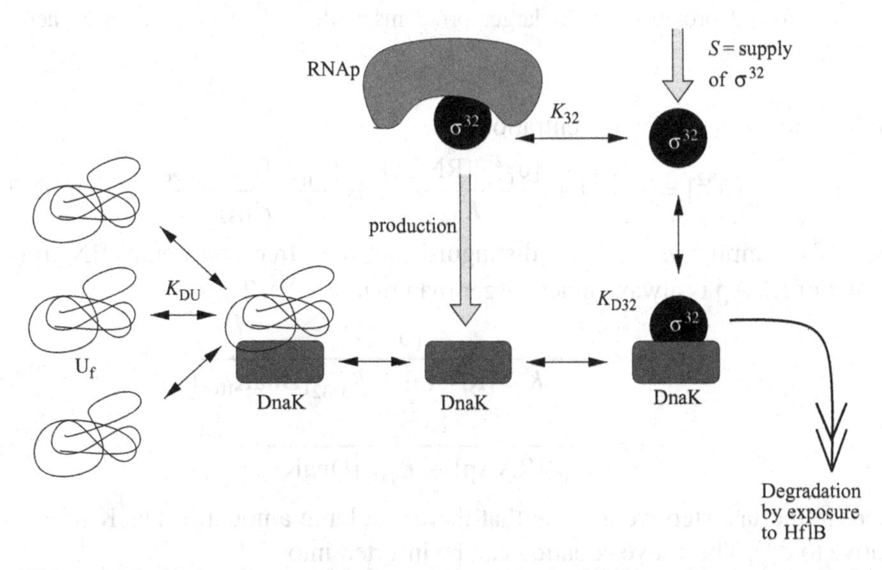


Figure 8.20. Molecular network for early heat shock. All lines with arrows at both ends are chemical reactions that may reach equilibrium within a few seconds (they represent the homeostatic response). The broad arrows are one-way reactions, with the production of DnaK through the $\sigma^{32}$–RNAp complex being the central one. The temperature dependence of the early heat shock is reproduced when the concentration of unfolded proteins ([U]) is much larger than the concentration of DnaK, which in turn should be much larger than the concentration of $\sigma^{32}$. Equilibrium feedback works from high to low concentrations! The DnaK–$\sigma^{32}$ complex exposes only $\sigma^{32}$ to degradation.

inhibitory link of DnaK to $\sigma^{32}$. This network allows us to build a dynamic model for the production of DnaK.

DnaK is produced owing to the presence of $\sigma^{32}$. If one assume that $\sigma^{32}$ are bound either to DnaK or to RNAp then there is a tight coupling between DnaK levels and DnaK production rates (see Fig. 8.20). Formally the total $\sigma^{32}$ concentration $[\sigma^{32}_{\text{total}}]$

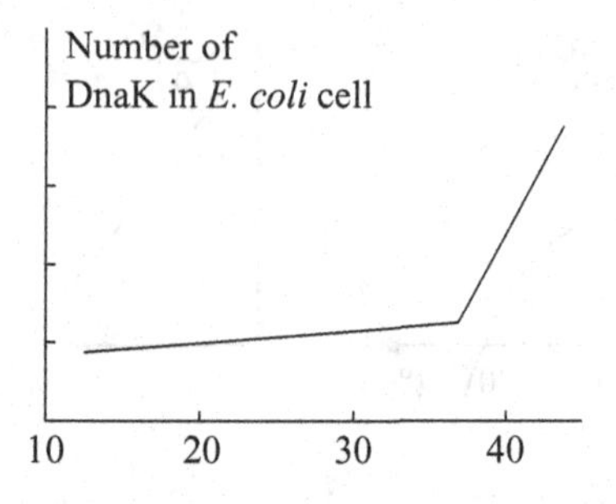


Figure 8.21. DnaK steady-state level in *E. coli* as a function of temperature. Steady-state levels of DnaK in an *E. coli* cell vary from $\sim 4000$ at $T = 13.5°C$ to $\sim 6000$ at 37°C, thereby remaining roughly proportional to the number of ribosomes (Herendeen *et al.*, 1979; Pedersen *et al.*, 1978). One observes a sharp rise above 37°C, reflecting either (1) decreased stabilization of already folded proteins, or (2) larger problems with folding proteins at higher temperatures.

is related to the free $\sigma^{32}$ concentration

$$[\sigma^{32}] = [\sigma^{32}_{\text{free}}] + \frac{[\sigma^{32}_{\text{free}}][\text{RNAp}]}{K_{32}} + \frac{[\sigma^{32}_{\text{free}}][\text{DnaK}_{\text{free}}]}{K_{\text{D32}}} \tag{8.30}$$

where, for simplicity, we don't distinguish between free and bound RNAp (the amount of RNAp is always much larger than that of $\sigma^{32}$). Thus

$$\begin{aligned}[\sigma^{32}_{\text{free}}] &= \frac{[\sigma^{32}]}{1 + K_{32}^{-1}[\text{RNAp}] + K_{\text{D32}}^{-1}[\text{DnaK}_{\text{free}}]} \\ &= \frac{[\sigma^{32}]}{K_{32}^{-1}[\text{RNAp}] + K_{\text{D32}}^{-1}[\text{DnaK}_{\text{free}}]}\end{aligned} \tag{8.31}$$

where in the last step we assume that there is a large amount of DnaK or RNAp relative to $\sigma^{32}$. The above equation can be inserted into

- $[\text{RNAp} \cdot \sigma^{32}] = [\text{RNAp}][\sigma^{32}_{\text{free}}]/K_{32}$
- $[\text{DnaK} \cdot \sigma^{32}] = [\text{DnaK}_{\text{free}}][\sigma^{32}_{\text{free}}]/K_{\text{D32}}$

which direct DnaK production and $\sigma^{32}$ degradation, respectively.

To obtain expressions for the feedback from the amount of unfolded proteins [U] we also need to relate the free DnaK to the total DnaK concentrations, given the amount of free unfolded proteins $[\text{U}_{\text{f}}]$

$$[\text{DnaK}] = [\text{DnaK}_{\text{free}}] + \frac{[\text{DnaK}_{\text{free}}][\text{U}_{\text{f}}]}{K_{\text{DU}}} + \frac{[\text{DnaK}_{\text{free}}][\sigma^{32}]}{K_{\text{D32}}} \tag{8.32}$$

We stress that the key inhibitory coupling between $\sigma^{32}$ and DnaK is not only inhibition by binding, but also expose $\sigma^{32}$ to proteases (Gottesman, 1996, see also Fig. 8.20). Thus we need both an equation for the DnaK production given total $\sigma^{32}$

concentration, $[\sigma^{32}]$, and an equation for changes in $[\sigma^{32}]$. With a little algebra the DnaK production rate may be shown to depend on the changed amount of unfolded proteins through

$$\frac{\mathrm{d}[\mathrm{DnaK}]}{\mathrm{d}t} \propto a[\mathrm{RNAp}\cdot\sigma^{32}] - \frac{[\mathrm{DnaK}]}{\tau_{\mathrm{DnaK}}} \approx \frac{a\cdot[\sigma^{32}]}{1+([\mathrm{DnaK}]/u)} - \frac{[\mathrm{DnaK}]}{\tau_{\mathrm{DnaK}}}$$
$$\frac{\mathrm{d}[\sigma_{32}]}{\mathrm{d}t} \propto S - b[\mathrm{DnaK}\cdot\sigma^{32}] \approx S - \frac{b\cdot[\sigma^{32}]}{1+(u/[\mathrm{DnaK}])} \tag{8.33}$$

where $\tau_{\mathrm{DnaK}}$ is the effective decay of DnaK concentration in the growing and dividing *E. coli* cell, $S$ is the production rate of $\sigma^{32}$ and $a$, $b$ are two rate constants. The factor $u$ parametrizes the amount of unfolded proteins. In the approximation where we ignore free $\sigma^{32}$, and the fraction of DnaK bound by $\sigma^{32}$, we obtain the following equation:

$$u \propto K_{\mathrm{D32}} + \frac{K_{\mathrm{D32}}}{K_{\mathrm{DU}}}[U_{\mathrm{f}}] \tag{8.34}$$

when we take into account the fact that the free DnaK concentration in Eq. (8.31) primarily depends on the unfolded protein concentration. The heat shock is as an increased level of $\sigma^{32}$, which in turn is caused by its decreased degradation because DnaK instead binds to unfolded proteins. In this way DnaK provides a feedback that also has consequences for the about 40 other genes that are regulated through $\sigma^{32}$.

To summarize the dynamics of the network: when moving away from the temperature $T_{\mathrm{s}}$ where proteins are most stable, we increase $u$, and thereby the rate for DnaK production. For a unchanged "supply" $S$ of $\sigma^{32}$ the extreme in DnaK production occurs when $\mathrm{d}[\sigma^{32}]/\mathrm{d}t = 0$ to a value $\propto u$. One may thus identify $R = \max(r)$ with $u$ and thereby with the free energy difference $\Delta G$ between folded and unfolded proteins.

To summarize the network lesson of the heat shock, we note that its core motif shown in Fig. 8.22 is a common control element in molecular networks. A similar interplay between proteins is found, for example, in the apoptosis network controlled by proteins p53 and mdm2 (see Fig. 8.23 and Question (4)). Thus negative feedback between a transcription factor and its product often takes place through an inhibitory binding away from the DNA, a motif also emphasized in Fig. 8.15. In general, stress responses may often be an interplay between a slowly reacting transcription part, and feedbacks facilitated by fast chemical bindings; for example, a feedback that goes from highly abundant proteins (DNAK) to the scarce regulatory protein $[\sigma^{32}]$.

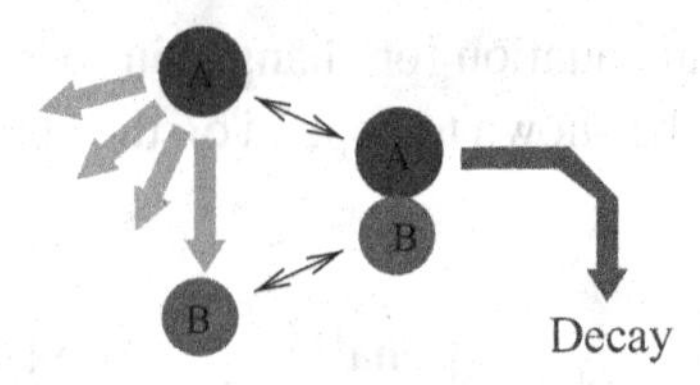


Figure 8.22. The core of the heat shock network, consisting of a motif where a protein A directs transcription of protein B, and where protein B directs degradation of A by binding to it. This motif is common in several molecular networks that exhibit negative feedback. To secure a fast response, the complex AB must not activate transcription of B. Further, the concentration [B] > [A], and the binding between A and B should be saturated ($K_{AB}$ < [A]). Then production of B is sensitive to free concentration of A (equal [A] – [AB]), and thereby to its own concentration [B].

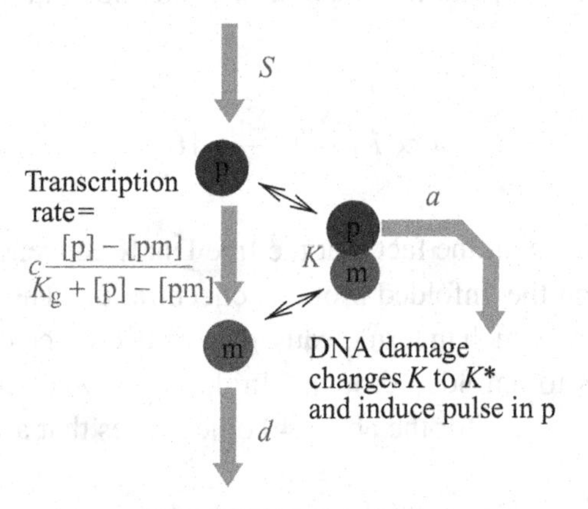


Figure 8.23. The p53 induction network exhibiting the same core motif as found in the heat shock. Induction of p53 pulse is done through phosphorylation of p = p53, that increase (see Tiana *et al.*, 2002) binding between p53 and m = mdm2.

## *Questions*

(1) Assume that a typical protein has a maximum stability of 15 kcal/mol at $T_s = 23$°C, and that the protein melts at 73°C. Assume parabolic behavior of $\Delta G = G(\text{unfolded}) - G(\text{folded})$ with $T$. Calculate the corresponding $\alpha$ (in Eq. (8.25)) that would govern the heat shock dependence with temperature for temperature quenches of 7°C.

(2) Arnvig *et al.* (2000) measured the heat shock in a strain where the $\sigma^{32}$ gene is located on a high copy number plasmid. In this strain where the synthesis rate for $\sigma^{32}$ may approach that of DnaK, the HS was smaller and also remained positive down to temperature jumps from $T$ well below $T_s = 23$. Explain these findings in terms of the model in the text.

(3) Tilly *et al.* (1983) found that over-expression of DnaK through a $\sigma^{32}$ dependent pathway represses HS. Explain this in terms of the above model. Notice that DnaK–$\sigma^{32}$ binding still plays a crucial role here, and discuss why the heat shock disappears.

(4) Feedback loops such as the one of the HS are seen in other systems. Consider the network that controls induced cell death in eukaryotes, through the p = p53 protein level (Fig. 8.23). Discuss the equations (Tiana *et al.*, 2002) $\mathrm{d}p/\mathrm{d}t = S - a \cdot (\mathrm{pm})$ and $\mathrm{d}m/\mathrm{d}t = c(\mathrm{p} - (\mathrm{pm}))/(K_\mathrm{g} + \mathrm{p} - (\mathrm{pm})) - \mathrm{dm}$ for the network with $K_\mathrm{g}$ being the binding constant for protein p53 to operator for p53, $c$ the production rate when p53 is bound, and the rest of the parameters identified from Fig. 8.23. Typical values are $a = 0.03\ \mathrm{s}^{-1}$, $K = 180$ nM, $K_\mathrm{g} = 28$ nM, $S \sim 1\ \mathrm{s}^{-1}$, $c = 1\ \mathrm{s}^{-1}$ and $d = 0.01\ \mathrm{s}^{-1}$ (see Tiana *et al.*, 2002). Notice that the concentration of p53–mdm2 complex (pm) is given by $K = (\mathrm{p} - (\mathrm{pm}))(\mathrm{m} - (\mathrm{pm}))/(\mathrm{pm})$. Simulate the equations and, after reaching steady state, simulate the response of $K \to 15K$ and $K \to K/15$, respectively.

(5) Consider the damped equation $\mathrm{d}y/\mathrm{d}t = -y(t)$ and compare this with the time-delayed version $\mathrm{d}y/\mathrm{d}t = -y(t - \tau)$. For which values of $\tau$ does the time-delayed equation start to oscillate? (Hint: make the guess $y = \exp(rt)$ with $r$ complex.)

(6) Consider the p53 network from Question (4) but with a time delay where mdm2 production is delayed by 1200 s relative to the p53 concentration. After reaching steady state, simulate the response to the same $K$ changes as in Question (4).

## Bacterial chemotaxis: robust pathways and scale invariant greed

Molecular networks are used for more than just changing concentrations of proteins. They are used to send signals, mediated through protein interactions where one protein modifies another protein. That is, the acting protein is an enzyme. A classical example of such a protein signaling system is the chemotaxis network in *E. coli*.

*E. coli* can chemotax: it can find food by sensing a gradient in food concentration. The sensing happens through receptors on the surface of the bacterial outer membrane (see Fig. 8.24) that through a sequence of events changes the phosphorylation status of a protein CheY and thereby the direction of motors placed on other positions in the bacterial membrane.

An example of such a receptor is the maltose receptor used by the λ-phage. Many other receptors exist. Each receptor can have a food molecule bound to it. When this is the case, the receptor properties on the inside of the membrane change, thus facilitating a signal. Upon processing the signal the bacterium decides whether to change direction or continue where it is heading, through a so-called tumbling process where its flagella motor changes direction (see Fig. 8.25). When the food concentration drops, it tends to shift direction (tumble). When the food concentration increases the bacterium continues straight by diminishing the frequency of tumbling. Thereby the bacterium makes a directed random walk towards larger concentrations of food.

A key element in chemotaxis is that the bacterium measures the change in food concentration over some distance. Thus it measures the difference in food occupation of the receptor, and does that sensitively for a huge range (factor $10^4$)

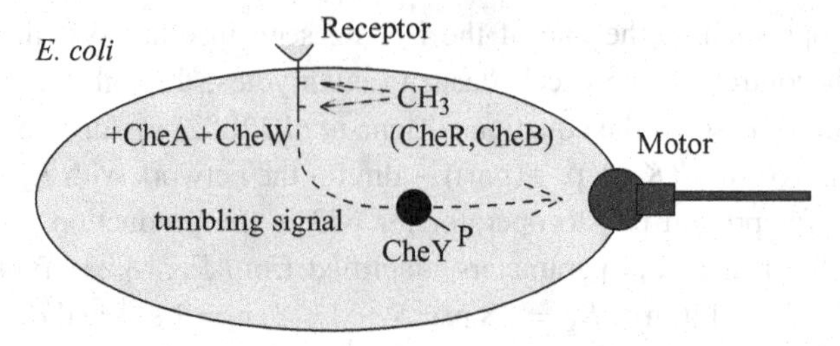


Figure 8.24. The decision to tumble or not is mediated through the receptor that modulate phosphorylation status of CheY, which thereby signals which way the motor should run: clockwise or counterclockwise (→ tumbling or moving straight). Proteins involved in signaling are Tar (the *lac* receptor), CheR methylation (adding $CH_3$ group to receptor), CheB demethylation, CheY messenger, CheAW kinase and CheZ phosphatase. CheZ adds a phosphate group to CheY, and deletion of CheZ is found to increase the tumbling rate by a factor of 5. Other mutant examples are a 12-fold increase in CheB, which decreases tumbling rate by a factor of 4, and increases adaptation time by a factor of 2.5.

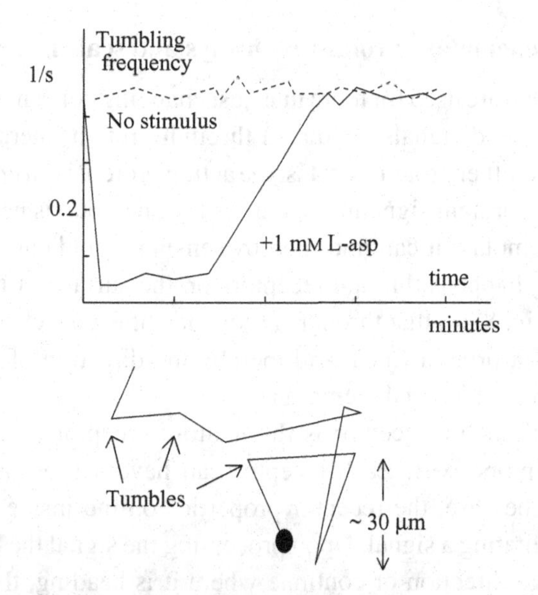


Figure 8.25. Chemotaxis of *E. coli*. The lower panel shows a trajectory of an *E. coli* bacterium consistent with a smooth path interrupted by sudden changes. Each sudden direction change is associated with a reversal of the bacterial movement engine, and is called a tumble. The upper figure illustrates an experiment where tumbling rate is shown to depend on addition of food that the bacteria are attracted to. At early times the tumbling frequency drops suddenly, allowing the bacteria to run longer. Later the bacteria adjust to a new level of attracting molecules, an adjustment that remarkably raises to exactly the same level as before the attracting molecules were added.

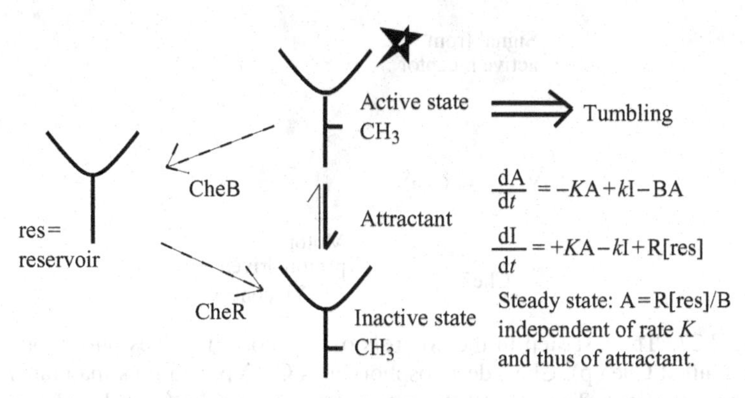


Figure 8.26. The Barkai–Leibler model for adaptation. Receptors in the membrane of *E. coli* control chemotaxis. When "attractants" are bound to a receptor it tends to stay in normal mode (low level of active state A). When "attractant" concentration increases, reaction rate $K$ is increased and thus transiently occupation of state A is decreased. Thus there is less tumbling, until a new steady state is reached. The steady state is the same with any value of $K$, so there is perfect recovery from any change in $K$. One obtains perfect recovery for a wide range of parameters. This is denoted robustness, because the adaptation does not change when one perturbs or alters the system.

of absolute food concentrations. Figure 8.26 demonstrates how Barkai & Leibler (1997) imagined this to be possible. In the figure the reaction rate $K$ determines the rate at which the active form of the receptor becomes deactivated, that is $A \rightarrow I$. The active form of the receptor (A) determines the tumbling rate. Upon changing outside food occupation of receptor, the rate $K$ changes, thereby giving a transient change in the number of active receptors, and thus in the tumbling rate. However, when one assumes a reservoir of receptors indicated on the left, then a one-way flow of the active receptors to this reservoir makes the steady-state number of active receptors independent of $K$. The equations governing the amount of A are

$$dA/dt = -K \cdot A + k \cdot I - B \cdot A \quad (8.35)$$

$$dI/dt = +K \cdot A - k \cdot I + R \cdot [res] \quad (8.36)$$

The steady-state level of $A = R \cdot [res]/B$ is independent of $K$ (the outside food level). This set of equations implicitly assumes that the flow from the reservoir, $R \cdot [res]$, is independent of the flow to the reservoir $B \cdot A$. Then the tumbling frequency will be independent of the absolute level of the external amount of food; see also Fig. 8.27.

After a sudden increase in food concentration, $K$ increases, and one first observes a decrease in A and thus in the tumbling rate. Subsequently, after some minutes, the rate recovers **exactly** to the same rate as with the lower food concentration, see Fig. 8.25. This is robustness toward external conditions. Further, when one changes

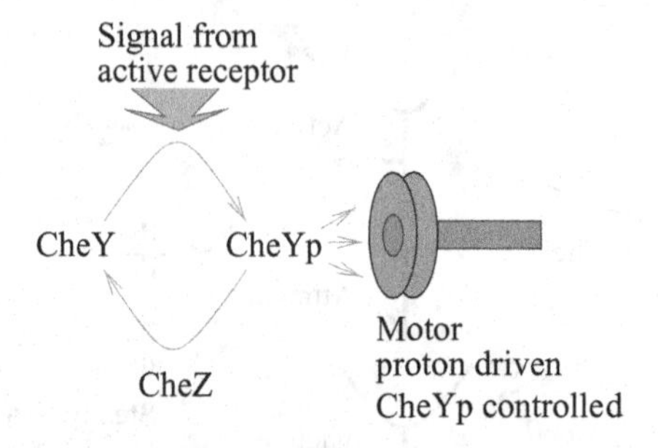


Figure 8.27. The last step in the controls of the motor involves phosphorylated CheY (called CheYp). CheZ de-phosphorylates CheYp, and thus maintains it at low concentrations. The state of the motor depends on CheYp to the 11th power (very large Hill coefficient). The motor uses a proton gradient as an energy source, not GTP or ATP as usually used by other motors in the cells. Notice also that the motor is cylindrical, probably with eight cylinders: the motor is an eight-cylindrical electro-motor!

the chemistry inside the cell through mutation, one may change the tumbling rate as a function of A. However, when A always recovers to the same level, the tumbling rate will also recover to the same level after a change in food concentration. Thus different bacteria may have different intrinsic tumbling rates, but each bacterium will always maintain its own food level independent tumbling rate. Chemotaxis is also robust against details of internal signaling chemistry that comes after the receptor (Alon *et al.*, 1998).

In summary, bacterial chemotaxis is a remarkable example of scale-invariant greed: at (nearly) any food concentration, the bacteria respond equally well to a relative change in external food concentration.

## *Questions*

(1) Consider the Barkai–Leibler (1997) model at some value $K_0$. Set $k = K_0$ and $R = 1$, $B = 1$ and res $= 1$ and find expression for A and I. Simulate the rate of tumbling as a function of a constantly increasing $K$, $K = \alpha t + K_0$, over a time interval where $K$ changes by a factor of 10.

(2) Consider a random walker along a one-dimensional line, performing simple chemotaxis. After each step in some direction it may change direction. If it moves left it changes direction with probability $p > 0.5$, say $p = 0.55$. If it moves right it changes direction with probability 0.5. If it does not change direction, it continues one step in same direction as before. This resembles some increased food gradients towards the left. Simulate the average displacement per step as a function of $p$.

(3) Simulate the push–pull reaction (Stadtman & Chock, 1977; Koshland *et al.*, 1978), where A converts B to C, and C spontaneously converts back to B with rate $r$ (assume that [B]+[C] is constant). Plot concentration of C as a function of concentration of A for two different values of $r$.

## References


Alon, U., Surette, M. G., Barkai, N. & Leibler, S. (1998). Robustness in bacterial chemotaxis. *Nature* **397**, 168–171.

Arnvig, K. B., Pedersen, S. & Sneppen, K. (2000). Thermodynamics of heat shock response. *Phys. Rev. Lett.* **84**, 3005 (cond-mat/9912402).

Barabasi, A.-L. & Alberts, R. (1999). Emergence of scaling in random networks. *Science* **286**, 509.

Barkai, N. & Leibler, S. (1997). Robustness in simple biochemical networks. *Nature* **387**, 913–917.

Bock, J. & Sneppen, K. (2004). Quantifying the benefits of translation regulation in the unfolded protein response. *Phys. Biol.* (in press).

Bornholdt, S. & Ebel, K. H. (2001). World Wide Web scaling from Simon's 1955 model. *Phys. Rev. E.*, 035104(R).

Broder, A. *et al.* (2000). Graph structure in the Web. *Computer Networks* **33**, 309–320.

Caldarelli, G., Capocci, A., De Los Rios, P. & Muoz, M. A. (2002). Scale-free networks from varying vertex intrinsic fitness. *Phys. Rev. Lett.* **89**, 258702.

Cox, J. S. & Walter, P. (1996). A novel mechanism for regulating activity of a transcription factor that controls the unfolded protein response. *Cell* **87**, 391–404.

Faloutsos, M., Faloutsos, P. & Faloutsos, C. (1999). On power-law relationships of the internet topology. *Comput. Commun. Rev.* **29**, 251.

Field, G. B. & Saslow, W. C. (1965). A statistical model of the formation of stars and interstellar clouds. *Astrophys. J.* **142**, 568.

Giot, L. *et al.* (2003). A protein interaction map of *Drosophila melanogaster*. *Science* **302**, 1727.

Gottesman, S. (1996). Proteases and their targets is *Escherichia coli. Ann. Rev. Genet.* **30**, 465–506.

Gross, C. A. (1996). *Cellular and Molecular Biology*. ASM Press.

Herendeen, S. L. *et al.* (1979). Levels of major proteins of *Escherichia coli* during growth at different temperatures. *J. Bacteriol.* **139**, 185.

Ito, T. *et al.* (2001). A comprehensive two-hybrid analysis to explore the yeast protein interactome. *Proc. Natl Acad. Sci. USA* **98**, 4569.

Jeong, H., Mason, S., Barabasi, A.-L. & Oltvai, Z. N. (2001). Centrality and lethality of protein networks. *Nature* **411**, 41–42.

Kim, B. J., Trusina, A., Minhagen, P. & Sneppen, K. (2004). Scale free networks from merging and creation. *Eur. Phys. J.* B (in press.)

Koshland Jr., D. E., Goldbeter, A. & Stock, J. B. (1978). Amplification and adaptation in regulatory and sensory systems. *Science* **217**, 220.

Maslov, S. & Sneppen, K. (2002a). Specificity and stability in topology of protein networks. *Science* **296**, 910.

(2002b). Pattern detection in complex networks: correlation profile of the Internet. Submitted to *Phys. Rev. Lett.*, cond-mat/0205379.

(2004). Computational architecture of the yeast regulatory network (preprint).

Newman, M. E. J., Strogatz, S. H. & Watts, D. J. (2001). Random graphs with arbitrary degree distributions and their applications. *Phys. Rev. E* **64**, 026118, 1.

van Nimwegen, E. (2003). Scaling laws in the functional content of genomes. *Trends Genet.* **19**, 479.

Norel, R. & Agur, Z. (1991). A model for the adjustment of the mitotic clock by cyclin and MPF levels. *Science* **251**, 1076–1078.

Pastor-Satorras, R., Smith, E. & Sole, R. V. (2003). Evolving protein interaction networks through gene duplication. *J. Theor. Biology* **222**, 199–210.

Pedersen, S., Block, P. L., Reeh, S. & Neidhardt, F. C. (1978). Patterns of protein synthesis in *E. coli*: a catalog of the amount of 140 individual proteins at different growth rates. *Cell* **14**, 179.

Price, D. J. de S. (1976). A general theory of bibliometric and other cumulative advantage processes. *J. Am. Soc. Inform. Sci.* **27**, 292.

Rosenfeld, N., Elowitz, M. & Alon, U. (2002). Negative autoregulation speeds the response times of transcription networks. *JMB* **323**, 785–793.

Rosvall, M. & Sneppen, K. (2003). Modeling dynamics of information networks. *Phys. Rev. Lett.* **91**, 178701.

Rosvall, M., Trusina, A., Minnhagen, P. & Sneppen, K. (2004). Networks and cities: an information perspective. *Phys. Rev. Lett.* **94**, 028701.

Shen-Orr, S. S., Milo, R., Mangan, S. & Alon, U. (2002). Network motifs in the transcriptional regulation of *Escherichia coli*. *Nature Genetics*, published online: 22 April 2002, DOI:10. 1038/ng881.

Sidrauski, C., Cox, J. & Walter, P. (1997). tRNA Ligase is required for regulated mRNA splicing in the unfolded protein response. *Cell* **87**, 405–413.

Simon, H. (1955). On a class of skew distribution functions. *Biometrika* **42**, 425.

Sneppen, K., Rosvall, M., Trusina, A. & Minhagen, P. (2004a). A simple model for self-organization of bipartite networks. *Europhys. Lett.* **67**, 349.

Sneppen, K., Trusina, A. & Rosvall, M. (2004b). Hide and seek on complex networks. Cond-mat/0407055.

Stadtman, E. R. & Chock, P. B. (1977). Superiority of inter convertible enzyme cascades in metabolic regulation: analysis of monocyclic systems. *Proc. Natl Acad. Sci. USA.* **74**, 2761–2765.

Stover, C. K. *et al.* (2000). Complete genome sequence of *Pseudomonas Aeruginosa* PA01, an opportunistic pathogen. *Nature* **406**, 959.

Tiana, G., Jensen, M. H. & Sneppen, K. (2002). Time delay as a key to apoptosis induction in the p53 network. *Eur. Phys. J.* B**29**, 135.

Tilly, K., Mckittrick, N., Zylicz, M. & Georgopoulos, C. (1983). The dnak protein modulates the heat-shock response of *Escherichia coli*. *Cell* **34**, 641.

Uetz, P. *et al.* (2000). A comprehensive analysis of protein–protein interactions in *Saccharomyces cerevisia*. *Nature* **403**, 623.

Watts, D. J. & Strogatz, S. H. (1998). Collective dynamics of "small-world" networks. *Nature* **393**, 440.

## Further reading

Bornholdt, S. & Sneppen, K. (2000). Robustness as an evolutionary principle. *Proc. R. Soc. Lond.* B**267**, 2281.

Bæk, K. T., Svenningsen, S., Eisen, H., Sneppen, K. & Brown, S. (2003). Single-cell analysis of lambda immunity regulation. *J. Mol. Biol.* **334**, 363.

Chalfie, M. (ed.) (1998). *Green Fluorescent Protein: Properties, Applications, and Protocols*. NewYork: Wiley-Liss.

Costanzo, M. C. *et al.* (2001). YPD, PombePD, and WormPD: model organism volumes of the BioKnowledge library, an integrated resource for protein information. *Nucl. Acids Res.* **29**, 75–79.

Davidson, X. *et al.* (2002). A genomic regulatory network for development. *Science* **295** (5560), 1669–2002.

Davidson, E. H. (2001). *Genomic Regulatory Systems. Development and Evolution.* San Diego: Academic Press.
Derrida, B. & Pomeau, Y. (1986). Random networks of autometa: a simple annealed approximation. *Europhys. Lett.* **1**, 45.
Detwiler, P. B., Ramanathan, S., Sengupta, A. & Shraiman, B. I. (2000). Engineering aspects of enzymatic signal transduction: photoreceptors in the retina. *Biophys. J.* **79**, 2801–2817.
Elena, S. F. & Lenski, R. E. (1999). Test of synergetic interactions among deleterious mutations in bacteria. *Nature* **390**, 395.
Elowitz, M. B. & Leibler, S. (2000). Synthetic oscillatory network of transcriptional regulators. *Nature* **403**, 335.
Elowitz, M. B., Levine, A. J., Siggia, E. D. & Swain, P. S. (2002). Stochastic gene expression in a single cell. *Science* **297**, 1183.
Erdös, P. & Rényi, A. (1960). On the evolution of random graphs. *Publ. Math. Inst. Hung. Acad. Sci.* **5**, 1760.
O'Farrell, P. H. (1975). High-resolution two-dimensional electrophoresis of proteins. *J. Biol. Chem* **250**, 4007.
Ferrell Jr., J. E. (2000). *What do Scaffold Proteins Really Do*, Science STKE, www.stke.sciencemag.org/cgi/content/full/sigtrans;2000/52/pe1.
Gavin, A.-C. *et al.* (2002). Functional organization of the yeast proteome by systematic analysis of protein complexes. *Nature* **415**, 141–147.
Hartwell, L. H., Hopfield, J. J., Leibler, S. & Murray, A. W. (1999). From molecular to modular cell biology. *Nature* **402** (6761 Suppl.), C47.
Horwich, A. (2002). Protein aggregation in disease: a role for folding intermediates forming specific multimeric interactions. *J. Clin. Invest.* **110**, 1221–1232.
Huerta, A. M., Salgado, H., Thieffry, D. & Collado-Vides, J. (1998). RegulonDB: a database on transcriptional regulation in *Escherichia coli. Nucl. Acid Res.* **26**, 55 (see also http://www.smi.stanford.edu/projects/helix/psb98/thieffry.pdf).
Jacob, F. & Monod, J. (1961). Genetic regulatory mechanisms of the synthesis of proteins. *J. Mol. Biol.* **3**, 318.
Jeong, H., Tombor, B., Albert, R., Oltvai, Z. N. & Barabasi, A.-L. (2000). The large scale organization of metabolic networks. *Nature* **407**, 651–654.
Kalir, S. *et al.* (2001). Ordering genes in a flagella pathway by analysis of expression kinetics from living bacteria. *Science* **292**, 2080.
Kauffman, S. (1969). Metabolic stability and epigenesis is randomly constructed genetic nets. *J. Theor. Biol.* **22**, 437.
Krapivsky, P. L. (1993). Aggregation-annihilation processes with injection. *Physica A* **198**, 157.
Minnhagen, P., Rosvall, M., Sneppen, K. & Trusina, A. (2004). Self-organization of structures and networks from merging and small-scale fluctuations. Cond-mat/0406752.
Neidhardt, F. C. *et al.* (1984). The genetics and regulation of heat-shock proteins. *Ann. Rev. Genet.* **18**, 295.
Peterson, K. J. & Davidson, E. H. (2000). Regulatory evolution and the origin of the bilaterians. *Proc. Natl Acad. Sci. USA* **97**, 4430–4433.
Price, D. J. de S. (1965). Networks of scientific papers. *Science* **149**, 510.
Ptashne, M. & Gann, A. (1997). Transcriptional activation by recruitment. *Nature* **386**, 569.
Spellman, P. T. *et al.* (2000). Genomic expression programs in the response of yeast cells to environmental changes. *Mol. Biol. Cell.* **11**, 4241–4257.

Takayasu, H., Nishikawa, I. & Tasaki, H. (1988). Power-law mass distribution of aggregation systems with injection. *Phys. Rev. A* **37**, 3110–3117.

Thieffry, D., Huerta, A. M., Prez-Rueda, E. & Collado-Vides, J. (1998). From specific gene regulation to global regulatory networks: a characterisation of *Escherichia coli* transcriptional network. *BioEssays* **20**, 433–440.

Trusina, A., Maslov, S., Minnhagen, P. & Sneppen, K. (2004). Hierarchy and anti-hierarchy in real and scale free networks. *Phys. Rev. Lett.* **92**, 178702.

Vazquez, A., Flammini, A., Maritan, A. & Vespignani, A. (2001). Modelling of protein interaction networks *arXiv* (cond-mat/0108043).

Vogelstein, B., Lane, D. & Levine, A. J. (2000). Surfing the p53 network. *Nature* **408**, 307.

Wagner, A. (2001). The yeast protein interaction network evolves rapidly and contains few redundant duplicate genes. *Mol. Biol. Evol.* **18**, 1283–1292.

Zhou, Y. *et al.* (1993). CAP. *Cell* **73**, 375.

Yi, T.-M., Huang, Y., Simon, M. I. & Doyle, J. (2000). Robust perfect adaptation in bacterial chemotaxis through integral feedback control. *Proc. Natl Acad. Sci. USA* **97**, 4649–4653.

# 9

# Evolution

Kim Sneppen

## Evolution and evolvability

Evolution, and the ability to evolve, is basic to life. In fact, it separates life from non-life. The process of evolution has brought us from inorganic material in ~3.5 billion years, our bacterial ancestors in ~2 billion years, from primitive sea-dwelling chordates in ~500 million years and from common ancestors to mice in only ~100 million years. Evolution has inspired our way of looking at life processes throughout modern biology, including the idea that evolvability is an evolving property in itself.

Most mutations are either without any consequences, i.e. neutral, or deleterious. Thus evolution is a costly process where most attempts are futile. That it works anyway reflects the capacity of life to copy itself so abundantly that it can sustain the costly evolutionary attempts.

Evolution is the necessary consequence of:

- heredity (memory);
- variability (mutations);
- each generation providing more individuals than can survive (surplus).

The last of these points implies selection, and thus the principle of "survival of the fittest". In fact, the surplus provided by growth of successful organisms is so powerful that it provides a direct connection between the scale of the molecule and the worldwide ecological system: a successful bacterium with a superior protein for universal food consumption may in principle fill all available space on Earth within a few days. In fact, bacteria on Earth number about $5 \times 10^{30}$ (Whitman *et al.*, 1998), a biomass that dominates all other animal groups together. The bacteria have already won! The dynamics of this interconnected microbiological ecosystem thus challenges any model built on a separation of phenomena at small and larger length scales.

Another important fact of evolution is that it always works on a number of individuals, $N$, that are much smaller that the combinatorial possibilities in their genome. There are maybe $5 \times 10^{30}$ bacteria on Earth, but a bacterial genome provides a number of codes of order $4^L$, where a typical genome length of a prokaryote is $L \sim 10^6$–$10^7$. The implication of this is that **evolution is a historical process**, where new variants arise by rather small variations of what is already present. Or, stated in another way, if one imagined starting life again, the outcome would in all likelihood be entirely different, and this difference would probably be on all scales, from strategies of molecular interactions to design of animal bodies.

An example of widely different body designs is found during Cambrian times (530 million years ago), where there was a wealth of organisms with body designs that are very different from what is found among phyla alive today (for a beautiful review see Gould, 1989). Each of these now-extinct life designs existed for millions of years, with no apparent inferiority to the survivors. They subsequently became extinct, leaving only the well-separated phyla (= body plans) that we know today. Figure 9.1 shows life history since Cambrian times in a way that illustrates the coherent emergence and collapse of whole ecosystems, as is evident by the nearly block diagonal form of the origination–extinction "matrix".

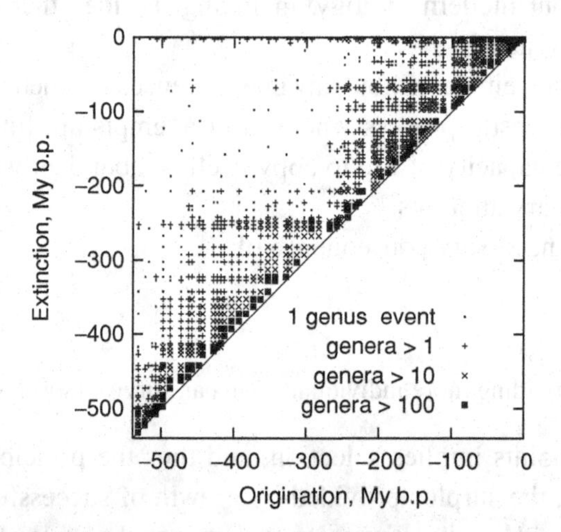


Figure 9.1. Origination and extinction of 35 000 genera in the Phanerozoic, from Bornholdt & Sneppen (2004). Data from Sepkoski (1993). Every event is quantified by the number of genera. A genera is a group of related animal species. For example, all types of humans, like Neanderthals, homo sapiens, cro-magnon, etc., form one genera. The vertical distance from a point to the diagonal measures the lifetime of the group of genera with that particular origination and extinction. Notice the collapse of many points close to the diagonal, reflecting the fact that most genera exist for less than the overall genera average of about 30 My (million years). Notice also the division of life before and after the Permian extinction 250 My b.p. (million years before present).

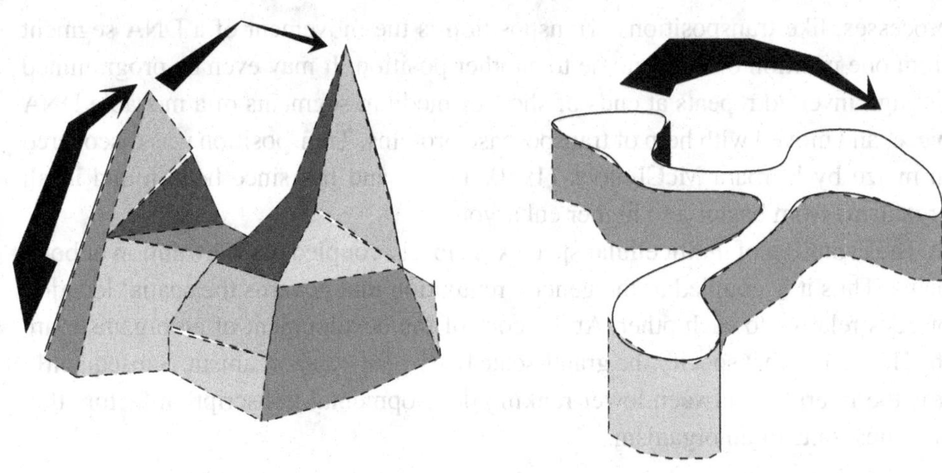

Walking in landscapes with seperated peaks? Walking in neutral landscapes?

Figure 9.2. Possible fitness landscapes: does evolution proceed as a steady increase in fitness, eventually interrupted by barriers, as Wright envisioned (1982), or is evolution a random walk on large fitness plateaus, as in the neutral evolution scenario of Kimura (1983)?

There are some concepts of evolution that are useful to keep in mind when thinking about life: for example, the genetic drift model (Kimura, 1980, 1983) or the adaptive trait concept (Fisher, 1930; Wright, 1982). Or, reformulated (see Fig. 9.2):

- the neutral scenario where every mutant is equal, thus allowing for large drift in genome space;
- fitness climbers – evolution viewed as an optimization process, where some individuals/species are better than others.

The neutral theory has experimental support in molecular evolution, where the sequence of most proteins/RNAs varies substantially between species. Probably all variants of the same macromolecule (homologs) do their job well, and evolution on a larger scale is seldom associated with developing better molecules. Instead, evolution of species is associated with reallocation of functions between cells. The reallocation is done through rewiring of the underlying genetic networks, which are responsible for cell differentiation and subsequent positioning of the large variety of organs in multicellular organisms. Presumably such network rewirings are not always neutral: occasionally they lead to mutant organisms that are better suited to deal with particular environments or competitors.

Genetic network rewirings are fast, and probably faster than single-point mutations. Mechanistically they may be done through a number of DNA edition processes, like transpositions. Transposition is the movement of a DNA segment from one position of the genome to another position. It may even be programmed through inverted repeats at ends of short or medium segments of a movable DNA piece, and moved with help of transposease proteins. Transposition was discovered in maize by Barbara McClintock (1950, 1953), and has since been found in all organisms from bacteria to higher eukaryotes.

The evolution of multicellular species is closely coupled to the evolution of body plans. Thus it is coupled to the genetic regulation that governs the spatial location of cells relative to each other. At the core of the development of an organism are the Hox genes that specify the grand-scale body plan of an organism. Subsequently it is the interplay between lower ranking developmental transcription factors that specifies traits of an organism.

### *Questions*

(1) The K12 strain of bacteria doubles itself every 20 min (at 37 °C), and every 2 h at 20 °C. If this bacterium could utilize all material on Earth, how long would it take to convert all of the Earth into bacterial material at the two temperatures?

(2) (a) How many bacterial generations have passed since life started on Earth? (Assume near optimal growth, with 1 h doubling time.)
(b) Assume that, on average, there is one base pair mutation per bacterium per generation. Assume that there have always been about $5 \times 10^{30}$ bacteria on Earth. Give an estimate of how many of all possible bacterial mutants have been tested.
(c) With 5 000 000 million base pairs in the genome, what fraction of all mutants have been tested until now?

## Adaptive walks

The conceptually simplest model of evolution is the one based on local fitness optimization, as illustrated in Fig. 9.3. In this one follows a species consisting of individuals with some variation in genotype. One describes the population as approximately Gaussian in a one-dimensional genome space. With a fitness gradient as shown in Fig. 9.3 the fittest part of the population tends to be more represented in the next generation. For a simple linear increasing fitness landscape this model predicts a steady increase in fitness with a rate proportional to the spread of the Gaussian distribution.

If a local fitness maximum is reached, the evolution is arrested, and the population remains at the maximum until, by random fluctuation, a sub-part of the population appears with a higher fitness. The chance of such a fluctuation depends both on the size of the fitness barrier and on the effective size of the population $N_e$. The

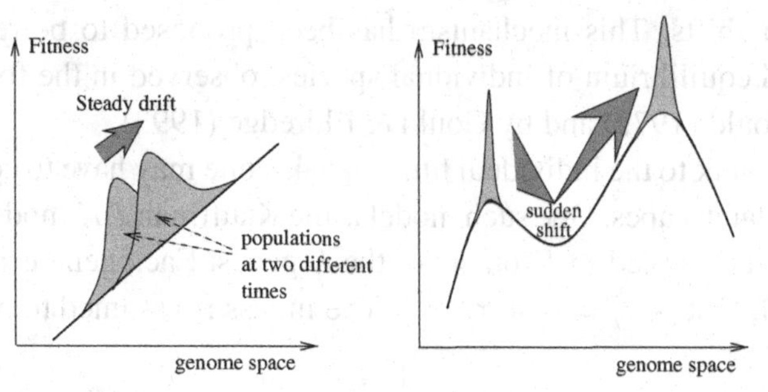


Figure 9.3. Walks in fitness landscapes. The left-hand figure illustrates the Fisher theory of evolution, where the population indicated by the Gaussian moves steadily to higher fitness by selecting the individuals with highest fitness to dominate subsequent generations. Evolutionary drift proportional to genetic variation, which in turn depends on mutation rate $\mu$. The right-hand figure illustrates the trapping of a population on a local fitness peak for long periods followed by a sudden jump when the fitness barrier is punctuated by a rare event.

effective population is given by

$$\frac{1}{N_e} = \langle \frac{1}{N} \rangle_{\text{time}} \tag{9.1}$$

The effective size may be the whole population $N$, but typically it will be smaller, as only a fraction of individuals may contribute directly to the next generation.

If we estimate the probability that a population of size $N_e$ escapes from a fitness peak by a random fluctuation (mutation), we need to know the minimal fitness "barriers" $\Delta F$ that separate the peak from a neighboring higher peak. Given that, one may consider the escape as a Kramer's escape problem (see Appendix) with a barrier $\Delta F$ and an effective "temperature" set by the random mutations (Lande, 1985):

$$\mathcal{P} \propto N_e \cdot \exp\left(-\frac{N_e \Delta F}{\mu}\right) \tag{9.2}$$

Here $\mu$ is the individual mutation rate and $\mu/N_e$ is the effective mutation rate for a well-mixed population consistent of $N_e$ individuals. Thus when the population grows, the escape probability diminishes exceedingly quickly! This is behind the belief that speciation is connected with the isolation of small sub-populations, for example owing to localization on small islands. Also this type of behavior is behind the "hopeful monster" proposal of Wright (1945), who suggested that evolutionary jumps are associated with individuals taking large evolutionary leaps by rare accidents. In any case, the walks between subsequent fitness maxima will be intermittent, with large waiting times on peaks, and rather fast adaptations when the population shifts. This mechanism has been proposed to be responsible for the punctuated equilibrium of individual species, observed in the fossil record by Eldredge & Gould (1972) and by Gould & Eldredge (1993).

Beyond the walk to the individual fitness peaks, one may have to consider multi-peaked fitness landscapes. One such model is the Kauffman $LK$ model, where each of the $L$ genes is coupled to $K$ others of the $L$ genes. Each gene is assumed to be either on or off, that is $\sigma_j = 0$ or $\sigma_j = 1$. The fitness is assumed to be of the form

$$F = \frac{1}{L} \sum F_i \tag{9.3}$$

where $F_i = F_i(\sigma_i, \sigma_{i,1}, \ldots \sigma_{i,K-1})$ is a random number between 0 and 1 for each of the possible $K$ inputs to gene $i$. For $K = 1$, the "$F$" landscape is a simple Mount Fuji landscape with one fitness peak. For $K = L - 1$ the landscape is multi-peaked; any change in any gene affects all random numbers in the sum $\sum F_i$. Thus the energy landscape resembles the random energy model of Derrida (1980). As proven by Bak *et al.* (1992), in these extremely rough landscapes we can say the following.

- The probability that a given $\sigma$ is at a fitness peak is equal to $1/(L+1)$, because this is the probability that $\sigma$ is the largest of itself plus all $L$ neighboring variants. Note also that the fitness of a local maximum will be of order $1 - 1/(L+1) \sim 1 - 1/L$.
- The number of upwards steps from a random $\sigma$ to a local fitness maximum is, on average, $\log_2(L)$. This is because at each step from a fitness value $F$ one selects a new random fitness in the interval $1 - F$. Thus the interval $1 - F$ will be a half length at each consecutive step, and when it reaches $1/(L+1)$ we should have reached a maximum. If $l$ is the number of upwards steps, then $1 - 2^{-l} = 1 - 1/(L+1)$, giving a typical walk length of $l = \log(L)/\log(2)$.

Looking at the history of life on large scales as in Fig. 9.1, there is no particular sign of any overall increase in fitness with time. In fact, when one compares existence times of various species groups, they show no sign of an overall increase with time (Van Valen, 1973). On the other hand, the survival time of a given species has a characteristic length that can be estimated from the average survival time for its "cousins" (see Fig. 9.4). Thus species evolve and inherit their basic robustness, but they typically do not improve this robustness against extinction.

In summary, adaptive walks are fine for local exploration but hopeless when large-scale jumps are needed. Reasonable population sizes will prohibit these jumps. As we will see now, neutral landscapes, on the other hand, open the way for large-scale meanderings and thus facilitate evolution. Another increase of evolution possibilities may occur by changing fitness landscapes dynamically. Such dynamics may be induced by the ever-changing ecosystem, maybe because other species evolve and thereby open new opportunities for each other.

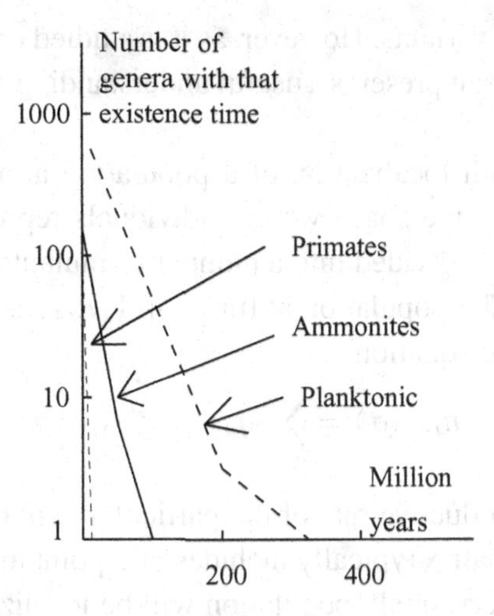


Figure 9.4. Sketch of typical existence time distributions for genera in the fossil record. If evolution was a steady hill climbing to an ever improved robustness, one should expect that species are better able to survive with time. This is not the case; distribution of survivorship in a given class of species is close to exponential, with similar time constants for early and late members of the class. This implies that the extinction risk for a species is a constant, independent of time. This led Van Valen (1973) to propose the "Red Queen" hypothesis: one has to run to remain at the same relative fitness to its environment. Further, the figure indicates that stability against extinction is inherited, whereas the actual extinction is a random event.

### *Questions*

(1) Draw random numbers from a continuous distribution, and remember the largest you have had until now. Simulate the times at which the largest number increases, and show that the times for such changes get subsequently larger as time passes. What is the distribution of these times? The numbers may be fitness values, and each newly drawn number may be a mutation attempt of the species. Only a fitter species will "outrun" the current species.
(2) The sex ratio of all species with two sexes is 1:1. Given the fact that each child has both a father and mother, argue that when there are fewer males than females in the population, it is favorable to produce males for the propagation of your genes.
(3) This is a computer exercise: study numerically the space-time trajectory of a particle in a double-well potential $V(x) = -x^2 + x^4$ by using overdamped Langevin dynamics (see Appendix) and different strengths of the noise term. Discuss the results in terms of evolution in a double-peaked fitness landscape $F = -V$.

## Quasi-species model

The quasi-species model (Eigen, 1971; Eigen *et al.*, 1989) is an illustrative study of the non-biological limit where $N \gg 2^L$. One assumes that there is an infinite supply of all genotypic variants. However, as it is studied only in the neighborhood of some optimal species, it presents a useful understanding of the role of mutational load.

The model deals with localization of a population around an optimal fitness, which is a point in genome space where individuals reproduce faster than other places. The population is divided into a number of mutants, each characterized by a certain genome, $\sigma$. The population at time $t+1$, $n_{t+1}$ is given from population at time $t$, $n_t$ through the equation

$$n_{t+1}(\sigma) = \sum_{\sigma'} q_{\sigma\sigma'} w(\sigma') n_t(\sigma') \tag{9.4}$$

where $w(\sigma)$ is the reproductive rate of the particular genome, and $q$ is a mutation matrix. The mutation matrix typically includes one-point mutations with some low probability, $\mu$. For $\mu$ very small, population will be localized in the area where $w$ is largest. For larger $\mu$ the system undergoes a phase transition where suddenly the individuals lose contact to an optimal point.

Consider a single-peaked fitness landscape, with replication rate

$$R(\sigma) = R_0 \text{ if } \sigma = \sigma_0 = (1, 1, 1, 1, 1, 1, 1 \ldots) \tag{9.5}$$

$$R(\sigma) = R < R_0 \quad \text{for all other points} \tag{9.6}$$

For genome length $L$ the growth of biomass at the master sequence is proportional to

$$\frac{dN_{11111\ldots}}{dt} \propto (1-\mu)^L R_0 \, N_{1111\ldots} \tag{9.7}$$

per generation. The net growth of species in the immediate neighborhood is

$$\frac{dN_{\text{others}}}{dt} \propto RN_{\text{others}} + \cdots \tag{9.8}$$

per generation. The condition for maintaining the population at the fitness peak is that the main contribution in any generation comes from the master sequence. To first order this means:

$$(1-\mu)^L R_0 > R \tag{9.9}$$

thus the critical $\mu = \mu^*$ is given by

$$(1-\mu^*)^L = R/R_0 \tag{9.10}$$

For $\mu > \mu^*$ the population de-localizes. This de-localization transition is called the error catastrophe: when errors become larger, the species collapses as an entity. Not surprisingly one obtains a critical $\mu = \mu^*$, which is of order $1/L$; see examples of $\mu$ tabulated below.

| | | |
|---|---|---|
| Replication of RNA without enzymes | $\mu = 0.05$ | $L_{\max} \sim 20$ |
| Viruses | $\mu = 0.0005$ | $L_{\max} \sim 1000$–$10000$ |
| Bacteria | $\mu = 1.0 \times 10^{-6}$ | $L_{\max} \sim 10^6$ |
| Vertebrates | $\mu = 1.0 \times 10^{-9}$ | $L_{\max} \sim 10^9$ |

In fact most organisms appear to work close to the error threshold: if one plots the average error rate per replication of stem cells and per genome, one obtains a fairly constant number close to 1! An extreme example is the HIV virus, which has a very small genome and by operating close to its error threshold has a very high mutation rate. This allows many mutations, which help the virus to escape the host immune system, as well as to escape any single drug with which we attack the virus. An interesting applicaton of host–parasite co-evolution of quasispecies is discussed by Kemp *et al.* (2003) and Anderson (2004).

## Neutral evolution

Mutation rates of living species are quite high, for humans $\mu \sim 2 \times 10^{-8}$ base pairs/generation. If most mutations were deleterious, there would be a substantial mutational load on the species: most offspring would die. Accordingly Kimura (1980, 1983) suggested that most mutations are without noticeable effects, i.e. that they are neutral. A few mutations are presumably deleterious. If the chance that a mutation leads to death is denoted $p_{\text{kill}}$, then survival of the offspring requires

$$p_{\text{kill}} \cdot \mu \cdot L < 1 \tag{9.11}$$

where $L$ is the genome length. Thus the chance that one mutation is deleterious has to be less than $(\mu \cdot L)^{-1}$. The deleterious mutations are instantly removed from the gene pool, and can thus be ignored. Finally an even smaller fraction of mutations may be favorable, a fraction that one may ignore for the first analysis. In summary, Kimura argues that most mutations are without effects, and accordingly one can count time by measuring changes in the genome. This molecular clock idea is a good working hypothesis, in spite of the variability of mutation rates between various parts of the genome, between species and maybe also over time.

One remarkable effect of neutral evolution is the fact that any new neutral mutation has a chance to take over the population. And that the probability of new mutations doing so is independent of population size. Say the probability that a neutral mutation arises in an individual at generation $t$ is $r \cdot N_e$, where $N_e$ is the effective population size and $r = \mu \cdot L$ is the probability to mutate per generation

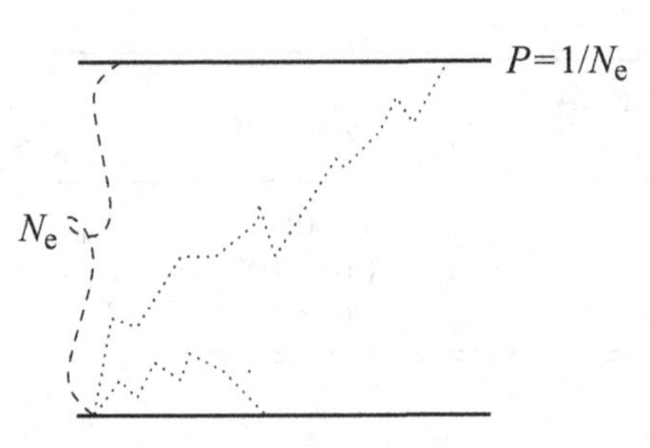


Figure 9.5. Probability of fixation of a single mutation in a population of size $N_e$ is proportional to $1/N_e$. The figure shows two random walks: one where the initial mutation is eliminated, and one where it is fixated (i.e. the mutation spreads to the total population $N_e$). The $x$-axis is not in real time, but in the number of events where the number of mutants increases or decreases in the population.

per individual. In the next generation this mutation may spread such that two individuals may carry it, or it may be annihilated. Since the mutation was neutral, and the total population size was constant, these two events are equally likely. Thus at each time step the number of surviving individuals with this mutation follows a random walk. If this reaches the absorbing state, 0, the mutation vanishes. If it reaches the absorbing state, $N_e$, the mutation takes over the population, and it is *fixated*. As random walks represent a fair game, the probability $\mathcal{P}$ to win (become fixated), multiplied by the amount one wins ($N_e$) has to be equal to the initial investment (one mutation). As the probability to obtain the mutation in the first place was proportional to $N_e$, then the overall rate of fixations becomes

$$\mathcal{P}(\text{fixation}) = r\mathcal{P}N_e = r \tag{9.12}$$

independent of effective population size (see Fig. 9.5). For an alternative approach using random walker theory, see the Questions below.

## *Questions*

(1) Consider an initial single-point mutation in one member of a population of size $N$. At each event, the mutation may duplicate with the duplication of the member, or die out with the member. In a steady-state population the numbers of members that carry the mutation thus make a random walk. What is the chance that the mutation becomes fixated, defined as being present in all $N$ members of the species?
(2) Consider a gambler's ruin version of species evolution, where at each time step each individual in the species has a fifty-fifty chance to double, or be eliminated. If species always start with one individual, what is the distribution of lifetime of the species?
(3) In a computer model of sympatic speciation, consider a population of $N$ individuals each with $N_g$ binary genes $(1, 0, 0, 1, \ldots)$. One individual is chosen, and another is selected randomly until its Hamming distance to the first is less than $k$. Then an offspring of the two is created, by randomly selecting genes from the two parents and mutating

each gene with probability $\mu$. Finally one selects a random individual and removes it (constant population assumption). Demonstrate that this may lead to speciation of the population (when the population fills only a small fraction of genomic space).

## Molecular evolution of species

Evolution on the molecular level can be quantified by comparing genomes of different species that live today. To do this reliably, one must know something about mutation rates of the genes, and between the respective species. For a start one may assume that all mutation rates are the same, independent of species, and are the same for all single base-pair changes. From this one can construct phylogenetic trees, which illustrates who are closest to whom. Thus distances between two species may be quantified by the number of million years of evolution that separate the two species.

Let us first address the evolution on the smallest molecular scale, where the different types of events occur with widely different frequency. For example, insertions/deletions (indels) are a factor of 10–20 less likely than single base-pair mutations. And, further, different single base-pair mutations depend on which base is exchanged by which, with G–C substitutions being more frequent than A–T substitutions, as seen in Table 9.1. More striking, however, is that the relative frequencies of these events differ between species. Deletions, for example, are more prominent in *Drosophila* than in humans (Petrov & Hartl, 1999).

Single-point mutations are typically either completely neutral, or confer a rather small change to a given protein (if they lead to a change in amino acid). Thus they are associated with the fine-tuning of properties of given proteins, not to the overall properties defining an organism. Thus species are presumably not evolved through single-point mutations, or simple deletions, but rather with an intricate set of genomic re-arrangements. When comparing genomes of different life forms, one often finds similar genes, but at different chromosomes and arranged in a different order.

Table 9.1. *Nucleotide mutation matrix $q_{ij}$ on the so-called PAM 1 level, defined as one substitution per 100 nucleotides: of the 1% substitutions, a fraction 0.6 will be an A replaced with G (or vice versa)*

| | A | G | T | C |
|---|---|---|---|---|
| A | 0.990 | 0.006 | 0.002 | 0.002 |
| G | 0.006 | 0.990 | 0.002 | 0.002 |
| T | 0.002 | 0.002 | 0.990 | 0.006 |
| C | 0.002 | 0.002 | 0.006 | 0.990 |

From Dayhoff *et al.* (1978).

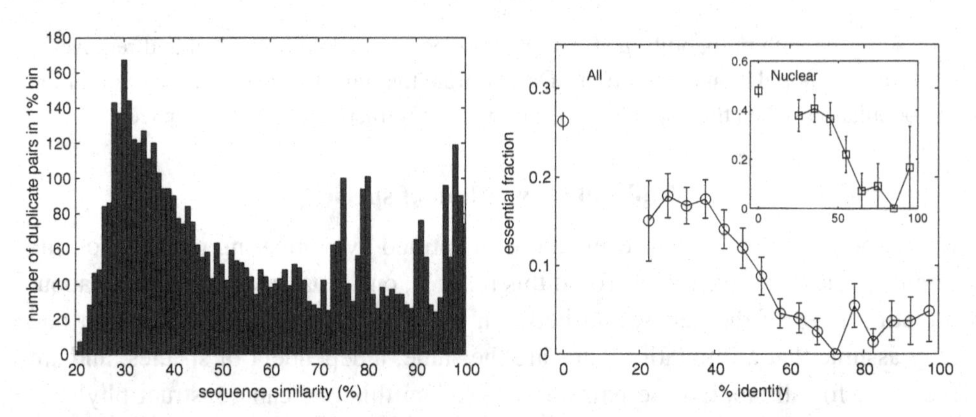


Figure 9.6. Distribution of protein paralogs in *S. Cerevisiae*. Two proteins are paralogs if they have a common origin. This is the case for about one third of all the proteins in yeast. The left-hand panel shows the number of paralog pairs as a function of their sequence similarity. The right-hand panel shows the fraction of essential proteins in single gene knockout experiments, as a function of similarity to the most similar paralog in the yeast genome. It demonstrates that proteins with very similar paralogs can be eliminated from the organism, because the duplicate can take over its basic functions. Notice also the insert, illustrating that proteins in the yeast nucleus tend to have a much larger probability to be essential than the "workforce" proteins of the cytoplasm (from Maslov *et al.*, 2004).

Even more strikingly, different groups of species differ hugely in their junk DNA, in particular by their repeat sequences (see Shapiro, 1997). Primates are, for example, characterized by huge sections of the so-called ALU repeat family. Thus evolution is random, but some events are self-propelled, even on the scale of the full genome.

There are many mechanisms that can rearrange or copy genes in molecular biology: mechanisms that are found even in quite primitive bacterial phages. This fact in itself makes one wonder to what extent genetic material can be transferred between species (horizontal transfer). An example of gene rearrangements is retroviruses, which are able to read an RNA message into DNA and insert it into the genome. Within a given cell, an mRNA → DNA mechanism thus copies genes into other places in the genome. However, it typically copies a working gene into a silent gene (a pseudogene), because it will not have the appropriate start signals for new transcription and translation. In any case many of the naturally occurring DNA engineering mechanisms can in principle be coupled to the environment of the organism, and be triggered by, for example, external stresses like osmotic stress, toxins, etc.

Genomic rearrangements can be traced by comparing protein sequences in an organism. Genes duplicate in the chromosome of any organism (Ohno, 1970). This provides a raw material of proteins, from which new functions can be developed. A pair of duplicated proteins within the same organism are called paralogs. Paralogs

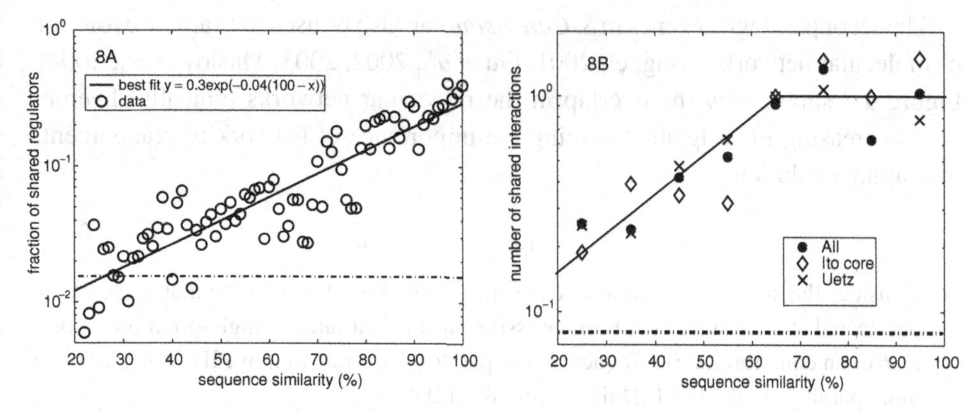


Figure 9.7. In the left-hand panel we show the decay of overlap of upstream transcription regulation (for yeast), with decreasing similarity of proteins, from Maslov *et al.* (2004). As the time since duplication increases, similarity decreases, and similarity in who regulates the two proteins decreases. In the right-hand panel we show the decay in overlap between the paralog binding partners, with binding partners determined from the two hybrid experiments of Ito *et al.* (2001) and Uetz *et al.* (2000).

can be identified with statistical methods that identify related proteins by examining their sequence similarity. For the yeast *Saccaromyces Cerevisiae* one finds that about 2000 of 6200 proteins have more than 20% similarity with at least one other yeast protein (Gilbert, 2002).

Obviously, gene duplication starts as a single organism event that only rarely spreads to the whole species. However, for the few duplications that actually take over the whole population, subsequent mutations will change the two paralogs independently. Thereby the paralogous proteins diverge in sequence similarity with time. As a result we see today a wide variety of protein similarities in any living organism. The left-hand panel of Fig. 9.6 shows the numbers of paralog pairs in yeast, distributed according to their amino acid similarity (PID = percent identity). Here, the duplicated genes consist of 4443 pairs of paralogous yeast proteins (data from Gilbert (2002); plot from Maslov *et al.* (2004)).

Recent duplicated proteins will be close to identical, whereas ancient duplications will have small PID. To set a very rough timescale, paralogs with a PID of about 80% typically were duplicated about 100 My ago. Following Gu *et al.* (2003) and Maslov *et al.* (2004) the right-hand panel of Fig. 9.6 shows divergence/redundancy of a pair of paralogs by investigating viability of a null-mutant lacking one of them. Notice that lethality first increases for PID$<$ 60%, indicating that paralogs with a higher level of similarity can substitute for each other. This shows substantial room for neutral evolution, also on the amino acid sequence level.

The set of paralogous genes in *S. Cerevisiae* can also be used to examine evolution of molecular networks (Wagner, 2001; Gu *et al.*, 2002, 2003; Maslov *et al.*, 2004). Figure 9.7 shows how the overlap in the molecular networks typically diverges with decreasing PID, again indicating the importance of network rearrangements in shaping evolution.

### *Question*

(1) Consider the spectrum of paralog frequencies; see Fig. 9.6. Assume that proteins are duplicated at a constant rate. Further assume a constant rate of single-point mutations. Derive an equation for the frequencies of paralogs as a function of PID. Why are there more paralogs with 20% PID than with 98% PID?

## Punctuated equilibrium and co-evolution

Life on our scale, and also the evolution of species during the past 540 million years, present some intriguing evidence for cooperative behavior: Often during the history of life there have been major "revolutions" where many species have been replaced simultaneously. This is illustrated in Fig. 9.1, and also in Fig. 9.8. Spectacular examples are the Cambrian explosion 540 million years ago where a huge variety

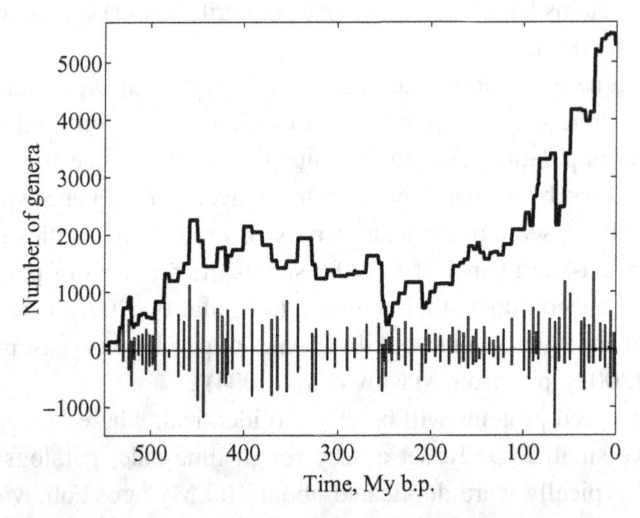


Figure 9.8. Extinction and origination as well as total number of observed genera throughout the Phanerozoic (the period of multicellular life on Earth). The two intermittent signals along the time axis are the horizontal and vertical projections of the activity in Fig. 9.1. Extinction is negative, origination is counted positively, and the total diversity of course is positive. Notice that extinction and origination appear synchronously, presumably all happening in the instants that divide the geological stages. Data from Sepkoski database on genera level (1993).

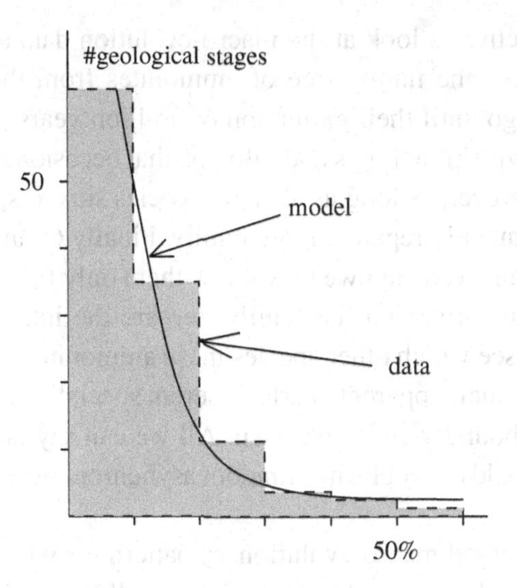


Figure 9.9. Histogram of family extinctions in the fossil record as recorded by Raup (1991, 1994) and Raup & Sepkoski (1992). For comparison we show the prediction of the random-neighbor version of the BS model.

of life arose in a short time interval, and the Cretaceous–Tertiary boundary where mammals took over large parts of the ecosystem.

In between these major revolutions there have been periods of quiescence, where often all species seemed to live in "the best of all worlds", and only a few species suffered extinctions. However, the pattern of life is more subtle than completely on–off transitions of macro-evolutionary revolutions. When inspecting the extinction record of Fig. 9.8 one observes that there was often extinction of smaller size in the quieter periods. In fact if one plots the size distribution of ecological events (see Fig. 9.9), one observes all sizes of extinctions. That is, the large events are becoming gradually less frequent than smaller events. There is no "bump" or enhanced frequency for the large-scale extinctions. However, the distribution is broader than a Gaussian, in fact it is close to scale-free, as indicated by the fitted $1/s^{1.5}$ curve. This overall gradual decline of event size distributions indicates the following.

- The probability distribution for extinction events are non-Gaussian, implying that the probability for obtaining large events is relatively large. This shows that the species in the ecosystem do not suffer extinction independently of each other. Thus the overall macro-evolutionary pattern supports cooperativity, even on the scale of the global ecosystem.
- Large and small events may be associated with similar types of underlying dynamics. If extinctions were external, because of asteroid impacts, for example (Alvarez, 1987), one would expect a peak at large events.

In may be instructive to look at the macro-evolution data a little more closely. In Fig. 9.10 we show the family tree of ammonites from their origination over 350 million years ago until their extinction 66 million years ago. One observes a tree-like structure, with branchings and killings, that occasionally undergoes a near total extinction. However, as long as a single species survives, one sees that it can diversify and subsequently regain a large family. Finally 66 million years ago, not a single ammonite survived, and we now know them only from their very common fossils. What we cannot see on the family tree are the interactions between the species, nor can we see which other species these ammonites interacted with. Thus the dynamics behind their apparent nearly coherent speciation and extinction events far beyond species boundaries is unknown. All we can say is that the behavior is far from what we would expect from a random asynchronous extinction/origination tree.

To model the observed macro-evolutionary pattern we will start with objects of the size of the main players on this scale; let us call them species. A species, of course, consists of many individual organisms, and dynamics of species represent the coarse-grained view of the dynamics of these entire populations. Thus, whereas population dynamics may be governed by some sort of fitness, we propose that species dynamics is governed by stabilities. In the language of fitness landscapes, we picture the species dynamics as erratic jumps over fitness barriers.

| Population dynamics | $\xrightarrow{\text{large time scale}}$ | Barrier dynamics |
|---|---|---|
| Survival of fittest | | Evolve the least stable |

Obviously, stability is a relative concept that depends on the particular environment and type of fluctuations at hand. It is also important to realize that evolution is not an equilibrium process. If one really had a fixed fitness landscape one may be misled to view movements in this landscape as part of an equilibrium dynamics. However, even for a fixed landscape, its sheer size would prevent any equilibration. Further, there is no reason to consider a fixed landscape. Changes in environment would change the landscape (Kauffman, 1990). And genomic rearrangement in itself would invalidate the concept of an underlying metric for such a landscape. Overall, fitness is ill defined in large-scale evolution. However, stability still has meaning, in particular in a given environment.

We characterize the basic evolutionary entity, called a species, with one number $B_i$. This number is the stability of the species on some rather long evolutionary timescale (a timescale much longer that the organism reproduction timescale). An ecosystem of species consists of selecting $N$ numbers $B_i$, each representing a

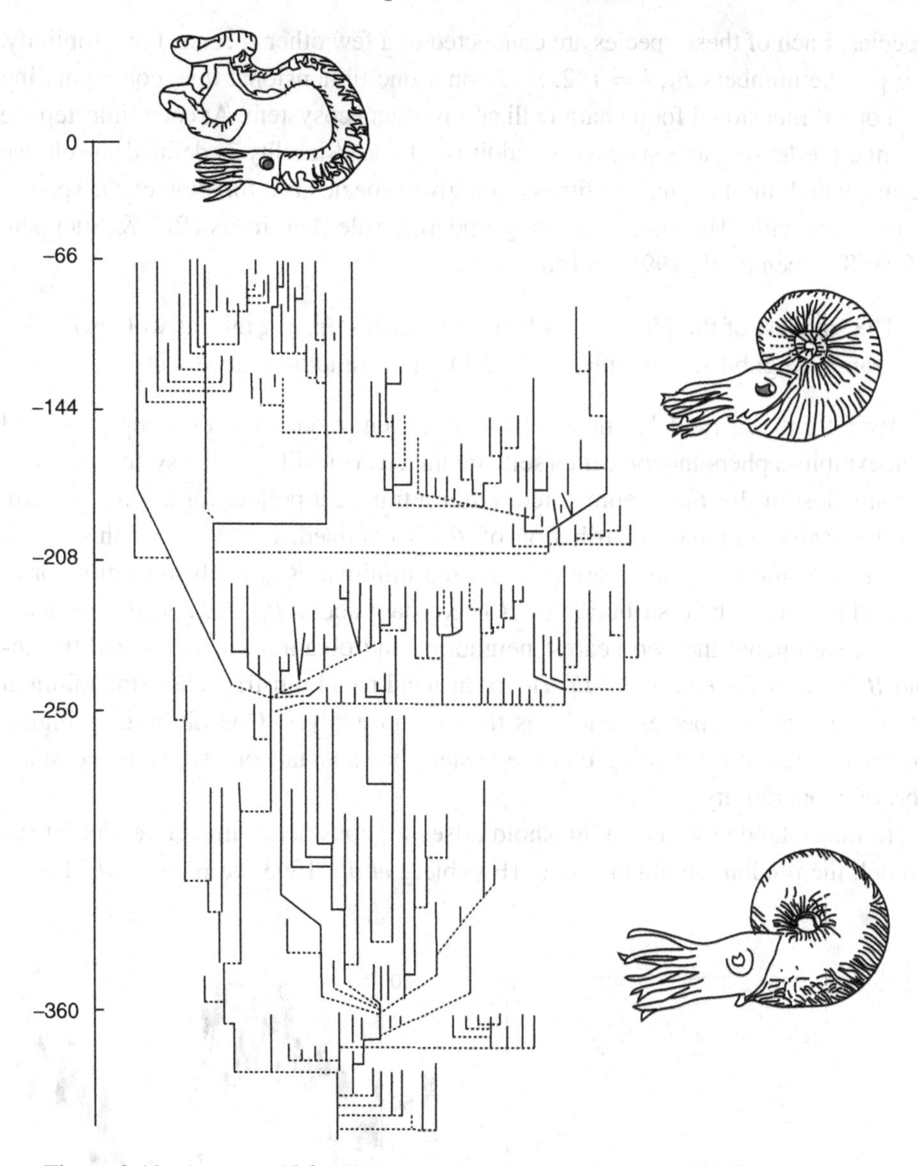


Figure 9.10. Ammonoid family tree, redrawn from Eldredge (1987). Ammonites lived in water on the continental shelves. They fossilized well and there are about 7000 different ammonoid species in the fossil record. During their evolutionary history (from 400 million to 66 million years ago) we observe times with fast speciation into many species and also of simultaneous extinction of many different species. Today there are no living ammonites.

species. Each of these species are connected to a few other species. For simplicity, we put the numbers $B_i$, $i = 1, 2, \ldots N$ on a one-dimensional line, corresponding to a one-dimensional food chain realization of an ecosystem. At each timestep we change the least stable species. In addition, fitness/stability is defined in relative terms, which implies that the fitness of a given species is a function of the species it interacts with. The co-evolutionary updating rule then reads (Bak & Sneppen, 1993; Sneppen *et al.*, 1995) as follows.

> The smallest of the $\{B_i\}_{i=1,N}$ is located at each step. For this, as well as its nearest neighbors, one replaces the $B$s by new random numbers in [0, 1].

We refer to this model as the BS (Bak–Sneppen) model. It is the simplest model that exhibits a phenomenon called self-organized criticality. As the system evolves, the smallest of the $B_i$s is eliminated. After a transient period, for a finite system, a statistically stationary distribution of $B$s is obtained. For $N \to \infty$ this distribution is a step function where the selected minimal $B_{\min}$ is always below or at $B_c$ and therefore the distribution of $B$ is constant above $B_c$. For a grid dimension $d = 1$, we update the two nearest neighbors, and obtain a self-organized threshold $B_c = 0.6670$ (see Fig. 9.11). The right-hand panel illustrates how the minimal $B$ sites move in "species space" as the system evolves. One observes a highly correlated activity, signaling that the system has spontaneously developed some sort of cooperativity.

To understand how such a threshold arises we consider a simpler version of the model, the random neighbor model (Flyvbjerg *et al.*, 1993; de Boer *et al.*, 1994),

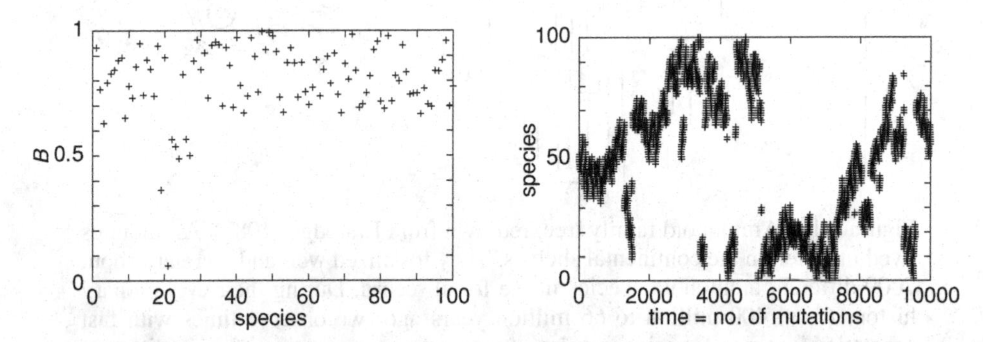


Figure 9.11. Left: Example of snapshot distribution of barriers $B_i$ in space. One observes that $B > B_c = 0.6670$ are distributed randomly in space. In contrast, sites with $B < B_c$ are highly correlated and tend to remain in a small region. Right: Space-time plot of the activity where "time" is counted by the number of updates. At any timestep the site with the minimum barrier is shown with a large dot.

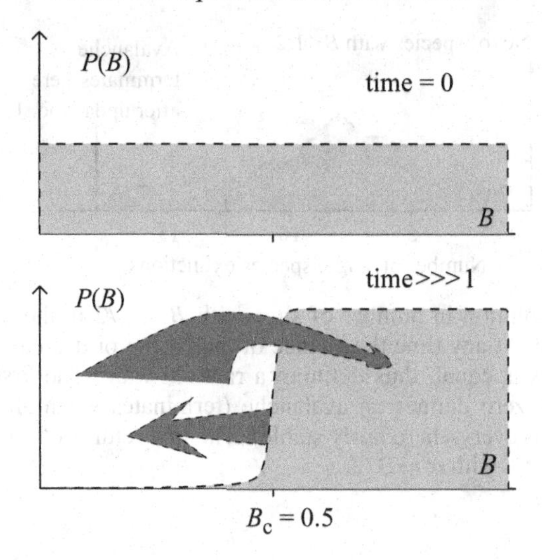


Figure 9.12. Distribution of barriers/fitnesses in the BS model, random neighbor version, where one takes a minimal $B$ and one random $B$, and replaces them with new random $B \in [0, 1]$. The $B$ distribution reaches a steady state where the one minimal selected below $B_c = 1/2$ and distributed everywhere in $[0, 1]$ provides the same net transport into the interval $[B_c, 1]$, as the one selected at random does for the opposite transport away from $[B_c, 1]$ to $[0, B_c]$.

where at each timestep we take the minimal $B$ as well as one other random $B$ (see Fig. 9.12). Because of the lack of spatial correlations between the $B$ values it can be solved, as we will now demonstrate. Starting with a random distribution of $B$s, as the system evolves, the smallest of the $B_i$s is eliminated. After a transient period a statistically stationary distribution of $B$s is obtained. For $N \to \infty$ this distribution is a step function where the selected minimal $B_{\min}$ is always below or at $B_c$. This means that species with $B > B_c$ cannot be selected as the minimum, and thus are selected only as the random neighbor, irrespective of their actual $B$ value. Therefore, the distribution of $B$ is constant above $B_c$. At each step the dynamics selects one $B$ below $B_c$ and the other $B$ above $B_c$. As the two newly assigned $B$s are assigned uniform random values in $[0, 1]$, the condition for a statistically stationary distribution of number of species in the interval $[0, B_c]$ is

$$-1 + 2\,B_c = 0 \quad \text{or} \quad B_c = 1/2 \tag{9.13}$$

The timeseries of selected minimal $B$ exhibits correlations. Let us now define an avalanche as the number of steps between two subsequent selections of minimal $B > B_t$. The number $n_c$ of $B$s below $B_t = B_c$ exhibits a random walk and the size of the avalanche is determined by the first return of this random walk: $P(s) \propto s^{-3/2}$, the distribution of waiting times in the gambler's ruin problem.

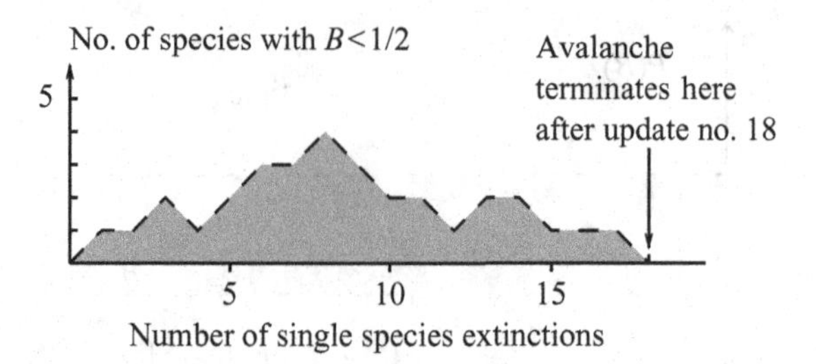


Figure 9.13. Evolution in number of sites with $B < 1/2$ in the random version of the BS model. At any time the chance of increasing or decreasing the number of active species is equal, thus defining a random walk. The first return of the random walk to zero defines an avalanche (terminates when all $B > 1/2$, and thus the system is everywhere fairly stable). The first returns of random walks are distributed as $1/t^\sigma$, with $\sigma = 3/2$.

To prove this we consider first returns of a random walk in one dimension. The random walk can be assumed to be scale invariant, because it repeats itself on all scales. Thus the distribution of first returns of the random walk is assumed to be distributed as $P_{\text{first-return}}(t) \propto 1/t^\tau$, and we want to determine $\tau$ (see Fig. 9.13). Let the total time interval be $T$. Now consider the division of the long time interval $T$ into first returns (Maslov *et al.*, 1994):

$$T = \langle t \rangle_T \cdot (\text{no. of returns in } T) = \langle t \rangle_T \cdot T^{1/2} \tag{9.14}$$

where the number of returns in time $T$ is counted by noting that the chance that the random walker is at $x = 0$ at any time $t < T$ is $\propto 1/\sqrt{t}$. When this is integrated from $t = 0$ to $t = T$ we obtain the above scaling. When we insert $\langle t \rangle_T = \int^T t/t^\tau \,\mathrm{d}t$ (the definition of average return withing the interval $[0, T]$) we obtain the equation

$$T = T^{1/2} T^{2-\tau} \tag{9.15}$$

giving the scaling

$$P_{\text{first-return}}(t) \propto 1/t^{3/2} \tag{9.16}$$

which is then also the distribution of avalanche sizes in the random neighbor version of the BS model.

We now return to the model with fixed geometry, in the sense that neighbors are put on the one-dimensional line. In right-hand panel of Fig. 9.11 we showed a "space-time" map of those sites on which species change barrier values in the

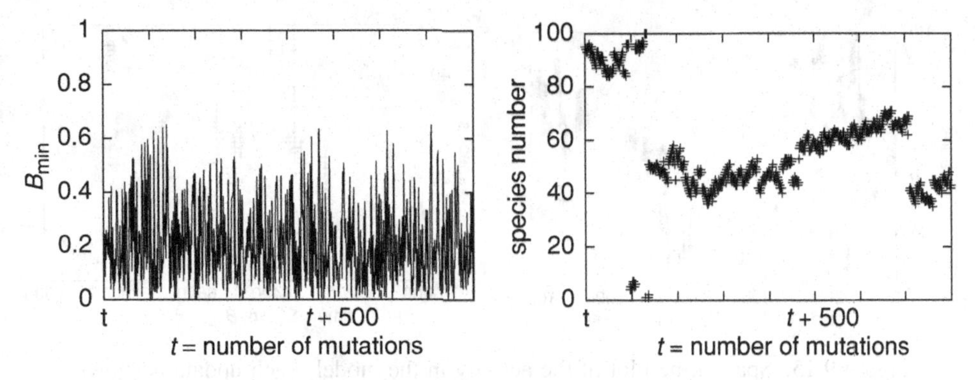


Figure 9.14. Left: development of minimal $B$ with the number of updates in the BS model. Notice that the minimum never (nearly) exceeds the threshold $B_c = 0.67$. Any horizontal line with $B = B_t < B_c$ will divide the dynamics into localized avalanches. For smaller $B_t$ one confines activity to sub-avalanches of the larger ones. Right: location of the minimal $B$ from the left-hand panel. Notice that the larger jumps in position at time $\sim t + 120$ and time $t + 710$ are associated with the largest values of $B_{min}$ on the left-hand panel.

time interval covered. Whenever the lowest barrier value is found among the three last renewed, the site of lowest barrier value performs a random walk, because those three sites have equal probability of being the one with smallest barrier value. Figure 9.11 shows that this is what happens most frequently. When the site of lowest barrier value moves by more than one lattice spacing (jumps), it most frequently backtracks by two lattice spacings to a site that was updated in the next-to-last timestep. But longer jumps occur, too; actually jumps of any length occur, the longer jumps typically to a less recently updated site.

In fact, as we evolve the system it self-organizes towards a state where there are power-law correlations in both space and time. This state is therefore critical, and the algorithm is one of a class of models that lets systems self-organize toward such critical conditions.

To compare the outcome of such a simplified model with real macro-evolutionary data is speculative. However, it is possible if we identify the replacement of one $B_i$ with a new one with an extinction/origination event. That is, let us assume that the species with the smallest $B$ gets eaten and the ecological space becomes occupied by a new species. Further one should of course not use a one-dimensional food chain model, but rather a ramified random network that represents the topology of a real ecological system. Finally, one should identify the observable to look for. In fact, the selection of least stable species to mutate next implicitly assumes a separation of timescales in the dynamics. Thus the selection of the least fit of all species to change next is the natural outcome of the updating model.

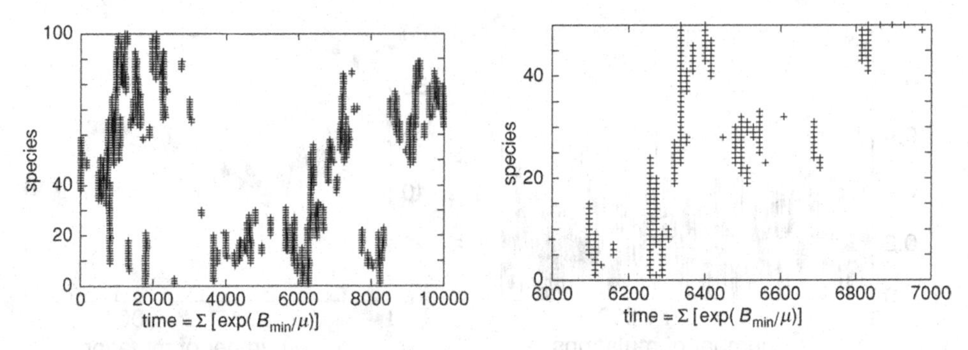


Figure 9.15. Space-time plot of the activity in the model. Each update is shown as a black dot. On the time resolution of the plot, the avalanches appear as almost horizontal lines. The increased magnification on the right shows that there are avalanches within avalanches. The calculations were done at a mutation rate set by $\mu = 0.02$.

At each step: select each of the $\{B_i\}_{i=1,N}$ with probability $\propto e^{-B_i/\mu}\, dt$. This selection defines a list of active sites. Replace members in this list, as well as their nearest neighbors, by new random numbers in [0, 1].

Here $\mu$ represents an attempt rate for microscopic evolutionary changes, and is proportional to the mutation rate per generation. In order to avoid too many iterations without any active sites, one in practice applies this algorithm with varying timestep $dt \sim e^{B_{min}/\mu}$. For low enough $\mu$ (mutation rate $\ll B_c$) this model degenerates into the one where the minimal $B_i$ is always selected. The value of $B_{min}$ represents a timescale (in fact $\propto$ log(time)). Thus if $B_{min}$ is large, exponentially long times pass without any activity at all. If $B_{min}$ is small it is selected practically instantly (see Fig. 9.15).

For each barrier $B_t$ below the self-organized critical threshold $B_c$, a "$B_t$ avalanche" starts when a selected $B_{min}$ is below $B_t$ and terminates when a selected $B_{min}$ is above $B_t$. In the *local* formulation, all activity within a $B_t$ avalanche occurs practically instantly when seen on a timescale of order $\exp(B_t/\mu)$. And this statement may be reiterated for the larger avalanches associated with $B_{t2} > B_t$, thereby defining a hierarchy of avalanches within avalanches. Thus one may view the avalanche-within-avalanche picture as burst-like activity on widely different timescales. This is illustrated in the space-time plot where each step of the algorithm is associated with a time interval

$$\Delta t \propto \frac{1}{\sum_i e^{-B_i/\mu}} \sim e^{B_{min}/\mu} \quad \text{for } B_c - B_{min} \gg \mu \tag{9.17}$$

The last approximation uses the fact that the distribution of barriers below $B_c$ is scarce, thus for low $\mu$ only the smallest barrier counts.

For $\mu \to 0$ the probability of having an avalanche of size $s$ associated with a punctuation of $B_t = B_c - \epsilon$ can be expressed as

$$P_0(s, \epsilon) \propto s^{-\tau} F\left(\frac{s}{\epsilon^{-1/\sigma}}\right) \tag{9.18}$$

where avalanches $s \ll \epsilon^{-1/\sigma}$ are power-law distributed with exponent $\tau$ ($F(x) \approx$ const for $x \ll 1$), whereas large avalanches $s > \epsilon^{-1/\sigma}$ are suppressed by $F$. The exponent $\tau$ depends on the dimension; $\tau = 1.07$ for $d = 1$, whereas $\tau = 3/2$ for high dimension $d$ of the ecosystem network. The value of $\tau = 3/2$ compares well with the histogram of extinction events on the family level (see Fig. 9.9).

Apart from the technical similarity to the overall macro-evolutionary dynamics, the above model exhibits a number of worthwhile lessons, of potential relevance for analysis of paleontological data.

(1) Each evolutionary avalanche consists of sub-avalanches on smaller scales. Thus when we analyze the fossil data on finer-grained time (and space) levels, we should expect to find each extinction event subdivided into smaller extinction events.
(2) Macro-evolutionary extinction is closely connected to micro-evolutionary changes in phenotype of selected species.
(3) Time separation between evolutionary events of a given lineage will be power-law distributed, with long periods of stasis, sometimes broken by a sequence of multiple small jumps. Punctuated equilibrium thus allows for large evolutionary meanderings, where barriers that would seem impossible at a stasis period are circumvented by changed fitness landscapes due to co-evolution adaptations.

In summary, we have discussed a model that extends the concept of punctuated equilibrium known from single fossil data. This demonstrates that both stasis and global extinction could be a natural part of co-evolution of the discrete life forms that we know from our surrounding biological world.

### *Questions*

(1) Simulate the BS model for 100 species placed along a line. Plot the selected $B_{\min}$ as a function of time. Change the local neighborhood to four nearest neighbors at each update, and repeat the simulation. How does the minimum of $B$ change as time progresses toward steady state (look at the envelope defined as maximum over all $B_{\min}$ at earlier times)?
(2) Consider the random neighbor version of the BS model, and update $K$ neighbors at each time step. What is the self-organized threshold?
(3) Generate a scale-free network with $N = 200$ nodes, and degree distribution $\mathrm{d}P/\mathrm{d}K \propto 1/K^3$ (see Chapter 8). Assign a number $B$ between 0 and 1 to each node and simulate the BS model on it. Use a version where, at each timestep, one updates the site with the smallest $B$ and ONE of its neighbors. Plot the accumulated activity versus degree of the different nodes.

(4) Calculate the distribution of time intervals between changes in the finite mutation rate version of the BS model. (Hint: consider $P(t) \propto \int P(B_{\min})P(t|B_{\min})\mathrm{d}B_{\min}$ with $P(t|B_{\min}) = \mathrm{e}^{-t_{\min}/\tau}$ and $\tau_{\min} = \mathrm{e}^{B_{\min}/\mu}$.)

(5) One may consider a globally driven version of evolution based on stability of species (Newman & Sneppen, 1996). Consider a system of $N$, say 1000, species. Assign a random number $B$ in [0, 1] to each species. At each time step select an external noise $x$ from a narrow distribution $p(x) \propto \exp(-x/\sigma)$, $\sigma \ll 1$. At each time step: replace all $B < \sigma$ with a new random number between 0 and 1 and, in addition, select one random species and replace its stability $B$ with a new random number. Simulate this model and explain why it predicts a power law of type $p(s) \propto 1/s^2$ for extinction sizes (Sneppen & Newman, 1997).

## Evolution of autocatalytic networks

One of the main problems in the origin of life is understanding the emergence of structures that can replicate/copy themselves. A main problem in this regard is that the size of the replication machinery in even the simplest living organisms must have arisen from spontaneous fluctuations. In fact the requirement of sufficient exact replications is so complicated that it seems impossible for it to have arisen by spontaneous fluctuations. Formally this problem has been addressed by Eigen & Schuster (1977, 1978) who suggested a hyper-cycle to circumvent it (see Fig. 9.16). A hyper-cycle is a group of interconnected molecular species that each help other members of the set to replicate. None of the molecules can replicate themselves, and thus there is no competition between them. However, acting together they should be able to replicate the whole. The basic trick is that if any single molecule makes

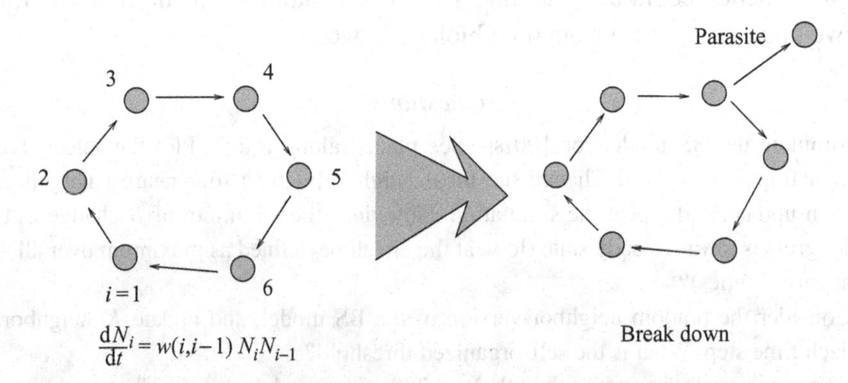


Figure 9.16. A hyper-cycle, an inspiring possibility suggesting that a system may use mutual dependence to avoid the winner taking all. This was proposed in order to circumvent the huge error rates associated with spontaneous copying of larger molecules. However, the proposed mechanism is unstable against emergence of parasites that drain but do not contribute to the cycle.

$b$ is made from $a$:

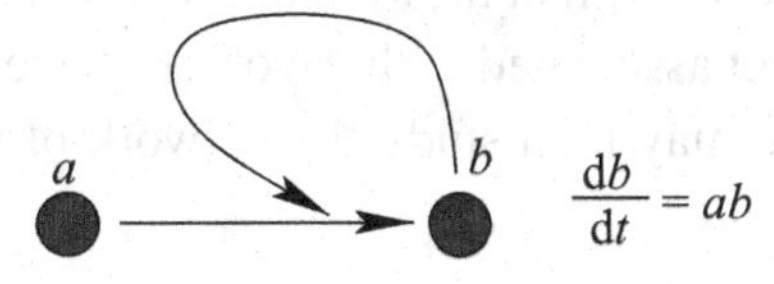


$b$ eats $a$ to replicate (catalyze its own formation):

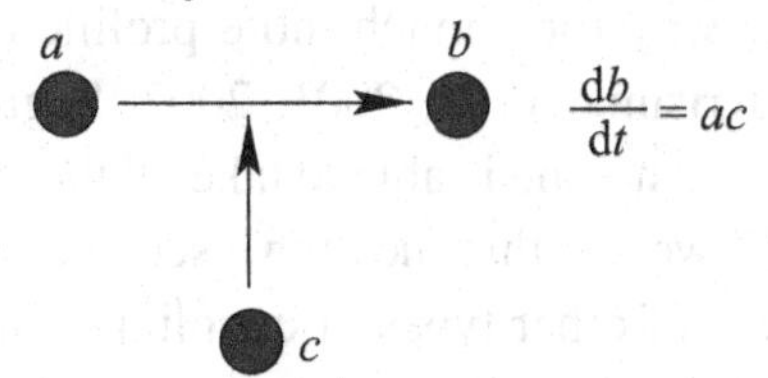


$c$ catalyzes the conversion of $a$ to $b$:

$a$ $b$

$$\frac{db}{dt} = ac$$

$c$

Figure 9.17. Catalytic reactions, as in the Jain–Krishna model (top), with predator–prey like autocatalysis (middle), and finally as in standard catalytic reaction as they typically occur in for example metabolic networks (bottom).

a faulty replication it is discarded and replaced by the next exact copy. However, the Eigen & Schuster hyper-cycle is unstable against parasites, in the sense that it dissolves when there are mutated molecules that take resources from the cycle but do not give back to it. This is a serious fault, first realized by Smith (1979).

The basic equation in Eigen & Schuster's hyper-cycle is of the form

$$\frac{dN_i}{dt} = \sum_{i,j} w(i, j) N_i N_j \tag{9.19}$$

a type of catalytic reaction that we visualize in the middle of Fig. 9.17. It represents a production of species $i$ that is proportional both to itself and to some other species (the populations $N_j$ where $w(i, j) > 0$). In addition there is an overall constraint that fixes the total amount of biomass $\sum N_i$ in the system. This is modeling an ecology based on self-replicating entities that each catalyze self-replication of some other species. The couplings are also known from the predator–prey model. The evolutionary consequence of this type of equation is exponential amplification of the best replicator, to the cost of all others. Thus only very few species survive with this type of model, and larger networks tends to break up owing to parasites.

The instability can be removed if one instead uses

$$\frac{dN_i}{dt} = \sum_{i,j} w(i, j) \frac{N_i N_j}{\text{Max} + N_i} \tag{9.20}$$

where "Max" sets an upper limit of $N_i$ that the system can sustain (a maximum that may be set by factors that are not associated with the other species). For a saturated system, where $N_i \gg \max$, one may then study the network of species described by

$$\frac{dN_i}{dt} = \sum_{j} w(i, j) N_j \tag{9.21}$$

This is a system that opens the way for a much more prolific evolution than the earlier nonlinear one (Jain & Krishna, 1998, 2001, 2002; Segre *et al.*, 2000). It allows many species to co-exist, as no one is able to take all the resources by itself. When comparing with Fig. 9.17 we see that the above set of couplings could also be obtained as saturated versions of other types of couplings, and as such may be a useful starting point for analyses.

The Jain–Krishna model for evolving networks of the type governed by Eq. (9.21) is a model for evolving the coupling matrix $w(i, j)$ for a number of $i = 1, 2, \ldots, \mathcal{N}$ species. The evolution is again done by separating the problem into two timescales. At the fast timescale one runs the equations until a steady state is reached (all $N_i$ constant). This state can in principle be analytically calculated as the eigenvector of the largest eigenvalue for a set of linear equations.

Once a steady state is reached (on a fairly fast timescale of population dynamics) one selects the species with the lowest (minimal) $N_i$, say with label $i_{\text{min}}$, and replaces all couplings $w(i_{\text{min}}, j)$ and $w(j, i_{\text{min}})$ with new random couplings with an a priori set probability $p < 1/\mathcal{N}$ for being non-zero. This defines a new network with a new steady state, and the evolution is repeated.

On the longest timescale one records how the network evolves. Starting from no network (disconnected species) one obtains a large autocatalytic network that remains stable for quite long periods (see Fig. 9.18). When the network breaks down, on a timescale $1/p^2$ it replaces itself with a new network. The Jain–Krishna model may thus be seen as an elegant network extension of the punctuated evolution scenario outlined in the previous section.

## *Questions*

(1) Simulate the dynamics of the two species cycle: $dx/dt = y$ and $dy/dt = -x$.

(2) Consider dynamics of the two species cycle: $dx/dt = x + ax \cdot y$ and $dy/dt = y + bx \cdot y$. Simulate for $(a, b) = (1, 1)$, $(a, b) = (10, 10)$ and $(a, b) = (1, -1)$.

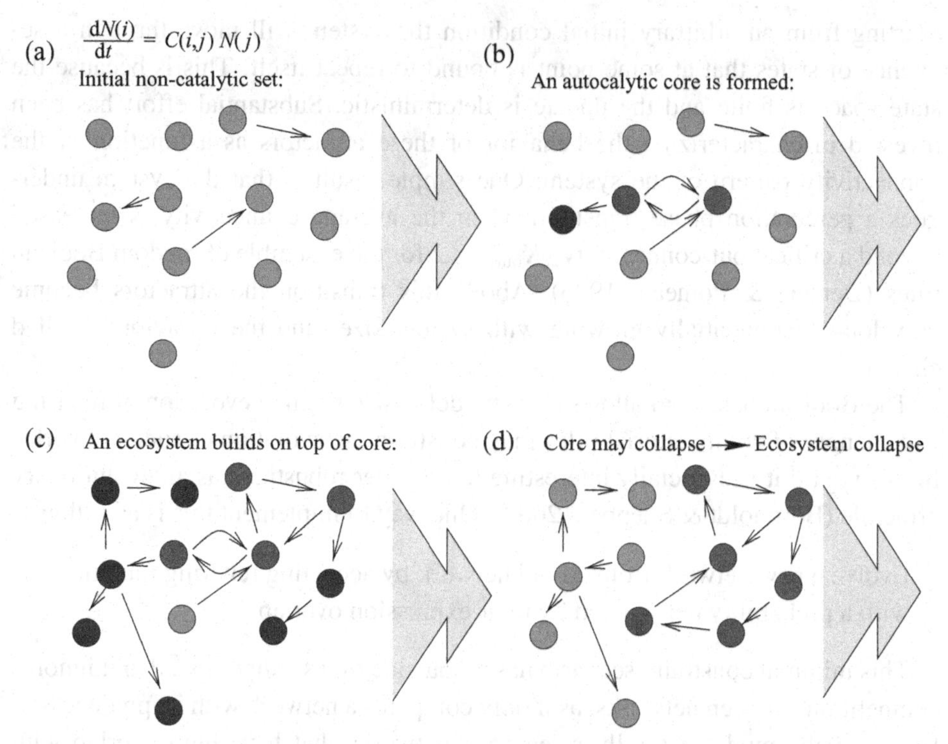


Figure 9.18. Evolving networks as suggested by Jain & Krishna. Initially (a) no species replicate, but after some time a self-catalyzing core (b) is formed. Subsequently a larger network is developed on this basis (c). Occasionally new parts of the networks develop that out-compete the original core, and without these "key-stone" species a larger part of the ecosystem may collapse (d).

## Evolution of evolvability

In addition to the topological properties discussed in Chapter 8, the regulation of genes also has a combinatorial part (Davidson *et al.*, 2002). In other words, the regulation of a given gene may depend on the combination of the concentration of its regulators. An inspiring way to model such combinatorial regulation was proposed already in 1969 by Kauffman. In this very simplified approach, each gene is assigned a binary number, 0 or 1, that counts whether the gene is off or on. Each gene $i$ is assumed to be on or off, depending on some logical function of the state of the genes that regulate it. A simple version of this Boolean rule is the threshold network where the state of gene $i$ at time $t$ is given by

$$\sigma(i,t) = \Theta\left(\sum A_{ij}\sigma(j,t-1)\right) \tag{9.22}$$

where $\Theta(x)$ is zero for $x < 0$ and $= 1$ else. The $A_{ij}$ is non-zero for genes $j$ that regulate gene $i$. $A_{ij}$ is positive for activation, and negative for repression.

Starting from an arbitrary initial condition the system will move through a sequence of states that at some point is bound to repeat itself. This is because the state space is finite and the update is deterministic. Substantial effort has been invested in characterizing the behavior of these attractors as a function of the connectivity pattern of the system. One simple result is that the system undergoes a percolation phase transition when the average connectivity is increased beyond a critical out-connectivity $\langle K_{\text{out}} \rangle \sim 2$ for the ensemble of random Boolean rules (Derrida & Pomeau, 1986). Above this transition the attractors become very long (exponentially growing with system size) and the behavior is called chaotic.

The Boolean paradigm allows us to model how long time evolution of rewiring and changes of combinatorial rules may constrain the evolution of such networks. In this regard it is especially interesting to consider robustness as an evolutionary principle (Bornholdt & Sneppen, 2000). One way to implement this is as follows.

> Evolve a new network from an old network by accepting rewiring mutations with a probability determined by their expression overlap.

This minimal constraint scenario has no outside fitness imposed. Also it ignores competition between networks, as it only compares a network with its predecessor. However, the model naturally selects for networks that have high overlap with neighboring mutant networks. This feature is associated with robustness, defined as the requirement that, not only should the present network work, but also mutations of the networks should work. In terms of network topology this means a change in the wiring $\{A_{ij}\} \rightarrow \{A'_{ij}\}$ that takes place on a much slower timescale than the $\{\sigma_j\}$ updating using the Boolean dynamics itself.

The system that is evolved is the set of couplings $A_{ij}$ in a single network. One evolutionary timestep of the network is as follows.

(1) Create a daughter network by (a) adding, (b) removing, or (c) adding and removing a weight in the coupling matrix $A_{ij}$ at random, each option occurring with probability $p = 1/3$. This means turning $A_{ij} = 0$ to a randomly chosen $\pm 1$, or vice versa.
(2) Select a random input state $\{\sigma_i\}$. Iterate simultaneously both the mother and daughter systems from this state until they have reached and completed the same attractor cycle, or until a time where $\{\sigma_i\}$ differs between the two networks. In case their dynamics is identical then replace the mother with the daughter network. In case their dynamics differs, keep the mother network.

This dynamics favors mutations that are phenotypically silent, in the sense that they are neutral under at least some external condition. Iterating these steps represents an evolution that proceeds by checking overlap in expression pattern

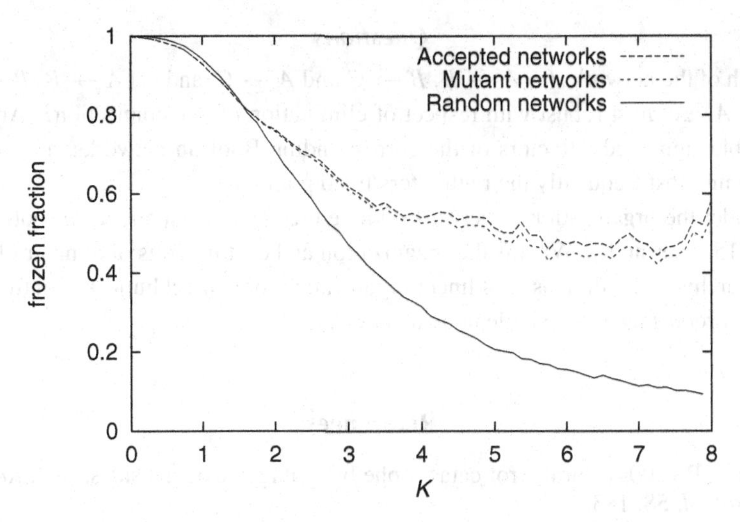


Figure 9.19. Evolving networks with local overlap automatically generate networks with simpler expression patterns. Here the expression is quantified in terms of size of the frozen component, which for any value of average connectivity $\langle K \rangle$ is much larger than the one found in a randomized network with the same overall connectivity. The frozen component consists of nodes that always have the same output state. The more frozen components, the simpler the dynamics of the system. Figure from Bornholdt & Sneppen (2000).

between networks. If there are many states $\{\sigma_i\}$ that give the same expression of the two networks, then transitions between them are fast. On the other hand, if there are only very few states $\{\sigma_i\}$ that result in the same expression for the two networks, then the transition rate from one network to the other is small. If this is true for nearly all neighbors of a network, the evolution of this network will be hugely slowed down.

The result of this type of modeling is an intermittent evolutionary pattern: occasionally, the evolved network is trapped in states with few evolution paths away from it, sometimes it is instead in an evolutionary active region of "network space", which allows for fast readjustments of genome architecture. Further, the obtained networks exhibit a less chaotic behavior than random networks. A computational structure has emerged; see Fig. 9.19. The system self-organizes into robust logical structures, characterized by simplicity. This is reminiscent of the simplification in topology that one sees in the yeast molecular network with its highly connected proteins placed on the periphery of the network. In the Boolean model networks explored here, the attractors for the networks are shorter, and there are more frozen regulators than in a random network at the same average connectivity.

### Questions

(1) Which of the networks (a) $A \to B$, $B \to C$ and $A \to C$, and (b) $A \to B$, $B \to C$ and $C \to A$, are most robust with respect of elimination of one connection? (Analyze all possible signs and attractors of the corresponding Boolean networks, and see which maintain most frequently their attractors/fixed points.)

(2) Consider the organization of networks where the signal from one node should control $N = 15$ other nodes. What is the organization and cost in terms of number of links by a hierarchy, a one-dimensional linear organization, or a direct hub-like control? Which is most robust against a single node deletion?

## References

Anderson, J. P. (2004). Viral error catastrophe by mutagenic nucleosides. *Ann. Rev. Microbiol.* **58**, 183.

Alvarez, L. W. (1987). Mass extinctions caused by large solid impacts. *Physics Today* (July), 24–33.

Bak, P. & Sneppen, K. (1993). Punctuated equilibrium and criticality in a simple model of evolution. *Phys. Rev. Lett.* **71**, 4083.

Bak, P., Flyvbjerg, H. & Lautrup, B. (1992). Co-evolution in a rugged fitness landscape. *Phys. Rev.* A **46**, 6724.

de Boer, J., Derrida, B., Flyvbjerg, H., Jackson, A. D. & Wettig, T. (1994). Simple model of self-organized biological evolution. *Phys. Rev. Lett.* **73**, 906.

Bornholdt, S. & Sneppen, K. (2000). Robustness as an evolutionary principle. *Proc. R. Soc. Lond.* B**267**, 2281–2286.

(2004). Hierarchy of life times in the fossil record. *Paleobiology* (in press).

Davidson, E. H. *et al.* (2002). A genomic regulatory network for development. *Science* **295** (5560), 1669–2002.

Dayhoff, M. O., Schwartz, R. M. & Orcutt, B. C. (1978). A model of evolutionary change in proteins. In *Atlas of Protein Sequence and Structure*, Vol. 5, Suppl. 3, ed. M. O. Dayhoff. Washington, DC: National Biomedical Research Foundation, pp. 345–352.

Derrida, B. (1980). Random energy model: limit of a family of disordered models. *Phys. Rev. Lett.* **45**, 2.

Derrida, B. & Pomeau, Y. (1986). Random networks of automata: a simple annealed approximation. *Europhys. Lett.* **1**, 45–49.

Eigen, M. (1971). Selforganization of matter and the evolution of biological macromolecules. *Naturwissenshaften* **58**, 465.

Eigen, M., McCaskill, J. & Schuster, P. (1989). The molecular quasispecies. *Adv. Chem. Phys.* **75**, 149.

Eigen, M. & Schuster, P. (1977). The hypercycle. A principle of natural self-organization. Emergence of the hypercycle. *Naturwissenschaften* **64**, 541.

(1978). The hypercycle. A principle of natural self-organization. Part A: The abstract hypercycle. *Naturwissenschaften* **65**, 7.

Eldredge, N. (1987). *Life Pulse, Episodes in the History of Life*. New York: Facts on File. Pelican edition (Great Britain).

Eldredge, N. & Gould, S. J. (1972). Punctuated equilibrium: an alternative to phyletic gradualism. In *Models in Paleobiology*, ed. T. J. M. Schopf & J. M. Thomas. San Francisco: Freeman and Cooper.

Fisher, R. A. (1930). *The Genetical Theory of Natural Selection*. New York: Dover.
Flyvbjerg, H., Sneppen, K. & Bak, P. (1993). Mean field model for a simple model of evolution. *Phys. Rev. Lett.* **71**, 4087.
Gilbert, D. G. (2002). euGenes: a eucaryote genome information system. *Nucl. Acids Res.* **30**, 145.
Gould, S. J. (1989). *Wonderful Life*. London: Penguin.
Gould, S. J. & Eldredge, N. (1993). Punctuated equilibrium comes of age. *Nature* **366**, 223–227.
Gu, Z., Nicolae, D., Lu, H. H.-S. & Li, W.-H. (2002). Rapid divergence in expression between duplicate genes inferred from microarray data. *Trends Genet.* **18**, 609.
Gu, Z., Steinmetz, L. M., Gu, X., Scharfe, C., Davis, R. W. & Li, W.-H. (2003). Role of duplicate genes in genetic robustness against null mutations. *Nature* **421**, 63.
Ito, T. *et al.* (2001). A comprehensive two-hybrid analysis to explore the yeast protein interactome. *Proc. Natl Acad. Sci. USA* **98**, 4569.
Jain, S. & Krishna, S. (1998). *Phys. Rev. Lett.* **81**, 5684.
(2001). A model for the emergence of cooperation, interdependence, and structure in evolving networks. *Proc. Natl Acad. Sci. USA* **98**(2), 543–547.
(2002). Large extinctions is an evolutionary model: the role of innovation and keystone species. *Proc. Natl Acad. Sci. USA* **99**(4), 2055–2060.
Kauffman, S. A. (1969). Metabolic stability and epigenesis in randomly constructed genetic nets. *J. Theor. Biol.* **22**, 437–467.
(1990). Requirement for evolvability is complex systems: orderly dynamics and frozen components. *Physica D* **42**, 135–152.
Kemp, C., Wilke, C. O. & Bornholdt, S. (2003). Viral evolution under the pressure of an adaptive immune system: optimal mutation rates for viral escape. *Complexity* **8**, 28.
Kimura, M. (1980). A simple model for estimating evolutionary rates of base substitutions through comparative studies of nucleotide sequences. *J. Mol. Evol.* **16**, 111–120.
(1983). *The Neutral Theory of Molecular Evolution*. Cambridge: Cambridge University Press.
Lande, R. (1985). Expected time for random genetic drift of a population between stable phenotype states. *Proc. Natl Acad. Sci. USA* **82**, 7641–7644.
Maslov, S., Paczuski, M. & Bak, P. (1994). Avalanches and $1/f$ noise in evolution and growth models. *Phys. Rev. Lett.* **73**, 2162.
Maslov, S., Sneppen, K. & Eriksen, K. (2004). Upstream plasticity and downstream robustness in evolving molecular network. *BMC Evol. Biol.* **4**, 9.
McClintock, B. (1950). The origin and behavior of mutable loci in maize. *Proc. Natl Acad. Sci. USA* **36**, 344–355.
(1953). Induction of instability at selected loci in maize. *Genetics* **38**, 579–599.
Newman, M. & Sneppen, K. (1996). Avalanches, scaling and coherent noise. *Phys. Rev.* E**54**, 6226.
Ohno, S. (1970). *Evolution by Gene Duplication*. Berlin: Springer-Verlag.
Petrov, D. A. & Hartl, D. L. (1999). Patterns of nucleotide substitutions in drosophila and mammalian genomes. *Proc. Natl Acad. Sci. USA* **96**, 1475–1479.
Raup, D. M. (1991). *Extinction, Bad Genes or Bad Luck?* Oxford: Oxford University Press.
(1994). The role of extinction in evolution. *Proc. Natl Acad. Sci. USA* **91**(15), 6758–6763.

Raup, D. M. & Sepkoski Jr., J. J. (1992). Mass extinctions and the marine fossil record. *Science* **215**, 1501.
Segre, D., Ben-Eli, D. & Lancet, D. (2000). Compositional genomes: prebiotic information transfer in mutually catalytic noncovalent assemblies. *Proc. Natl Acad. Sci. USA* **97**, 4112–4117.
Sepkoski Jr., J. J. (1993). Ten years in the library: new data confirm paleontological patterns. *Paleobiology* **19**, 43.
Shapiro, J. A. (1997). Genome organization, natural genetics engineering and adaptive mutation. *TIG* **13** (3), 98–104.
Smith, J. M. (1979). Hypercycles and the origin of life. *Nature* **280**, 445–446.
Sneppen, K., Bak, P., Flyvbjerg, H. & Jensen, M. H. (1995). Evolution as a self-organized critical phenomenon. *Proc. Natl Acad. Sci. USA* **92**, 5209–5213.
Sneppen, K. & Newman, M. (1997). Coherent noise, scaling and intermittency in large systems. *Physica D* **110**, 209.
Uetz, P. *et al.* (2000). A comprehensive analysis of protein–protein interactions in *Saccharomyces Cerevisiae. Nature* **403**, 623.
Van Valen, L. (1973). A new evolutionary law. *Evolutionary Theory* **1**, 1.
Wagner, A. (2001). The yeast protein interaction network evolves rapidly and contains few redundant duplicate genes. *Mol. Biol. Evol.* **18**, 1283.
Whitman, W. B., Coleman, D. C. & Wiebe, W. J. (1998). Procaryotes: the unseen majority. *Proc. Natl Acad. Sci. USA* **95**, 6578.
Wright, S. (1945). Tempo and mode in evolution: a critical review. *Ecology* **26**, 415–419.
(1982). Character change, speciation, and the higher taxa. *Evolution* **36**, 427–443.

## Further reading

Alon, U., Surette, M. G., Barkai, N. & Leibler, S. (1999). Robustness in simple biochemical networks. *Nature* **397**, 168–171.
Axelrod, R. *The Evolution of Cooperation*. New York: Basic Books.
Benton, M. J. (1995). Diversification and extinction in the history of life. *Science* **268**, 52.
Boerlijst, M. C. & Hogeweg, P. (1991). Spiral wave structure in pre-biotic evolution: hypercycles stable against parasites. *Physica* D**48**, 17–28.
Burlando, B. (1993). The fractal geometry of evolution. *J. Theor. Biol.* **163**, 161.
Cairns, J., Overbaugh, J. & Miller, S. (1988). The origin of mutants. *Nature* **335**, 142.
Castro, H. F., Williams, N. H. & Ogram, A. (2000). Phylogeny of sulfate-reducing bacteria. *FEMS Microbiol. Ecol.* **31**, 1–9.
Christensen, K., Donangelo, R., Koiler, B. & Sneppen, K. (1998). Evolution of random networks. *Phys. Rev. Lett.* **87**, 2380–2383.
Cronhjort, M. & Nybers, A. M. (1996). 3D hypercycles have no stable spatial structure. *Physica* D**90**, 79.
Darwin, C. (1872). *The Origin of Species*, 6th edition. London: John Murray.
Dawkins, R. (1982). *The Extended Phenotype: The Gene as the Unit of Selection.* Oxford, San Fransisco: W. H. Freeman and Co.
Derrida, B. & Peliti, L. (1991). Evolution in flat fitness landscapes. *Bull. Math. Biol.* **53**, 355.
Doring, H. P. & Starlinger, P. (1984). Barbara McClintock's controlling elements: now at the DNA level. *Cell* **39**, 253–259.
Dyson, F. (1985). *Origins of Life*. Cambridge: Cambridge University Press.

Ehrlich, P. R. & Raven, P. H. (1964). Butterflies and plants: a study in coevolution. *Evolution* **18**, 586–608.

Eickbush, T. H. (1997). Telomerase and retrotransposons: which came first? *Science* **277**, 911–912.

Eigen, M., Biebricher, C. K., Gebinoga, M. & Gardiner, W. C. (1991). The hypercycle. Coupling of RNA and protein biosynthesis in the infection cycle of an RNA bacteriophage. *Biochemistry* **30**(46), 11 005.

Eigen, M. & Schuster, P. (1979). *The Hypercycle: A Principle of Natural Self-organization.* New York: Springer-Verlag.

Eisen, J. A. (2000). Horizontal gene transfer among microbial genomes: new insights from complete genome analysis. *Curr. Opin. Genet. Devel.* **10**, 606–612.

Elena, S. F. & Lenski, R. E. (1997). Long-term experimental evolution in *Escherichia coli.* VII. Mechanisms maintaining genetic variability within populations. *Evolution* **51**, 1058.

Elena, S. F., Cooper, V. S. & Lenski, R. E. (1996a). Punctuated evolution caused by selection of rare beneficial mutations. *Science* **272**, 1802.

(1996b). Mechanisms of punctuated evolution (Technical Comment). *Science* **274**, 1749.

Federoff, N. V. (1999). The suppressor–mutator element and the evolutionary riddle of transposons. *Gene to Cells* **4**, 11–19.

Felsenstein, J. (1981). Evolutionary trees from DNA sequences: a maximum likelihood approach. *J. Mol. Evol.* **17**, 368–376.

Force, A., Lynch, M., Pickett, F. B., Amores, A., Yi-Lin Yan & Postlethwait, J. (1999). Preservation of duplicate genes by complementary, degenerative mutations. *Genetics* **151**, 1531.

Futyuma, D. (1990). Phylogeny and the evolution of host plant associations in the leaf beetle genus *Ophraella* (Coleoptera: Chrysomelidae). *Evolution* **44**, 1885–1913.

(1998). *Evolutionary Biology*, 3rd edition. Sunderland, MA: Sinauer.

Futuyma, D. J. & Mitter, C. (1996). Insect-plant interactions: the evolution of component communities. *Phil. Trans. R. Soc. Lond.* B**351**, 1361–1366.

Fox, G. E. *et al.* (1980). The phylogeny of prokaryotes. *Science* **209**, 457–463.

Giaever, G. *et al.* (2002). Functional profiling of the *Saccharomyces Cerevisiae* genome. *Nature* **418**, 387.

Glen, W. (1990). What killed the dinosaurs. *Am. Scientist* (July–Aug.), pp. 354–370.

Gojobori, T., Li, W. H. & Graur, D. (1982). Patterns of nucleotide substitution in pseudogenes and functional genes. *J. Mol. Evol.* **18**, 360–369.

Hendrix, R. W., Smith, M. C. M., Burns, R. N., Ford, M. E. & Hatfull, G. F. (1999). Evolutionary relationships among diverse bacteriophages and prophages: all the world's a phage. *Proc. Natl Acad. Sci. USA* **96**, 2192.

Jukes, T. H. & Cantor, C. R. (1969). Evolution of protein molecules. In *Mammalian Protein Metabolism*, ed. H. N. Munro. New York: Academic Press, p. 21.

Karlin, S., Campbell, A. M. & Mraze, J. (1998). Comparative DNA analysis across diverse genomes. *Ann. Rev. Genet.* **32**, 185–225.

Kashiwagi, A. *et al.* (2001). Plasticity of fitness and diversification process during an experimental molecular evolution. *J. Mol. Evol.* **52**, 502–509.

Kauffman, S. A. (1993). *The Origins of Order*. Oxford: Oxford University Press.

Kidwell, M. G. (1992). Horizontal transfer. *Curr. Opin. Genet. Dev.* **2**, 868–873.

Kimura, M. & Otha, T. (1971). *Theoretical Aspects of Population Genetics*. Princeton: Princeton University Press.

Kondrashov, F. A., Rogozin, I. B., Wolf, Y. I. & Koonin, E. V. (2002). Selection in the evolution of gene duplications. *Genome Biology* **3**, 2.
Kuhn, T. S. (1962). *The Structure of Scientific Revolutions.* Chicago: University of Chicago Press.
Kumar, S. & Hedges, S. B. (1998). A molecular timescale for vertebrate evolution. *Nature* **392**, 917–920.
Lawrence, J. G. & Ochman, H. (1998). Molecular archaeology of the *Escherichia coli* genome. *Proc. Natl Acad. Sci. USA* **95**, 9413–9417.
Lee, T. I. *et al.* (2002). Transcription regulatory networks in *Saccharomyces Cerevisiae. Science* **298**, 799.
Lenski, R. E. & Mongold, J. A. (2000). Cell size, shape, and fitness in evolving populations of bacteria. In *Scaling in Biology*, ed. J. H. Brown & G. B. West. Oxford: Oxford University Press, pp. 221–235.
Li, W.-H., Gu, Z., Wang, H. & Nekrutenko, A. (2001). Evolutionary analyses of the human genome. *Nature* **409**, 847–849.
Lindgren, K. & Nordahl, M. (1993). Artificial food webs. In *Artificial Life III*, ed. C. G. Langton, vol. XVII of the Santa Fe Institute Studies in the Science of Complexity. Addison-Wesley.
Lovelock, J. E. (1992). A numerical model for biodiversity. *Phil. Trans. R. Soc. Lond.* B**338**, 383–391.
Lovelock, J. E. & Margulis, L. (1974). Homeostatic tendencies of the earth's atmosphere. *Origin of Life* **1**, 12.
Lynch, M. & Conery, J. S. (2000). The evolutionary fate and consequences of duplicate genes. *Science* **290**, 1151.
MacArthur, R. M. & Wilson, E. O. (1967). *The Theory of Island Biogeography.* Princeton: Princeton University Press.
Manrubia, S. C. & Paczuski, M. (1996). A simple model for large scale organization in evolution. (cond-mat/9607066).
Maslov, S. & Sneppen, K. (2002). Specificity and stability in topology of protein networks. *Science* **296**, 910.
May, R. M. (1991). Hypercycles spring to life. *Nature* **353**, 607–608.
Mayer, E. (1961). Causes and effect in biology. *Science* **134**, 1501–1506.
McClintock, B. (1984). The significance of responses of the genome to challenge. *Science* **226**, 792.
Newell, N. D. (1967). Revolutions in the history of life. *Geol. Soc. Amer. Spec. Paper* **89**, 63–91.
Newman, C. M. & Kipnis, C. (1985). Neo-Darwinian evolution implies punctuated equilibria. *Nature* **315**, 400–402.
Newman, M. E. J. (1997). A model for mass extinction. (adap-org/9702003).
Newman, M. E. J. & Eble, G. J. (1999). Power spectra of extinction in the fossil record. *Proc. R. Soc. Lond.* B**266**, 1267–1270.
Novella, I. S., Duarte, E. A., Elena, S. F., Hoya, A., Domingo, E. & Holland, J. J. (1995). Exponential increase of RNA virus fitnesses during large population transmissions. *Proc. Natl Acad. Sci. USA* **92**, 5841.
Peliti, L. (2002). Quasispecies evolution in general mean-field landscapes. *Europhys. Lett.* **57**, 745–751.
Savageau, M. A. (1971). Parameter sensitivity as a criterion for evaluating and comparing the performance of biochemical systems. *Nature* **229**, 542–544.
Schluter, D. (1995). Uncertainty in ancient phylogenies. *Nature* **377**, 108–109.
Segre, D. & Lancet, D. (2000). Composing life. *EMBO Reports* **1**(3), 217–222.

Sigmund, K. (1995). *Games of Life*. London: Penguin Books.
Simon, H. A. (1962). The architecture of complexity. *Proc. Amer. Phil. Soc.* **106**, 467–482.
Simpson, G. G. (1944). *Tempo and Mode in Evolution*. New York: Colombia University Press.
Smith, J. M. (1989). *Evolutionary Genetics*. Oxford: Oxford University Press.
Sneppen, K. (1995). Extremal dynamics and punctuated co-evolution. *Physica A***221**, 168.
Sole, R. V. & Manrubia, S. C. (1996). Extinction and self organized criticality in a model of large scale evolution. *Phys. Rev. E***54**, R42.
Szathmary, E. & Maynard-Smith, J. (1995). The major evolutionary transitions. *Nature* **374**, 227–232.
Voytas, D. F. (1996). Retroelements in genome organization. *Science* **274**, 737.
Vrba, E. S. (1985). Environment and evolution: alternative causes of the temporal distribution of evolutionary events. *S. Afr. J. Sci.* **81**, 229.
Vulic, M., Lenski, R. E. & Radman, M. (1999). Mutation, recombination and incipient speciation of bacteria in the laboratory. *Proc. Natl Acad. Sci. USA* **96**, 7348–7351.
Wilson, D. S. (1992). Complex interactions in metacommunities with implications for biodiversity and higher levels of selection. *Ecology* **73**, 1984.
Woese, C. R. (1987). Bacterial evolution. *Microbiol. Rev.* **51**, 221.
Yule, G. U. (1924). A mathematical theory of evolution, based on the conclusions of Dr. J. C. Willis. *Phil. Trans. R. Soc. Lond.* B**213**, 2187.

# Appendix Concepts from statistical mechanics and damped dynamics

Kim Sneppen

## Thermodynamics from statistical mechanics

Here we introduce a few counting techniques associated with thermodynamic quantities. The most elementary quantity is the entropy $S$, defined as the logarithm of the number $\Omega$ of possible states the system could be in:

$$S = k_{\mathrm{B}} \ln(\Omega) \tag{A1}$$

$\Omega$ can be calculated by counting the number of microstates consistent with the constraints on the system. Thus the probability of finding a system in some macroscopic state characterized by an extropy $S$ is proportional to $\mathrm{e}^{S/k_{\mathrm{B}}}$, where $k_{\mathrm{B}}$ is the Boltzmann constant. A key property of $S$ is that the combined entropy of two independent systems equals the sum of the entropy of each system: the entropy is an extensive quantity.

Before quantum mechanics one did not know what a microstate was and accordingly one would know the entropy $S$ only up to an additive constant. However, quantum mechanics specify the size of the discretization of our world, thereby allowing us to count $S$ by counting phase-space cells $\mathrm{d}x \cdot \mathrm{d}p$ in units of Planck's constant $h$. Thus the $\Omega$ of Eq. (A1) is the phase space counted in units of $h$.

Entropy is closely associated with information. If we know a system is in one of $\Omega$ states we can specify exactly in which of the states it is by answering $\log_2(\Omega)$ yes–no questions. At one instant we may know its location to within one of the states $\Omega$. As times proceed, this certainty will diminish, owing to molecular chaos, for example, and at a later time we will only be able to locate the system to within a larger set of states $\Omega_2 > \Omega$. Thus the entropy has increased. This is a general phenomenon that is called the second law of thermodynamics.

In addition to its entropy, the other fundamental quantity for a system is its energy $E$. The first principle of thermodynamics says that the energy for a closed system never changes.

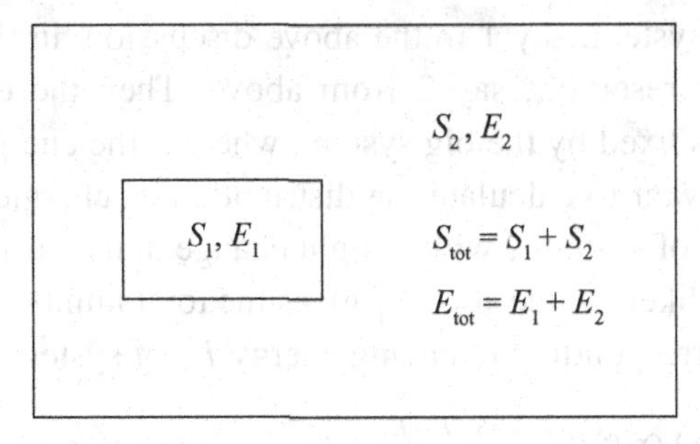


Figure A1. The total system, here $1 + 2$, isolated from its surroundings. The energy change in 1 must be compensated by an opposite energy change in 2. In equilibrium $S_1 + S_2$ maximal and thus $dS_1/dE_1 = dS_2/dE_2 = 1/T$. The Boltzmann distribution for energy $E_1$ of system 1, $P(E_1) \propto \exp(-E_1/k_B T)$ is caused by an entropy reduction of the large system 2 when the energy of system 1 increases (surrounding system 2 is assumed large because we use the linear approximation $S_2(E - E_1) = S_2(E) - (dS_2/dE)E_1$).

For a system with several/many degrees of freedom, a large $E$ implies that there are many ways to distribute energy between the variables (see Fig. A1). In contrast, for sufficiently low energy there will be large volumes of phase space that are excluded. Thus $S$ typically increases with $E$. The variation of $S$ with $E$ is written as

$$\frac{dS}{dE} = \frac{1}{T} \tag{A2}$$

where $T$ is the temperature. A nice property for this quantity is that when two systems exchange energy, their temperature approaches the same value. To see this, consider two systems with energy and entropy $E_1$, $S_1$ and $E_2$, $S_2$. The total entropy of the combined system is

$$S_{tot}(E) = S_1(E_1) + S_2(E_2) \tag{A3}$$

where $E = E_1 + E_2$ is fixed. The probability for any given partition of the energy between the two systems is proportional to the total number of microstates with that partition. Thus the probability that system 1 has energy $E_1$ is

$$P(E_1) \propto e^{(S_1(E_1)+S_2(E-E_1))/k_B} \tag{A4}$$

The most likely partition is accordingly given by $dP/dE_1 = 0$, or

$$\frac{dS_1}{dE_1} + \frac{dS_2}{dE_1} = 0 \quad \text{or} \quad \frac{dS_1}{dE_1} = \frac{dS_2}{dE_2} \tag{A5}$$

where we use $dS_2/dE_1 = -dS_2/dE_2$. Thus the most likely partition of energy is where the temperatures of the two systems are equal. When this is the case we say that they are in thermal equilibrium.

Now consider a system, say 1 in the above discussion, in thermal equilibrium with a much larger reservoir, say 2 from above. Then the overall temperature $T = (\mathrm{d}S_2/\mathrm{d}E_2)^{-1}$ is fixed by the big system, whereas the energy in system 1 may fluctuate. We now want to calculate the distribution of energies of system 1. Expanding the entropy of system 2 with a slight change in its intrinsic energy $E - E_1$ away from its most likely value $E - E_1^0$ gives the total number of states (counting in both systems) corresponding to having energy $E_1$ of system 1

$$P(E_1) \propto \mathrm{e}^{(S_1(E_1)+S_2(E-E_1))/k_\mathrm{B}} \tag{A6}$$

$$\propto \mathrm{e}^{S_1(E_1)/k_\mathrm{B}-(E_1-E_1^0)/k_\mathrm{B}T} = \mathrm{e}^{S_1(E_1)/k_\mathrm{B}}\,\mathrm{e}^{-E_1/k_\mathrm{B}T} \tag{A7}$$

where the first term counts the degeneracy of states of $E_1$ with energy $E_1$, whereas the second term counts the probability that one state at energy $E_1$ is selected. The second term reflects the reduction of phase space for the surroundings of system 1. Thus we have obtained the famous Boltzmann weight factor, stating that if a system has two states separated by energy $\Delta E$, then the probability $P(\text{upper})$ of being in the upper state is a fraction of the probability of being in the lower ($P(\text{lower})$) state

$$\frac{P(\text{upper})}{P(\text{lower})} = \mathrm{e}^{-\Delta E/k_\mathrm{B}T} \tag{A8}$$

This is a very useful equation. Most processes in the living world are regulated by Boltzmann weights. With $\Delta E$ of order of 10 $k_\mathrm{B}T$–20 $k_\mathrm{B}T$ these weights make biological processes take place at sufficiently slow rates (of order one per millisecond to one per minute) to allow large cooperative structures to function.

When system 1 can populate a number of states $\{i\}$, the total statistical weight for populating all these states is given by a simple sum of all the single-state weights:

$$Z = \sum_i \mathrm{e}^{-E_1(i)/k_\mathrm{B}T} \tag{A9}$$

This quantity is called the partition function. It simply counts the total phase-space volume of system 1 and its surroundings under the constraint that system 1 has energy $E_1$.

The partition function is a useful quantity. Setting $\beta = 1/k_\mathrm{B}T$ one can for example express

$$\frac{\mathrm{d}\ln(Z)}{\mathrm{d}\beta} = \langle E\rangle \tag{A10}$$

Here $\langle E\rangle$ is the average energy of the system at temperature $T$, defined as

$$\langle E\rangle = \sum_i p_i E_i = \frac{\sum_i E_i \mathrm{e}^{-E_i/k_\mathrm{B}T}}{\sum_i \mathrm{e}^{-E_i/k_\mathrm{B}T}} \tag{A11}$$

where the probability to be in a state of energy $E_i$ is

$$p_i = \frac{e^{-E_i/k_B T}}{Z} \tag{A12}$$

Also, $Z$ allows us to calculate the entropy of the system. This is in general given as the sum over all its intrinsic states weighted by their relative occurrences:

$$S = -\sum_i p_i \ln(p_i) \tag{A13}$$

That is, when all $p_i$ are equal, each of them $= 1/\Omega$ and the definition agrees with that given for an isolated system where all allowed states are equally allowed.

When the $p_i$ vary, as they will for a system that exchanges energy with its surroundings, the sum over $p_i \cdot \ln(p_i)$ is the function that preserves the additive property of counting two independent systems as one. With the thermal distribution of energy levels, $S$ is

$$S = -\sum_i p_i \ln(p_i) = -\sum_i \left(\frac{e^{-E(i)/k_B T}}{Z}\right) \cdot \ln\left(\frac{e^{-E(i)/k_B T}}{Z}\right) \tag{A14}$$

or

$$S = \frac{1}{Z}\sum_i \left(\frac{E(i)}{k_B T} + \ln(Z)\right) e^{-E(i)/k_B T} \tag{A15}$$

Using the definition of $Z$ and the expression for average energy $\langle E\rangle$, we obtain:

$$-k_B T \ln(Z) = \langle E\rangle - TS \tag{A16}$$

which is also called the free energy $F$ for the system. Thus

$$Z = \sum_i e^{-E(i)/k_B T} = e^{-F/k_B T} \tag{A17}$$

Thus the free energy $F = E - TS$ is treated similarly to the energy of the system. In addition to the energy it counts the size of the state space.

The free energy $F$ can be used for extracting work out of the system, meaning that a difference in free energy can be transformed into mechanical work. This is associated with the fact that, for a closed system, available phase space increases with time. Normally this is formulated as the theorem that the entropy never decreases in a closed system. This is a pure probabilistic argument, which nevertheless becomes close to an absolute truth for rather small systems. Thus if we have a system at a rather high intrinsic free energy, for example by having constrained it from occupying certain parts of phase space, then when releasing these constraints we can convert the decrease in free energy into work. To be specific, let system 1 have free energy $F_1(\mathrm{c})$ in the constrained state, and $F_1(\mathrm{u})$ in the unconstrained state with

$F_1(\mathrm{c}) > F_1(\mathrm{u})$. Now open the system by combining it with a secondary system, which is at energy 0 when system 1 is constrained and at energy $W$ when 1 is unconstrained. The total partition sum for the combined system is then

$$Z_{\text{combined}} = \mathrm{e}^{-F_\mathrm{c}/k_\mathrm{B}T} + \mathrm{e}^{-W/k_\mathrm{B}T}\mathrm{e}^{-F_\mathrm{u}/k_\mathrm{B}T} \tag{A18}$$

thus when $F_\mathrm{u} + W < F_\mathrm{c}$, i.e. when $W$ is not too large, the last term in the partition sum dominates, implying that the system most likely elects to change free energy $F$ of the system into energy $W$ in the secondary system. Further, the maximum work $W$ that can be extracted equals the free energy drop.

Finally, in biology we typically deal with systems where the energy and also the volume of the considered system may change. This means that we have to couple the possible states of the system not only to the changed energy of its surroundings, but also to the changed volumes of the surroundings. The calculation corresponding to this change of constants proceeds exactly as with going from fixed energy to fixed temperature. However, we now have to consider our system 1 at volume $V_1$ in contact with a larger system with volume $V_2$, where the total volume $V = V_1 + V_2$ is fixed. The partition function for the whole system at a given volume $V_1$ of system 1 is

$$Z(V_1, V_2) = Z_1(V_1) \cdot Z_2(V_2) \tag{A19}$$

and, using that $V = V_1 + V_2$ is constant, one finds that the most likely partition is the one where

$$\frac{\mathrm{d}\ln Z_1}{\mathrm{d}V_1} = \frac{\mathrm{d}\ln Z_2}{\mathrm{d}V_2} \tag{A20}$$

Thus the quantity $\mathrm{d}\ln(Z)\mathrm{d}V$ tends to equilibrium for large systems, justifying the definition of a quantity $P$ called the pressure

$$P = -\frac{\mathrm{d}F}{\mathrm{d}V} = \frac{k_\mathrm{B}T\,\mathrm{d}\ln(Z)}{\mathrm{d}V} \tag{A21}$$

and the corresponding statistical weight of a particular volume $V_1$ is

$$Z(V_1) = \mathrm{e}^{-(F_1(V_1)+F_2(V-V_1))/k_\mathrm{B}T} = \mathrm{e}^{-(F_1(V_1)+P(V_1-V_1^0))/k_\mathrm{B}T} \propto \mathrm{e}^{-(F_1+PV_1)/k_\mathrm{B}T} \tag{A22}$$

Thus the effective free energy in case of constant pressure is the so-called Gibbs free energy

$$G = E - T \cdot S + P \cdot V \tag{A23}$$

which is often expressed through the enthalpy $H$ defined as

$$H = E + P \cdot V \tag{A24}$$

The constant temperature and constant pressure condition is the one we deal with in life as well as in chemistry. Thus the Gibbs free energy $G$ counts the corresponding statistical weights, and thereby also the possible mechanical work that one can extract from various reactions.

Now let us return to the constant volume situation. The heat capacity is defined as the amount of heat needed to increase the temperature of the sample by 1 K

$$C = \frac{\mathrm{d}\langle E\rangle}{\mathrm{d}T} \tag{A25}$$

It is instructive to rewrite the heat capacity in terms of energy fluctuations. Thus, using that the average energy is given in terms of the partition function, and taking the derivative

$$Z = \sum_i \exp\left(-\frac{\epsilon_i}{k_B T}\right) \tag{A26}$$

$$\langle E\rangle = \frac{\sum_i \epsilon_i \exp(-\epsilon_i/k_B T)}{\sum_i \exp(-\epsilon_i/k_B T)} = -\frac{\mathrm{d}\ln(Z)}{\mathrm{d}(\beta)}, \quad \text{with} \quad \beta = \frac{1}{k_B T} \tag{A27}$$

$$C = -\frac{1}{k_B T^2}\frac{\mathrm{d}\langle E\rangle}{\mathrm{d}\beta} = \frac{\sum_i \epsilon_i^2 \exp(-\epsilon_i\beta) - (\sum_i \epsilon_i \exp(-\epsilon_i\beta))^2}{k_B T^2(\sum_i \exp(-\epsilon_i\beta))^2} \tag{A28}$$

$$= \frac{\langle E^2\rangle - \langle E\rangle^2}{k_B T^2} \tag{A29}$$

one sees that a high heat capacity thus implies large (energy) fluctuations. For constant pressure, one similarly obtains

$$C = C_p = \frac{\mathrm{d}H}{\mathrm{d}T} = \frac{\langle H^2\rangle - \langle H\rangle^2}{k_B T^2} \tag{A30}$$

where $H$ is the enthalpy.

Typically $C$ represents the number of degrees of freedom that may change energy levels. In particular, when the energy of a system depends only on temperature and not on pressure, overall energy conservations give the condition that all added heat must go to increase average internal energy $\langle E\rangle = E$. This is, for example, the case for an ideal gas where equipartition tell us that the energy per degree of freedom is $\frac{1}{2}k_B T$, and thus that the heat capacity is equal to $N/2$, where $N$ equals the number of degrees of freedom[1].

[1] To prove equipartition use $\langle E\rangle \cdot \int \mathrm{d}p \exp(-p^2/2mk_B T) = \int \mathrm{d}p (p^2/2m) \exp(-p^2/2mk_B T)$ and find $\langle E\rangle = k_B T/2$, which is valid for any quadratic variable in the energy function. In particular it is valid for each of the three spatial directions of motions for an ideal gas, giving a total of $3k_B T/2$. And for a crystal the harmonic potential in each of the three directions gives $3k_B T/2$ additional energy contributions, for a total of $3k_B T$ per atom in the crystal.

Although $F$ or $G$ are always continuous functions with temperature, because $Z$ is, then $E$ or $H$ do not need to be continuously varying with temperature. When it is not, we have a first-order phase transition.

Phase transitions have been studied in great detail in a number of physical systems. They are implicitly important because they allow us to classify behavior into a few important regimes, like solid, liquid and gas. They are explicitly important also because they reflect a certain class of cooperative phenomena: that large systems can sometimes select to be in one of two states with only a slight change of one control parameter. A first-order phase transition is a transition where the *first* derivative of the free energy develops a discontinuity as the system size increases to infinity. A second-order phase transition is a transition where the first derivative is continuous, but the *second* derivative develops a discontinuity as system size increases to infinity. The difference between first and second order becomes more striking when considering the change in energy (enthalpy at fixed pressure) as a function of the control parameter, say temperature $T$.

As energy or enthalpy are given by a first derivative of the free energy, then for a first-order transition the energy contents of the two phases are different; this difference is called the latent heat. The canonical example of a first-order transition is the liquid–gas transition, a transition between a liquid phase with large binding energy on one side, and a gas phase with large entropy on the other side. The latent heat is the energy released when the gas condenses to the liquid. The latent heat also gives rise to a divergence of the heat capacity with system size, a divergence that simply follows from the association of the heat capacity to the second derivative of the free energy.

For a second-order phase transition, the energy contents of the two phases are equal. However, the second derivative, i.e. the heat capacity $C$, is discontinuous, signaling that the degrees of freedom are different in the two phases. A second-order transition is less cooperative than a first-order transition; the transition is less sharp. In any case, for biological applications one often has a rather small system, for proteins $N \sim 100$, which replace divergences by softer functions and make discussion of the order of the transition less useful than characterizing the sharpness by other methods (see Chapter 5).

## Dynamics from statistical mechanics

Free energies are the thermodynamical variables that can be converted to mechanical work. Thus the effective equations of motion of macroscopic variables are governed by free energies, and not by energies or enthalpies. To prove this from a probabilistic argument, let us consider a system in a volume $V$ that can be changed continuously along a one-dimensional coordinate $x$. The canonical example is gas in a piston.

Also let the system be in thermal equilibrium with its surroundings. Now let the statistical weight (partition sum) of the system being at position $x$ be $Z(x)$, and at position $x + \mathrm{d}x$ let it be $Z(x + \mathrm{d}x)$. The transition rate of going from $x$ to $x + \mathrm{d}x$ must be proportional to the number of exit channel states $(Z(x + \mathrm{d}x))$ *per entrance channel state* (of which there are $Z(x)$):

$$\Gamma(x \rightarrow x + \mathrm{d}x) \propto \frac{Z(x + \mathrm{d}x)}{Z(x)} \tag{A31}$$

Averaging over forward and backward motions, the overall drift at position $x$ must be proportional to

$$\Gamma(x \rightarrow x + \mathrm{d}x) - \Gamma(x \rightarrow x - \mathrm{d}x) \propto \frac{\mathrm{d}\ln(Z)}{\mathrm{d}x}\mathrm{d}x \tag{A32}$$

Thus the drift velocity

$$\frac{\mathrm{d}x}{\mathrm{d}t} = (\mu k_{\mathrm{B}} T)\frac{\mathrm{d}\ln(Z)}{\mathrm{d}x} = -\mu \frac{\mathrm{d}F}{\mathrm{d}x} \tag{A33}$$

is proportional to the free energy gradient with a prefactor $\mu$ called the mobility. This is what one should expect for a slowly changing system where inertia plays no role (overdamped motion). The factor $\mu$ measures the friction of moving the $x$ coordinate of the system, and the derivative $\mathrm{d}F/\mathrm{d}x = A \cdot \mathrm{d}F/\mathrm{d}V = -P \cdot A$ is the area multiplied by the pressure.

In biology, systems are often not truly macroscopic, but rather in the range where fluctuations from case to case may play a role. Therefore we have to consider not only the mean trajectories, but also the effect of noise/randomness. In general, for a particle moving in a potential $V(x)$, its equation of motion is given by a Langevin equation

$$m\frac{\mathrm{d}^2 x}{\mathrm{d}t^2} = -\frac{\mathrm{d}V}{\mathrm{d}x} - \frac{1}{\tau}\frac{\mathrm{d}x}{\mathrm{d}t} + f(x, t) \tag{A34}$$

where $f$ is some randomly fluctuating force coming from degrees of freedom that are not explicitly included in the description. The term $(1/\tau)(\mathrm{d}x/\mathrm{d}t)$ is the damping due to friction, and $f$ is often Gaussian uncorrelated noise, obeying:

$$\langle f(x_1, t_1) f(x_2, t_2)\rangle = \Gamma\delta(x_1 - x_2)\delta(t_1 - t_2) \tag{A35}$$

where $\langle\,\rangle$ denotes an average over many realizations of the noise (normally called ensemble averaging).

For most molecular biological applications the friction coefficient $1/\tau$ is so large that motion is overdamped. Large $1/\tau$ means that $v/\tau \gg \mathrm{d}v/\mathrm{d}t$ and therefore that dissipation overrules any other acceleration. Ignoring the acceleration term, the equation of motion is a Langevin equation

$$\frac{\mathrm{d}x}{\mathrm{d}t} = -\mu \frac{\mathrm{d}V}{\mathrm{d}x} + \xi(x, t) \tag{A36}$$

which is similar to the equation derived from phase space dynamics apart from the noise term $\xi = \xi(x, t) = \tau f$. $V$ is thus the free energy associated to a coarse-grained variable $x$, and the noise term $\xi$ should be of a size that makes the distribution of $x$ in potential $V$ resemble the equilibrium distribution

$$P(x) \propto \mathrm{e}^{-V(x)/k_\mathrm{B}T} \tag{A37}$$

Thus the size of $\xi$ is dictated by the temperature and mobility $\mu$.

There are two ways to deduce the relation between $\xi$ and $T$; these are the fluctuation–dissipation theorem and the Einstein relation. We now go through both derivations, because they both give some insight into the description of random processes. But first we discuss noise, as manifested in the Brownian random walk and diffusion.

## Diffusion: microscopic model

Normal random walkers are walkers that always make small steps (length $\delta l$), and always forget everything about the past. In a one-dimensional discrete simulation of a random walker consider the walker at position $x(t)$ at time $t$. At time step $t + \delta t$:

$$x(t + \delta t) = x(t) \pm \delta l \tag{A38}$$

where each neighboring position is chosen with equal probability. An example of a random walk is the so-called Brownian motion. Consider initially a random walker at $x = 0$ to time $t = 0$. After $t$ steps the position

$$x = \sum_{i=1}^{t} (\pm 1)\delta l = \sum_{i=1}^{t} \varsigma(i)\delta l \tag{A39}$$

is the sum of $t$ random numbers times the step length $\delta l$. Each of the random numbers is from a distribution with zero mean and with finite second moment. For example, each of the random nunbers $\eta$ may be either $+1$ or $-1$ as indicated above, corresponding to a step left or a step right; see Fig. A2. The mean displacement of $x$ is

$$\langle x(t)\rangle = \sum_{i=1}^{t} \langle \varsigma_i \rangle \delta l = 0 \tag{A40}$$

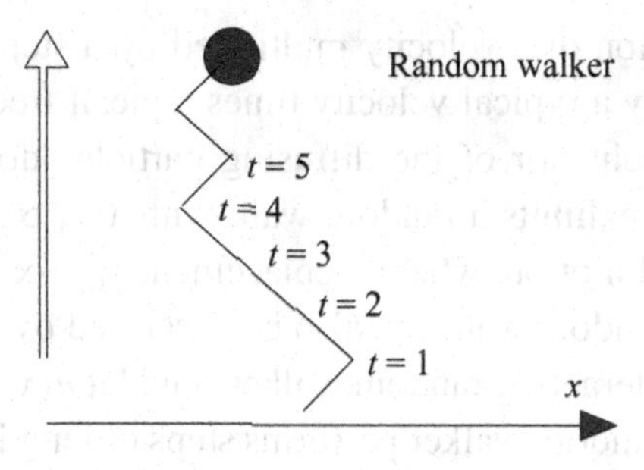


Figure A2. A random walk in one dimension; at each timestep it moves one step to the right or left with equal probability.

Because $x$ is the sum of $t$ independent numbers, the variance of $x$ is then the sum of the $t$ variances:

$$\text{var}(x) = \langle (x(t) - \langle x \rangle)^2 \rangle = \sum_{i=1}^{t} \text{var}(\varsigma_i)\, (\delta l)^2 = t\, \text{var}(\varsigma)\delta l^2 = t\ \delta l^2 \tag{A41}$$

where in the last identity we use the fact that the distribution of $\varsigma$ is given by $\varsigma = +1$ or $\varsigma = -1$, and then the variance of $\varsigma$ is 1 (the reader can prove that).

Further, according to the central limit theorem, when $t$ is large then the sum of the $t$ random numbers is Gaussian distributed. Therefore after $t$ timesteps the random walker is at position $x$ with probability

$$p(x)\,\mathrm{d}x = \frac{1}{\sqrt{2\pi t}\ \delta l_{\mathrm{p}}}\ \mathrm{e}^{-x^2/(2t\delta l^2)}\,\mathrm{d}x \tag{A42}$$

The average position after $t$ timesteps is thus $\langle x \rangle = 0$ whereas the root mean square position $x_{\mathrm{rms}}$ is

$$x_{\mathrm{rms}}(t) = \langle x^2 \rangle^{1/2} = \left( \int x^2 p(x)\mathrm{d}x \right)^{1/2}$$

$$= \left( \frac{\int x^2\ \mathrm{e}^{-x^2/(2t\delta l^2)}\,\mathrm{d}x}{\int \mathrm{e}^{-x^2/(2t\delta l^2)}\,\mathrm{d}x} \right)^{1/2} = \sqrt{2t\delta l^2} \tag{A43}$$

We can re-express this in terms of actual time, when we remember that $t$ was measured in units of the timestep $\delta t$ at which we take a step of length $\delta l$:

$$x_{\mathrm{rms}} = \sqrt{2t(\delta l^2/\delta t)} = \sqrt{2tD} \tag{A44}$$

where we introduce the diffusion constant

$$D = (\delta l)^2/\delta t \tag{A45}$$

which has the dimension of a velocity multiplied by a step length. For Brownian diffusion $D$ is given by a typical velocity times a mean free path. At scales much larger than this, the behavior of the diffusing particles does not depend on microscopic details, and exhibits a random walk with $x_{\text{rms}} \propto \sqrt{t}$. The random walk contrasts with directed motion, where displacement $x_{\text{rms}} \propto t$.

On large scales, a random walk can also be described by the diffusion equation. Consider many non-interacting random walkers and let $n(x, t)$ denote their density distribution. As each random walker performs steps of length $\delta l$ during time $\delta t$, the density distribution $n(x, t)$ of these walkers evolves according to:

$$n(x, t + \delta t) - n(x, t) \propto \frac{1}{2} n(x - \delta l, t) - \frac{1}{2} n(x, t) - \frac{1}{2} n(x, t) + \frac{1}{2} n(x + \delta l, t) \tag{A46}$$

where at position $x$ we add and subtract contributions according to the exchange of particles with neighboring positions. The proportionality factor would be the diffusion constant $D$. Thus the density distribution evolves as in the diffusion equation

$$\frac{\mathrm{d}n}{\mathrm{d}t} = D \frac{\mathrm{d}^2 n}{\mathrm{d}x^2} \tag{A47}$$

Notice that $n =$ constant is preserved by the diffusion equation. Also the Gaussian $n(x, t) = 1/\sqrt{4\pi D t} \exp(-x^2/4tD)$ solves the diffusion equation. Thus as time $t \to \infty$ then $n \to$ constant.

## Fluctuation–dissipation theorem

Consider the Langevin equation for a variable $x$ moving in a flat potential, $\mathrm{d}V/\mathrm{d}x = 0$

$$m \frac{\mathrm{d}^2 x}{\mathrm{d}t^2} = -\frac{1}{\tau} m \frac{\mathrm{d}x}{\mathrm{d}t} + \Gamma \xi(t) \tag{A48}$$

where the noise term fulfils $\langle \xi(x, t) \rangle = 0$, $\langle \xi(x_1, t_1) \xi(x_2, t_2) \rangle = \delta(t_1 - t_2)$ and where $\tau$ is the typical time it takes to dissipate velocity. Integration gives

$$m \frac{\mathrm{d}x}{\mathrm{d}t} = p = \int_{-\infty}^{t} \mathrm{d}t' \Gamma \xi(t') \mathrm{e}^{-(t-t')/\tau} \tag{A49}$$

reflecting a momentum determined by a series of random kicks at previous times, with an influence that decays exponentially because of the friction as illustrated in Fig. A3. This equation can also be proven backwards by differentiation. Try!

Consider the ensemble averaged (averaged over many trajectories)

$$\langle p^2 \rangle = \int_{-\infty}^{t} \int_{-\infty}^{t} \mathrm{d}t' \mathrm{d}t'' \; \Gamma^2 \langle \xi(t') \xi(t'') \rangle \; \mathrm{e}^{-(t-t')/\tau} \mathrm{e}^{-(t-t'')/\tau} \tag{A50}$$

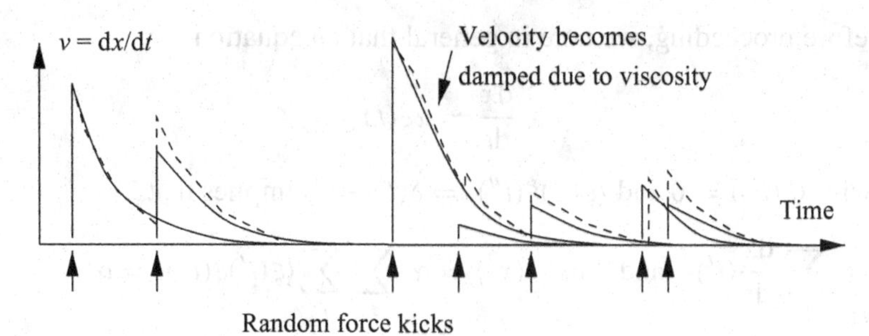


Figure A3. Newton equations with random forces and viscous damping transform into the Langevin equation. To secure that the resulting average velocity is such that there is $T/2$ energy per degree of freedom, there must be a relation between the size of the random kicks, and the damping due to viscosity. This is the fluctuation–dissipation theorem.

Because the noise has no temporal correlations, we obtain

$$\langle p^2 \rangle = \int_{-\infty}^{t} \mathrm{d}t' \; \Gamma^2 \, \mathrm{e}^{-2(t-t')/\tau} = \frac{1}{2}\Gamma^2 \tau \tag{A51}$$

On the other hand, the average kinetic energy for a one-dimensional free particle is $k_B T/2$, where $T$ is the temperature:

$$\left\langle \frac{p^2}{2m} \right\rangle = \frac{k_B T}{2} \tag{A52}$$

This follows from the Boltzmann distribution of kinetic energies $P(E)\mathrm{d}E \propto \exp(-E/k_B T)\mathrm{d}E$ with $E = p^2/(2m)$. Identifying the terms in Eqs. (A51) and (A52) one obtains

$$\Gamma = \sqrt{\frac{2mk_B T}{\tau}} \tag{A53}$$

an equation that relates the size of fluctuations to the damping time $\tau$ and thus to the friction. To simplify this further we return to Eq. (A48) in an external potential $V(x)$:

$$\frac{\mathrm{d}p}{\mathrm{d}t} = -\frac{1}{\tau}p - \frac{\mathrm{d}V}{\mathrm{d}x} + \sqrt{\frac{2mk_B T}{\tau}} \cdot \xi \tag{A54}$$

or, expressed in overdamped limit without the potential,

$$\frac{\mathrm{d}x}{\mathrm{d}t} = -\frac{\tau}{m}\frac{\mathrm{d}V}{\mathrm{d}x} + \sqrt{\frac{2\tau k_B T}{m}}\xi \tag{A55}$$

Now, before proceeding, we note in general that an equation

$$\frac{\mathrm{d}x}{\mathrm{d}t} = \alpha\xi(t) \tag{A56}$$

which with $\langle\xi(t')\rangle = 0$ and $\langle\xi(t')\xi(t'')\rangle = \delta(t' - t'')$ implies that

$$x = \sum_{t'} \frac{\mathrm{d}x}{\mathrm{d}t}(t') \quad \text{and thus} \quad \langle x^2\rangle = \alpha^2 \sum_{t'}\sum_{t''}\langle\xi(t')\xi(t'')\rangle = \alpha^2 t \tag{A57}$$

Thus the factor $\alpha^2$ is equal to $2D$; see Eq. (A44) where $D$ is the diffusion constant. Returning to Eq. (A55) we have

$$D = \frac{\tau k_{\mathrm{B}}T}{m} \tag{A58}$$

For large $1/\tau$, Eq. (A55) can be reformulated by ignoring the acceleration term:

$$\frac{\mathrm{d}x}{\mathrm{d}t} = -\mu\frac{\mathrm{d}V}{\mathrm{d}x} + \sqrt{2D}\cdot\xi \tag{A59}$$

Here $\langle\xi(t')\xi(t'')\rangle = \delta(t' - t'')$, and the mobility $\mu = \tau/m$. A larger mobility means that the particle moves faster. In this last equation the ratio between the mobility and the diffusion constant is

$$D/\mu = k_{\mathrm{B}}T \tag{A60}$$

This is often called the fluctuation dissipation theorem.

To iterate on the physics of the above derivation: the average thermal energy is $k_{\mathrm{B}}T/2$. The typical velocity is therefore given by $mv^2/2 = k_{\mathrm{B}}T/2$, or $v = \sqrt{k_{\mathrm{B}}T/m}$. The momentum is given a new independent kick every timestep $\tau$, thereby making any directed motion dissipate with rate $1/\tau$. Between each kick the particle moves a distance $l \approx v\cdot\tau = \sqrt{k_{\mathrm{B}}T/m}\cdot\tau$. The particle therefore performs a random walk quantified by a diffusion constant that is equal to the product of the velocity and the step length

$$D = v\cdot l = \frac{k_{\mathrm{B}}T\tau}{m} = k_{\mathrm{B}}T\mu \tag{A61}$$

To put the above equations in perspective we note that for a single amino acid in water, say glycine, the diffusion constant $D = 10^{-9}\ \mathrm{m^2/s} = 1000\ \mathrm{\mu m^2/s}$ and a timescale for dissipation

$$\tau = \frac{mD}{k_{\mathrm{B}}T} = 0.03\times10^{-12}\ \mathrm{s} \tag{A62}$$

On timescales longer than 0.03 ps, single amino acids have forgotten their momenta, and molecular dynamics will degenerate into first-order Langevin dynamics. For a single protein, say albumin in water, $D = 60\ \mu\text{m}^2/\text{s}$ (i.e. albumin diffuses 10 μm in 1 s) and $\tau = 1.7$ ps.

## The Fokker–Planck equation and the Einstein relation

The alternative approach to the relation between mobility and noise is through the Fokker–Planck equation. If $x$ fulfils a Langevin equation

$$\frac{\mathrm{d}x}{\mathrm{d}t} = -\mu\,\frac{\mathrm{d}V}{\mathrm{d}x} + \sqrt{2D}\eta(t) \tag{A63}$$

with $\langle \eta(t'')\eta(t')\rangle = \delta(t'' - t')$ then one can construct a probability distribution for $x$ at a given time $t$:

$$P(x,t)\mathrm{d}x = \text{probability for } x \in [x, x + \mathrm{d}x] \text{ at time } t \tag{A64}$$

By using the physical insights we have for the two terms in the Langevin equation in terms of loss and gain terms to a given $x$ coordinate, $P$ will evolve according to the Fokker–Planck equation

$$\frac{\mathrm{d}P(x,t)}{\mathrm{d}t} = -\frac{\mathrm{d}}{\mathrm{d}x}J \tag{A65}$$

where the current $J$ is given by

$$J = -\mu \cdot P\frac{\mathrm{d}V}{\mathrm{d}x} - D\frac{\mathrm{d}P(x,t)}{\mathrm{d}x} \tag{A66}$$

At equilibrium $P$ = constant, implying that $J = 0$ and thus

$$P(x,t) = P(x) \propto \mathrm{e}^{-\mu V(x)/D} \tag{A67}$$

which can only be $\propto \mathrm{e}^{-V(x)/k_\mathrm{B}T}$ if $\mu = D/k_\mathrm{B}T$.

Thus the Einstein relation states that mobility ($\mu$) is proportional to the diffusion constant $D$. The diffusion constant $D$ has dimension of a mean free path times a typical (thermal) velocity. The diffusive part of the Fokker–Planck equation describes how an initial localized particle spreads out in flat potentials to a Gaussian with spread $\sigma \propto \sqrt{Dt}$ that simply follows from the central limit theorem. The convective part of the Fokker–Planck equation states that the particle moves downhill, with a speed proportional to both $\mathrm{d}V/\mathrm{d}x$ and the mobility $\mu$.

## Kramers' formula

Escape from a potential well is an old and important problem in physics, as well as in chemistry. It has also recently proven important in a number of biological

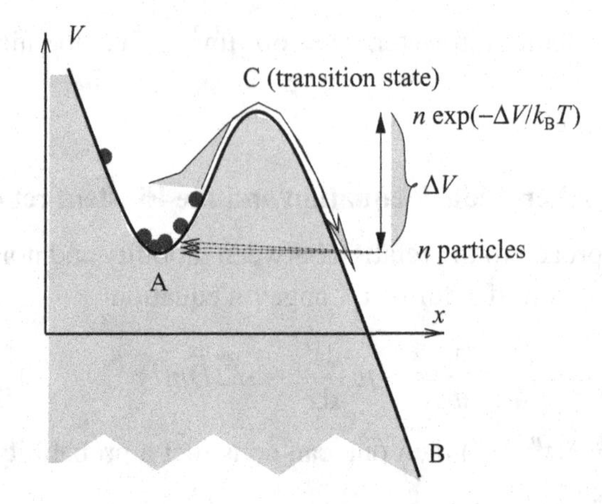


Figure A4. Escape over a one-dimensional potential, where a particle is confined at A, but is allowed to escape over barrier C to the outside of the well. In the figure we show a number of particles, to illustrate the approach by Kramers.

problems, such as stability of genetic switches in molecular biology and macroevolution in abstract fitness landscapes. Here we present the derivation proposed by H. A. Kramers (Brownian motion in a field of force and the diffusion model of chemical reactions. *Physica* **VII** (1940), 284–304) for escape from a potential well in the overdamped case.

Consider a one-dimensional potential well as in Fig. A4 where point A is at the bottom of the well and point B is somewhere outside the well. Following Kramers, we consider the stationary situation with the current leaking out from the well to be insignificant. Thus current $J$ in Eq. (A66) is an $x$-independent small number and the corresponding $P$ in Eq. (A65) is constant in time. Using $\mu = D/k_B T$ the $x$-independent current can be rewritten as

$$J = -\frac{D}{k_B T} P \frac{\mathrm{d}}{\mathrm{d}x} V - D \frac{\mathrm{d}P}{\mathrm{d}x} = -D \cdot \mathrm{e}^{-V/k_B T} \cdot \frac{\mathrm{d}}{\mathrm{d}x} \left( P \cdot \mathrm{e}^{V/k_B T} \right) \tag{A68}$$

When rewritten and integrated from point A to point B:

$$J \cdot \mathrm{e}^{V/k_B T} = -D \cdot \frac{\mathrm{d}}{\mathrm{d}x} \left( P \cdot \mathrm{e}^{V/k_B T} \right) \tag{A69}$$

or

$$J = -D \frac{\left[ P \mathrm{e}^{V/k_B T} \right]_{\mathrm{A}}^{\mathrm{B}}}{\int_{\mathrm{A}}^{\mathrm{B}} \mathrm{e}^{V/k_B T} \mathrm{d}x} \tag{A70}$$

where the quasi-stationary condition states that $P_B \sim 0$ and $P_A \approx$ local equilibrium value:

$$J = -D \frac{P_A e^{V_A/k_B T}}{\int_A^B e^{V/k_B T} dx} \tag{A71}$$

The value of $P_A$ can be estimated from using a harmonic approximation around the minimum A, and setting $P_A$ equal to the corresponding maximum of the corresponding Gaussian density profile. Thus around point A

$$V(x) \approx V_A + \frac{1}{2} \frac{d^2 V}{dx^2} (x - x_A)^2 = V_A + \frac{1}{2} k \, (\delta x)^2 \tag{A72}$$

For a particle with mass $m$ this is a harmonic oscillator with frequency $\omega = \sqrt{k/m}$. The peak density $P_A$ is given by normalization of $\exp(-k\delta x^2/(2k_B T))$ (everything is counted as if there is one particle in the potential well that can escape). The escape rate ($J$ per particle) is

$$r = \frac{\omega_A \tau}{m} \sqrt{\frac{m k_B T}{2\pi}} \frac{e^{V_A/k_B T}}{\int_A^B e^{V/k_B T} dx} \tag{A73}$$

The remaining integral is calculated by a saddle point around its maximum, i.e. around the barrier top at point C:

$$\int_A^B dx \; e^{V/k_B T} = e^{V_C/k_B T} \int_{-\infty}^{\infty} dx \; e^{-k_C \delta x^2/(2k_B T)} = \frac{\sqrt{2\pi k_B T}}{\sqrt{k_C}} e^{V_C/k_B T} \tag{A74}$$

which with $k_C = m\omega_C^2$ gives the final escape rate for overdamped motion:

$$r = \frac{\omega_A}{2\pi} \cdot \omega_C \tau \cdot \exp\left(-\frac{V_C - V_A}{k_B T}\right) \tag{A75}$$

This equation can be interpreted in terms of a product between a number of attempted climbs:

$$\text{number of attempts} = \frac{\omega_A}{2\pi} \tag{A76}$$

multiplied by the fraction of these climbs that can reach C, simply given by the Boltzmann weight $e^{-(V_C - V_A)/k_B T}$. Finally just because a climb reaches the saddle point, it is not certain that it will pass. The chance that it will pass is $\omega_C \cdot \tau$, which is equal to one divided by the width of the saddle, in units of the steps defined by the random kicking frequency $1/\tau$. That is, imagine that the saddle is replaced by a plateau of $w = 1/(\omega_C \tau)$ steps, and we enter the first (leftmost) of these steps. We then perform a random walk over the plateau, with absorbing boundaries on both sides. As this is equivalent to a fair game, the chance of escaping on the right-hand side is $1/w$. Thus one may interpret the overdamped escape as:

$$r = \text{(Attempts to climb)} \cdot \text{(chance to reach top, given it attempts)} \cdot \text{(chance to pass top, given it reached top)} \qquad \text{(A77)}$$

One notices that a higher viscosity, meaning a lower $\tau$, implies that the escape rate diminishes. This is not surprising, as a higher viscosity means that everything goes accordingly slower, and therefore this applies also to the escape.

Although all biology essentially takes place at this overdamped limit, it may for general interest be worthwhile to notice that Kramers also considered the general case and thus also treated the case of "underdamped" motion. In that case the motion is not limited by viscosity; instead the chance to get kicks to sufficiently high momenta will be proportional to the kicking frequency. In that case the escape rate $r \propto 1/\tau$, and when escape momentum is reached, no collisions will stop the particle from leaving the potential well.

# Glossary

**activator** Transcription factor that activates the production of a gene, typically through binding to a promoter upstream of the region for the gene and recruiting the RNAp.

**active site** The site on an enzyme where the substrate molecule binds and where its reaction is facilitated.

**amino acids** The building blocks of proteins; see Chapter 4.

**anticodon** The three nucleotide region of a tRNA that base pairs with the codon on mRNA during protein synthesis.

**Arrhenius law** Probability of escaping from a potential well of depth $E$ per unit time is $\propto e^{-E/k_BT}$, where $T$ is some temperature, and the relaxation probability at time $t$ is exponential in $t$. Non-Arrhenius behavior is found in glasses where the decay rate is $\propto e^{-(E/k_BT)^2}$. Signs of such transition-rate dependence is found for some supercooled liquids, for example glycerin. For an example of non-Arrhenius behavior found in flash photolysis of proteins at low temperatures, see R. H. Austin *et al.* (*Phys. Rev. Lett.* **32** (1974), 403–405).

**atoms in life** Biological molecules mostly consist of C, O, N and H atoms. This does not mean that other atoms are not important. Other atoms are used to facilitate special functions, e.g. Fe is used in hemoglobin and Zn is found in the large class of zinc-finger DNA binding proteins. In diminishing amount, living cells use Ca, K, Na, Mg, Fe, Zn, Cu, Mn, Mo and Co, where Ca is found in 15 g, Mg in 0.5 g and Cu in 2 mg per kg in a human.

**bases** Building blocks of DNA or RNA. For DNA they are divided into purines (A and G) and pyrimidines (T and C). In RNA, T is replaced by uracil (U).

**base pairs (bp)** Two bases linked by non-covalent forces (hydrogen bonds), that pair in double-stranded DNA.

**bending modulus** The bending of a polymer like dsDNA is similar to the bending of a beam, so it can also be described through a bending modulus. Consider a piece of DNA of length $L$ bent around a disk of radius $R$. The bending energy per unit length is quadratic in the curvature $E/L = B/2R^2$, where $B$ is the bending modulus. On the other hand, the persistence length $l_p$ is roughly the length that can be bent into a circle by a thermal fluctuation $k_BT \sim Bl_p/(l_p^2)$ or $l_p = B/k_BT$. For dsDNA, $l_p = 50\,\text{nm}$ corresponds to a bending modulus $B = 200\,\text{pN}\cdot\text{nm}$.

**Bethe lattice (or Cayley tree)** Hierarchically ordered network without ends and without any loops. Each unit is connected to one above and to two below, and there is only one line of connection between any two units. A Bethe lattice is effectively infinite dimensional.

**Boltzmann constant** $k_B$ is the conversion factor from temperature measured in Kelvin into energy per molecule, in biological terms into kcal/mol. The energy unit $1k_BT = 4\,\text{pN} \cdot 1\,\text{nm}$ in terms of a typical molecular force times a typical distance.

**Boolean automaton** An automaton that works with two outputs only (on or off). It can be used in a Boolean network. A candidate for simplified descriptions of single genes in genetic networks. See S. A. Kauffman (*J. Theor. Biol.* **22** (1969), 437).

**Boolean network** Random directed network where each node becomes either on or off, depending on inputs to the node. For $k$ input there are $n = 2^k$ input states. Each of these should be assigned either on or off, giving $2^n$ ($n = 2^k$) Boolean functions. A simple Boolean network is two states of cellular automata arranged on lattices with only nearest-neighbor interactions. A random Boolean network is critical if the average number of inputs is two per node. At this value a change in state will barely propagate along the network. For a review see pp. 182–235 in Kauffman's book, *The Origins of Order* (Oxford University Press, (1993)), and for possible connections to number of cell types in eukaryotic organisms, see pp. 460–462. See also R. Somogyi & C. A. Sniegoski (*Complexity* **1**(6) (1996), 45–63.

**cDNA** Complementary DNA.

**centromere** The site of attachment for spindle fibers during mitosis and meiosis.

**CD (or circular dicroism)** A way to detect secondary structure in proteins, using polarized light. It uses the effect that, for example, $\alpha$-helices absorb left- and right-handed polarized light differently, in a characteristic way that depends on wavelength. Similarly, $\beta$-sheets can be detected through a different wavelength-dependence absorbance of left- and right-handed polarized light. See the review by A. J. Alder *et al.* (*Method. Enzymology* **27** (1973), 675).

**cellular automata (CA)** Defined on a lattice, say on the integer position on a line. At each position assign a discrete variable, say 0 or 1. Parallel update; the variable on each position is changed deterministically according to the value it and its neighbors had at the previous timestep. If one considers only a binary variable and the nearest neighbors, there are $2^3 = 8$ possible neighborhoods to consider: 000, 001, 010, 011, 100, 101, 110, 111. For each of these one should define whether the outcome is 0 or 1. Thus any nearest-neighbor rule could be specified by a four-digit binary number.

*probabilistic CA.* Not deterministic, but with probabilities of having the outcome dependent on current values.

**chaperone** A protein that helps to fold other proteins. Most famous are GroEL and DnaK. Chaperones use energy to redirect the folding process, probably by unfolding misfolded proteins and thereby giving them a chance to fold correctly on a second try.

**coarse graining** Describes your system with larger-scale variables that each represent some effective description average of effects of smaller-scale variables. In equilibrium systems one can rewrite the partition sum $Z = \sum_{i,j} \exp(-E_{ij}/T) = \sum_i \exp(-F_i/T)$, where the free energy of the coarse-grained variable labeled by $i$ is given by $\exp(-F_i/T) = \sum_j \exp(-E_{ij}/T)$.

**codon** A sequence of three nucleotides (in a DNA of mRNA) that encodes a specific amino acid.

**complementary sequence** Sequence of bases that can form a double-stranded structure by matching base pairs. The complementary sequence to base pairs C–T–A–G is G–A–T–C.

**complexity** A loosely defined term used in many different contexts. High complexity implies a structure that is intermediate between a stage of perfect regularity (minimum entropy) and a stage of complete disorder (maximum entropy). One definition, by Hinegardner & Engleberg (*J. Theor. Biol.* **104** (1983), 7) is that an object's complexity is defined as the size of its minimal description. That a structure has large complexity roughly means that one needs either a long time or a complicated algorithm to develop it. The paleontolog J. W. Valentine proposed measuring complexity of an organism by the number of different cell types that it has. Valentine *et al.* (*Paleobiology* **20** (1994), 131) make the observation that the hereby defined complexity of the most complex species at time $t$ has grown as $\sqrt{t}$ since the Cambrian explosion, ending at human-like organisms, which have about 250 different cell types.

**consensus sequence** A sequence constructed by choosing at each position the residue that is found there most often in the group of sequences under consideration.

**convergence** The tendency of similar proteins placed close on the DNA to be "error" corrected during replication. For example many paralogs that have first diverged later converge owing to this error correction.

**cooperativity** More than the sum of its parts. Acting cooperatively means that one part helps another to build a better functioning system. Cooperative bindings include dimerization, tetramerization, and binding between transcription factors on adjacent DNA sites.

**cumulants** For a variable $x$ the first four cumulants are $K_1 = \langle x \rangle$, $K_2 = \langle (x - \langle x \rangle)^2 \rangle$, $K_3 = \langle (x - \langle x \rangle)^3 \rangle$ and $K_4 = \langle (x - \langle x \rangle)^4 \rangle - 3K_2^2$. The cumulant $K_n$ measures what new is to be said about the distribution, given all previous cumulants $K_i$, $i < n$. The cumulants can be obtained from the generating function $G(z) = \log(\langle \exp(zx) \rangle)$, by $K_n = \lim_{z \to 0}(\mathrm{d}^n G/\mathrm{d}z^n)$.

**dalton** Mass unit, 1 dalton (Da) is 1 gram per mole. Thus carbon weights 12 Da, and water weighs 18 Da.

**detailed balance** A dynamical algorithm that defines transition probabilities $T(1 \to 2)$ between states in some space has detailed balance if one can assign probabilities $P$ to all states such that any two states $a$ and $b$ fulfil $T(a \to b)P(a) = T(b \to a)P(b)$. A transition probability $T(1 \to 2)$ is the probability that in the next step we get state 2, given we are in state 1. Detailed balance is not a trivial requirement, and if it is violated then the algorithm typically will not sample an equilibrium distribution. In order to sample an equilibrium distribution there is the further requirement that the transition rules $T$ should be defined such that all phase space is accessible. The transition rule is then a Metropolis algorithm.

**diffusion coefficients** $\mathrm{d}f/\mathrm{d}t = D\mathrm{d}^2 f/\mathrm{d}x^2$. Large $D$ means fast dissipation. $D$ is a typical length times a typical velocity (i.e. closely connected to viscosity). For protein in water $D_\mathrm{p} \approx 100\ \mu\mathrm{m}^2/\mathrm{s}$; for sucrose in water $D = 520\ \mu\mathrm{m}^2/\mathrm{s}$; for oxygen in water $D = 1800\ \mu\mathrm{m}^2/\mathrm{s}$. A protein molecule diffuses a distance of order $x \approx \sqrt{2D_\mathrm{p}t} = 0.6$ mm in 1 h.

**distributions**

*exponential* $p(t) \propto \exp(-t/\tau)$. If $t$ is a waiting time this is the distribution for a random uncorrelated signal. In that case the expected waiting time for the next signal does not change as time passes since the last signal.

*power law* $p(t) \propto 1/t^\alpha$. For example, if $t$ is a waiting time, then expected waiting time for the next signal increases as time passes since the last signal.

*normal or Gaussian distribution* Obtained by sum of exponentially bounded random numbers that are uncorrelated. Distribution: $p(x) \propto e^{-x^2/\sigma^2}$.

*log normal* Obtained by product of exponentially bounded random numbers that are uncorrelated. If $x$ is normal distributed then $y = \exp(x)$ is log normal: $q(y)dy \propto \exp(-\log(y)^2/\sigma^2)\,dy/y \sim dy/y$ for $y$ within a limited interval.

*stretched exponentials* These are of the form $p(x) \propto \exp(-x^\alpha)$.

*Pareto–Levi* Obtained from the sum of numbers, each drawn from a distribution $\propto x^{-\alpha}$. A Pareto–Levi distribution has a typical behavior like a Gaussian, but its tail is completely dominated by the single largest event. Thus a Pareto–Levi distribution has a power-law tail.

**DNA chips** mRNA activity in a cell sample can be analyzed with DNA chips. These are large arrays of pixels. Each pixel represents part of a gene by having of the order of $10^6$–$10^9$ single-stranded DNA-mers, that are identical copies from the DNA of the gene. The chip size is of the order of 1 $cm^2$, and can (in 2001) contain 12 000 different pixels (each measuring one gene). The analysis consists of taking a cell sample, extracting all mRNA in this (hopefully) homogeneous sample, and translating it to cDNA (DNA that is complementary to the RNA, and thus identical to one of the strands on the original DNA). The cDNA is labeled with either red or green fluorescent marker. The solution of many cDNAs is now flushed over the DNA chip, and the cDNAs that are complementary to the attached single-stranded DNA-mers will bind to them. The DNA chip is washed and illuminated and the fluorescent light intensity thus measures the effective mRNA concentration.

**DNA forms** Besides the standard B-form for double-stranded DNA there are several other known conformations of DNA. In B-DNA the planes of the bases are perpendicular to the axis of the helix; when dehydrated, DNA assumes the A-form in which the planes of the bases are tilted by $\sim 15°$. Certain sequences (GCGCGC...) can form a left-handed helix, which is called Z-DNA; finally it is possible mechanically to stretch DNA into the S-form.

**DNAp** Short for DNA polymerase; this is the machinery that copies DNA.

**diploid** Cells have two copies of each chromosome. The human diploid number is 46. Sperm and egg cells are haploid, with one copy in each cell.

**epigenetics** Defined as "a heritable change in phenotype in the absence of any change in the nucleotide sequence of the genome."

**entropy/information** Information definition: $\mathcal{I} = \sum_i p_i \log_2(p_i)$, where the sum runs over all possible outcomes; $-\mathcal{I}$ counts the minimal number of yes/no questions needed to specify microstate completely. $S \propto -k_B\mathcal{I}$ is the corresponding entropy. Minimal information is equivalent to maximum entropy, and implies equilibrium distributions within an accessible state space $\{i\}$. For example, we minimize $\mathcal{I}$ with constraint $\sum_i p_i = 1$ to give all $p_i$ equal (micro-canonical ensemble). Minimization with constraint on average energy conservation gives the canonical ensemble $p_i \propto e^{-E_i/T}$, where $1/T$ is determined by average energy conservation.

**equilibrium/non-equilibrium** An isolated system is in equilibrium if all available states are visited equally often. This is secured by a detailed balance. Equilibrium distributions are obtained by minimizing information (maximizing entropy) under constraints given by the contacts between the system and its surroundings. An isolated system is out of equilibrium if some parts of state space (phase space) are more populated than others. Dynamic systems may be driven out of equilibrium by an externally imposed flux. For example, a convective fluid is driven out of equilibrium by its attempt to transport heat. Without an externally imposed free energy flow a system always decays towards equilibrium (second law of thermodynamics). To popularize: equilibrium describes dead systems; non-equilibrium opens the way for spontaneous organization and apparently even for organization as subtle as life.

**eukaryote** An organism whose genetic information is, in contrast to prokaryotes, contained in a separate cellular compartment, the nucleus. Eukaryotes are believed to have originated by a merging of normal bacteria with the Archea bacteria.

**evolutionary concepts**

*Macro-evolution* is the evolution of life on a large scale, including many (eventually, all) species.

A *phylogenetic tree* is a family tree of species. One lineage is the development line leading to one species.

*Taxa* are hierarchical subdivisions of species in groups according to familiarity. From low to high there are the following levels: species, genera, family, order (like primates), class (like mammals), phyla (like Chordata), kingdoms (like animals).

Some numbers: there are about $10^7$ species on Earth today. There have been about $10^{10}$ species on Earth in total during history. The largest extinction event (at the end of the Permian, *c.* 250 My ago) wiped out 60% of all genera. The largest origination event is the Cambrian explosion ($\approx$540 My ago, where multicellular life flourished in origination over a rather short timespan). Oxygen originated in atmosphere 1600 My ago. Eukaryotes appeared about 2000 My ago. The first life-forms originated about 3600 My ago. See S. Wright (*Evolution* **36** (1982), 427); and D. M. Raup (*Science* **231** (1986), 1528) and *Bad Genes or Bad Luck* (New York: W. W. Norton & Company, 1991).

**evolvability** This concept may be quantified by the probability of obtaining a new working and survivable property (phenotype) by a mutation. That means that evolvability is quantified by the fraction of small (likely) mutations that exhibit new viable phenotypes. Evolvability is thus related to robustness against mutations. Note, however, that whereas evolvability requires variability in adjacent phenotypes, robustness requires only that adjacent genotypes sustain basic functions. In any case adjacent mutants should be able to survive.

**evolution in the lab** For simple organisms, Lenski has recorded the evolution of 20 000 generations of *E. coli* in the lab, and observed a number of phenomena including history-dependent outcomes, stasis and punctuated equilibrium of fitness value (Elena *et al.*, *Science* **272** (1996), 1797), and coexistence of evolved strains that differ significantly in reproduction rate (Rozen & Lenski, *Am. Nat.* **155** (1) (2000), 24). Also the inter-dependence of fitness with environmental conditions has been recently explored in an interesting laboratory experiment of bacterial evolution by A. Kashiwagi *et al.* (*J. Mol. Evol.* **52** (2001), 502). In this work it was found that in evolving bacterial cultures starved of glutamine, the culture in which there was excessive crowding developed diversity in strain composition, whereas a culture grown under less crowded conditions tended to eliminate an already present diversity in strain composition. Thus, with crowding, the different strains develop symbiotic specializations that favor coexistence.

**exons** Segments of a eukaryotic gene that encode mRNA. Exons are adjacent to non-coding DNA segments called introns. After transcription, the non-coding sequences are removed.

**fractals** Scale-invariant structures that fill a measure zero part of space. Physical examples are the coastline of Norway, river networks, snowflakes, dielectric breakdown (from DLA-like processes) and snapshots of high dissipation regions in turbulent media. Fractals are also seen in biology, where there is a need to cover a big area with little material (in trees or in lungs). Self-organized critical models may be seen as an attempt to present a unified dynamical principle for the origination of fractals.

**frozen accidents** Concept of Brian Arthur that refers to decision points in history that freeze later decisions to conform to a frozen landscape. Frozen accidents confront an equilibrium view, by implying a non-ergodic evolution of the system in question.

**Gaia** The world viewed as one living organism. Paradigm of Lovelock that means that there are stabilizing biological feed-back mechanisms that keep conditions on Earth optimal for life. A simple example is the oxygen level, which is kept as high as possible given that forest fires should be self-terminating. Another is the fact that the average temperature on Earth has kept constant within a few degrees over billions of years. Biological mechanisms governing $CO_2$ concentration, surface albedo and amount of rainfall are suggested. Explicit modeling is through the Daisy world model (A. J. Watson & J. E. Lovelock, *Tellus* **35**B (1983), 384).

**gel electrophoresis (two-dimensional)** Two-dimensional gel electrophoresis is a way to sort proteins by both size and their surface charge (P. H. O'Farrel, *J. Biol. Chem.* **250** (1975), 4007). The proteins are first separated horizontally by their charge by meandering in a pH gradient until the point where they have no net charge (isoelectric focusing). This point is the isoelectric point for the corresponding protein. The pH gradient is made by a special polymer (an ampholyte), which migrates to form the gradient when an electric field is applied. In the perpendicular direction on the gel, the proteins are subsequently separated after size by electrophoresis. The velocity of migration of compact globular proteins is $\propto$ charge/radius $= q/M^{1/3}$ (this follows from Stokes' law). The size of the spot on the gel is a measure of the protein abundance. The two-dimensional gel technique thereby allows a determination of concentrations of many proteins in living cells. Quantitatively the counting involves radioactive labeling of the proteins.

**gene** A section of DNA that codes for a protein, a tRNA, or rRNA.

**genetic recombination** Transfer of genetic material between DNA molecules.

**GFP (green fluorescent protein)** A protein that one uses to measure ongoing activity of selected promoters. GFP was originally found in jellyfish. When irradiating the protein with some UV light, it emits light at some specific wavelength. The GFP proteins in a single cell can then be seen in a microscope. The fluorescent property of GFP is preserved in virtually any organism that it is expressed in, including prokaryotes (like *E. coli*). GFP can be genetically linked to other proteins (covalent bond along the peptide backbone). This allows microscopic tracking of this protein inside the living cell. Often this linking with GFP does not influence the properties of the particular protein. GFP comes in red and yellow variants. The most severe limitation with all these fluorescent proteins is that they typically obtain their maximum maturation after a sizable amount of time. The "maturation time" may be from 15 min to some hours.

**globular proteins** Proteins with a compact three-dimensional native state.

**haploid** The number of chromosomes in egg or sperm. It is half the diploid number.

**Hill coefficient** The number of molecules that must act simultaneously in order to make a given reaction. The higher the Hill coefficient, the sharper the transition.

**histones** DNA binding proteins that regulate the condensation of DNA in eukaryotes. The DNA makes two turns around each histone. Histones play a major role in gene silencing in eukaryotes, and a large fraction of transcription regulators in yeast, for example, is associated with histone modifications.

**Holiday junction** A DNA structure that is a precursor for recombination (see Chapter 3).

**homeostasis** A state where all gene products are maintained at a steady state, and is thus typically the result of a negative (stabilizing) regulative feedback.

**hybridization** Base pairing of two single strands of DNA or RNA.

**hydrophilic** A molecule that has an attractive interaction with water molecules. Molecules that engage in hydrogen bonding are typically hydrophilic.

**hydrophobic** The property of having no attractive interactions with water. Hydrophobic substances are non-polar, and are effectively repelled by water.

**intron** A non-coding sequence of DNA that initially is copied to mRNA, but later cut out before protein in produced from the mRNA.

**kinase** An enzymatic protein that transfers a phosphate group ($PO_4$) from a phosphate donor to an acceptor amino acid in a substrate protein. Kinases have been classified after acceptor amino acids, which can be serine, tyrosine, histedine, cytosine, aspartane or glutamate. All kinases share a common (homologous) catalytic core (T. Hunter, *Meth. Enzymol.* **200** (1991), 3–37).

**ligase** A enzyme that helps to put other macromolecules together.

**microarray** See DNA chips.

**microtubules** Very stiff parts of cytoskeleton that consist of tubulin sub-units arranged to form a hollow tube.

**mismatch repair** A system for the correction of mismatched nucleotides or single-base insertions or deletions produced during DNA replication.

**mitochondria** Semi-autonomous, self-reproducing organelles in eukaryotes. These organelles are responsible for the energy conversion into ATP.

**mitosis** The process of cell division in eukaryotic cells.

**mobility** Mobility of a particle, $\mu = v/F$, determines how fast ($v$) the particle moves, when exposed to force $F$.

**mRNA** (messenger RNA). A single-stranded RNA that acts as a template for the amino acid sequence of proteins.

**NKC models** These are models of interacting species, after Kauffman & Johnnson (*J. Theor. Biol.* **149** (1991), 467), where each species is assigned a "fitness" function. Interactions between species occur only through this fitness function. In the NKC models the fitness of a species is given by $F = \sum f_i$, where $i$ runs over $N$ internal spin variables (base pair in the genome), that are each coupled by $K$ internal couplings within the species and by $C$ external couplings to other species. In practice, for each spin variable $i$ the $f_i$ is assigned a random value between 0 and 1 for each of the $2^{K+1+C}$ configurations. For a small $C$, different species become effectively decoupled and the ecology evolves to a frozen state, whereas a large $C$ means that a change in one species typically implies a "fitness" change in other species and, subsequently, a reshuffling in some of their $N$ spin variables. For a fine-tuned value of $C$, the system is critical and one may say that the ecology evolves on the "edge of chaos".

**nucleosome** A structural unit made up of 146 bp of DNA wrapped 1.75 times around an octamer of histone proteins.

**nucleotide** Basic element of DNA and RNA. DNA nucleotides are A, T, C and G.

**open reading frames (ORF)** DNA stretches that potentially encode proteins. They always have a start codon in one end (ATG) and a translation terminating stop codon at the other end.

**operator** DNA site that regulates the activity of gene transcription. It is typically located behind the promoter for the gene.

**operon** A set of contiguous prokaryotic structural genes that are transcribed as a unit, plus the regulatory elements that control their transcription.

**paralogs** A pair of duplicated proteins within the same organism (homologs within the same species). The paralogs will be similar in sequence (PID); typically one can identify them with certainty only when PID > 20%.

**palindrome** A sequence of DNA that is the same on one strand read right to left as on the other strand read left to right.

**persistence length** Length $l_p$ that characterizes how stiff a polymer is: $\langle e(x) \cdot e(y) \rangle \propto \exp(-|x-y|/l_p)$, where $e$ denotes the local tangent vector to the polymer. Thus $\langle r(\text{end-to-end})^2 \rangle \approx 2l_p L$, where $L$ is the contour length of the polymer. Note that $l_p$ is half the Kuhn length.

**phage** Also known as a bacteriophage, this is a virus that attacks a bacteria.

**plasmid** A piece of double-stranded DNA that encodes some proteins, which is expressed in the host of the plasmid. It may be viewed as an extrachromosomal DNA element, and as such it can be transmitted from host to host. Plasmids are, for example, carriers of antibiotic resistance, and when transmitted between bacteria thereby help these to share survival strategies. Plasmids often occur in multi-copies in a given organism, and can thus be used to greatly overproduce certain proteins. This is often used for industrial mass production of proteins.

**prokaryotes** Single-celled organisms without a membrane around the nucleus. Whitmann *et al.* (*Proc. Natl Acad. Sci. USA* **95** (12) (1998), 6578) estimated that there are $(4\text{–}6) \times 10^{30}$ prokaryotes on Earth. The number of prokaryote divisions per year is $\sim 1.7 \times 10^{30}$. Prokaryotes are estimated to contain about the same amount of carbon as all plants on Earth ($5 \times 10^{14}$ kg). Some 5000 species have been described, but there are estimated to be more than $10^6$ species.

**promoter** Region on DNA where RNAp binds in order to initiate transcription of a gene. The gene is downstream of the promoter. The transcription factors mostly bind upstream of the promoter.

**protease** A protein that actively degrades other proteins. An example is HflB, which plays a role in both heat shock in *E. coli* and the initial decision of $\lambda$ to lysogenize.

**protein** A polymer composed of amino acids.

**punctuated equilibrium** Term from biological evolution meaning that evolution of species occur in jumps. It can be understood if each species is in a metastable state, and its evolution corresponds to escape from a metastable state (see R. Lande, *Proc. Natl Acad. Sci. USA* **82** (1985), 7641). The content of the Bak–Sneppen model is that local punctuated equilibrium implies global correlations. For data on local punctuations, see S. J. Gould & N. Eldredge, *Nature* **366** (1993), 223–227.

**random walk** This is a very important concept in physics, as it represents the microscopic mechanism for the phenomenon of diffusion. A diffusing particle moves a distance $\sqrt{2Dt}$ in the time $t$, where $D$ is the diffusion constant. For the random walk, we introduce a microscopic time $\tau$ as the time between subsequent randomization of velocity, and $l$ the distance moved in time $\tau$, then $R = \sqrt{2(l^2/\tau)t}$ . The diffusion constant $D = l^2/\tau$ = velocity × mean free path; with velocity $= l/\tau$ and mean free path $l$.

**receptor** A protein, usually in the cell wall, that initiates a biological response inside the cell upon binding a specific molecule on the outside of the cell.

**recombinant DNA technique** A method of inserting DNA segments into the chromosome of an organism, often an *E. coli*. The *E. coli* is put together with the DNA in pure water (no ions) and exposed to strong alternating electric fields (of order 10 kV/cm). The DNA segments enter the cell, and if the ends of the DNA are single-stranded and have some base pairs on these single-stranded ends that are identical to some of the chromosomal *E. coli* DNA, it is often inserted into this DNA on the matching sites. This insertion presumably happens when the replication fork passes the part of the chromosome where the injected DNA has matching segments. The DNA is inserted in the lagging strand of the replication fork, in competition with usual complementation with new base pairs. The DNA is successfully inserted in only a small fraction of the *E. coli* (maybe one in 10 000). Successful insertion is checked through reporter genes associated with successful inserts.

**repressor** Transcription factor that represses a gene, typically by blocking the promoter for the gene.

**reporter genes** Certain genes that can report the activity on some promoter system. Typically they are inserted into the chromosome after the promoter, possibly even after the protein that is normally encoded by the promoter. A much used reporter is green fluorescent protein (GFP), which emits green light after the cell is irradiated with UV light in some particular wavelength.

**ribosome** This is the machine that translates the mRNA into a protein.

**RNA (ribonucleic acid)** Polymer composed of a backbone of phosphate and sugar subunits to which different bases are attached: adenine (A), cytosine (C), guanine (G) and uracil (U).

**RNAi** RNA interference is the introduction of specific double-stranded RNA corresponding to specific target genes in an organism. When present in the cell this results in elimination of all gene products of the target gene, and thus is an efficient way of making a null-mutant of the corresponding gene. It is believed to act through activation of an ancient immune response inside a cell (against RNA viruses), an immune response that when induced removes all RNA that is homologous to the injected RNAi. Through its activation of an immune response it is found to be highly specific and very potent, and only a few molecules of RNAi induce complete silencing of a gene. It was first studied in *C. Elegans* by S. Guo & K. Kemphues (*Cell* **81** (1995), 611–20) and by Fire *et al.* (*Nature* **391** (1998), 806–811). It was subsequently studied in *Drosophila* by J. R. Kennerdell & R. W. Carthew (*Development* **95** (1998), 1017–1026) and by L. Misquitta & B. M. Paterson (*Proc. Natl Acad. Sci. USA* **96** (1999), 1451–1456).

**RNAp** RNA polymerase enzyme. The molecular machinery that translates DNA into RNA.

**rRNA** Ribosomal RNA.

**robustness** This refers to an overall non-sensitivity to some of the parameters/assumptions of a particular machinery/model. The robustness can be towards changed parameter values (different environments) or towards changed rules (mutational variants). Robustness is presumably coupled to evolvability.

**scRNA (small cytoplasmic RNA)** A set of RNAs that are typically smaller than 300 nucleotides. Also used in signalling.

**snRNA (small nucleic RNA)** A set of RNAs that are typically smaller than 300 nucleotides, used to regulate post-transcriptional RNA processing.

**spindle** A mechanical network of microtubules and associated macromolecules that is formed prior to mitosis in eukaryotic cells. It mediates the separation of the duplicated chromosomes prior to cell division.

**strain** A genetically well-defined sub-group of a given species. *E. coli* comes in many strains that can vary by up to 20% in genome length. Some *E. coli* are harmless, others lethal.

**stop codons** Triplets (UAG, UGA, and UAA) of nucleotides in RNA that signal a ribosome to stop translating an mRNA and release the translated polypeptide.

**suppressor** Protein that suppresses the effects of a mutation in another gene.

**Southern blot** A way to analyze genetic patterns. The technique involves isolating the DNA in question, cutting it into several pieces, sorting these pieces by size in a gel, and denaturing the DNA in the gel to obtain single-stranded segments in the gel. Denaturation can be done by NaOH. Finally the DNA is blotted in the gel through a sheet of nitrocellulose paper in a way that makes it stick permanently (by heating it). The DNA fragments will retain the same pattern as they had on the gel. Analysis is then done through hybridization with radioactive complementary ssDNA that can be detected on an X-ray film.

**Stokes' equation** The force needed to drag a sphere of radius $r$ through liquid with viscosity $\eta$ is $F = 6\pi\eta r v$, where $v$ is the velocity. For water $\eta = 0.001$ Pas $= 0.001$ kg s/m.

**sympatic speciation** A species separating into two species without geographical isolation. Contrast this with allopatic speciation.

**temperate phage** This is a phage that can live within the bacterium and be passively replicated with it, or can lyse the bacterium and use the host's raw material to generate a number (typically of order 100) progeny phages.

**terminator** Stop sign for transcription at the DNA. In *E. coli* it is typically a DNA sequence that codes for an mRNA sequence that forms a short hair-pin structure plus a sequence of subsequent Us. For example, the RNA sequence CCCGCCUAAUGAGCGGGCUUUUU-UUU terminates RNAp elongation in *E. coli*.

**transcription** The process of copying the DNA template to an RNA.

**transcription factors** Proteins that bind upstream to a gene and regulate how much it is transcribed.

**transduction** If a phage packs host DNA instead of phage it forms what are called transducing particles. These can inject their packed DNA into another bacterium, and thereby allow for horizontal transfer of DNA. Examples are the P22 and P1 phages of *E. coli*, where about 1/1000 phages carry host DNA. P22 also infects *Salmonella typhimurium*, and can thus carry DNA across species boundaries.

**translation** The process of copying RNA to protein. It is done in the ribosome with the help of tRNA.

**transposable element (transposon)** A mobile DNA sequence that can move from one site in a chromosome to another, or between different chromosomes.

**tRNA** This is transfer RNA – small RNA molecules that are recruited to match the triplet codons on the mRNA with the corresponding amino acid. This matching takes place inside the ribosome. For each amino acid there is at least one tRNA.

**two-hybrid method** The two-hybrid system is a way of recording whether two proteins bind to each other inside the cell (in the nucleus, as is presently mostly done for eukaryotes). It uses the mechanism of recruitment of RNA polymerase to the promoter for an inserted reported gene. The typical setup is to use the GAL protein and separate its DNA binding domain (GAL4-BD) and its RNAp activating domain (GAL4-AD). The genes for each of these domains are linked to their selected gene. Each such link is inserted in a plasmid, and two different plasmids are inserted into the same organism (usually yeast). Only the cells where there is an attraction between the bait and prey protein can transcribe the reporter gene. Thus one can test whether there is attraction or not between the two tested proteins. The two-hybrid method allows for large-scale screening of all pairs of proteins in the organism's genome.

**van't Hoff enthalpy** The $\Delta H_{\text{vH}}$ enthalphy deduced from plotting $\ln(K) = \ln(N_{\text{u}}) - \ln(N_{\text{f}})$ vs $1/T$ and measuring the slope. For a two-state system where $N_{\text{f}}$ and $N_{\text{u}}$ are well determined, this enthalpy would be identical to the calorimetrically measured enthalpy difference $\Delta H_{\text{cal}} = H_{\text{u}} - H_{\text{f}} = Q$, where $Q$ is the latent heat of folding. If $\Delta H_{\text{vH}} < \Delta H_{\text{cal}}$ there are folding intermediates.

**vector** A virus or a plasmid that carries a modified gene into a cell.

**Western blot** A technique to identify and locate proteins based on their binding to specific antibodies.

**Zipf law** See G. K. Zipf, *Human Behaviour and the Principle of Least Effort* (Reading, MA: Adldison-Wesley, 1949). This law is especially famous for the distribution of frequencies of words, and defines the rank of a word according to how frequently it appears. The most frequent word has rank 1, the next rank 2, etc. The actual frequency of the word is plotted against its rank, both on a logarithmic scale. One observes a power law with exponent of about 1. Zipf did similar plotting for frequencies of cities, and because larger cities are less frequent, the map directly translates into a plot of frequency of cities larger than a given size $m$ versus $m$. The exponent, observed to be about $z = 0.8$, indicates that the frequency of city of given size $m$ is $\propto m^{-\alpha}$ with $\alpha = 1 + 1/z \approx 2.5$. Recent data of D. H. Zanette and S. Manrubia (*Phys. Rev. Lett.* **79** (1997), 523) give $\alpha = 2.0 \pm 0.1$ for city size distributions. See also H. A. Simon (*Biometrika* **42** (1955), 425–440), G. U. Yule (*Phil. Trans. R. Soc. Lond.* B**213** (1924), 21–87) and B. M. Hill & M. Woodroofe (*J. Statist. Assoc.* **70** (349) (1975), 212).

# Index